New Insights into Antithrombotic Therapy for Cardio- and Cerebrovascular Disease: From Molecular Mechanisms to Clinical Application

New Insights into Antithrombotic Therapy for Cardio- and Cerebrovascular Disease: From Molecular Mechanisms to Clinical Application

Editors

Giovanni Cimmino
Plinio Cirillo

MDPI • Basel • Beijing • Wuhan • Barcelona • Belgrade • Manchester • Tokyo • Cluj • Tianjin

Editors
Giovanni Cimmino
University of Campania Luigi Vanvitelli
Naples, Italy

Plinio Cirillo
University of Naples "Federico II"
Naples, Italy

Editorial Office
MDPI
St. Alban-Anlage 66
4052 Basel, Switzerland

This is a reprint of articles from the Special Issue published online in the open access journal *Journal of Cardiovascular Development and Disease* (ISSN 2308-3425) (available at: https://www.mdpi.com/journal/jcdd/special_issues/Antithrombotic_Therapy_Application).

For citation purposes, cite each article independently as indicated on the article page online and as indicated below:

LastName, A.A.; LastName, B.B.; LastName, C.C. Article Title. *Journal Name* **Year**, *Volume Number*, Page Range.

ISBN 978-3-0365-8116-3 (Hbk)
ISBN 978-3-0365-8117-0 (PDF)

Contents

Preface to "New Insights into Antithrombotic Therapy for Cardio- and Cerebrovascular Disease: From Molecular Mechanisms to Clinical Application"

Thrombosis represents a pathophysiological phenomenon that may be observed in several clinical conditions, such as acute and chronic coronary syndrome, stroke, and peripheral artery disease. The activation of coagulation cascade, as well as platelet aggregation, are key steps of the thrombotic process. Pharmacological modulation of both components of thrombosis, the coagulation cascade and platelet activation, is of great clinical importance. Several clinical trials have clearly shown the efficacy of anticoagulation and/or anti-platelet aggregation in different thrombotic disorders. However, real-world practice clearly indicates that antithrombotic strategies need to be personalized according to patient characteristics, such as age, concomitant diseases already requiring antithrombotic drugs, or risk for bleeding. In this regard, the combination of multiple antithrombotic drugs represents a challenging scenario and was therefore the focus of multiple recent randomized controlled trials. However, several gray areas still persist.

This Special Issue, by collecting the points of view of authoritative international research groups, gives an updated overview of the state of the art, as well as of the most promising future prospects for mechanisms of thrombosis and antithrombotic therapy.

Giovanni Cimmino and Plinio Cirillo
Editors

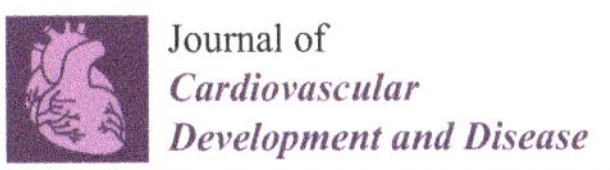
Journal of
Cardiovascular
Development and Disease

MDPI

Editorial

New Insights into Antithrombotic Therapy for Cardio- and Cerebrovascular Disease: From Molecular Mechanisms to Clinical Application

Plinio Cirillo [1,*] and Giovanni Cimmino [2]

1 Department of Advanced Biomedical Sciences, University of Naples Federico II, Via S. Pansini 5, 80131 Naples, Italy
2 Department of Translational Medical Sciences, University of Campania Luigi Vanvitelli, Via L. Bianchi, 80131 Naples, Italy; giovanni.cimmino@unicampania.it
* Correspondence: pcirillo@unina.it

Citation: Cirillo, P.; Cimmino, G. New Insights into Antithrombotic Therapy for Cardio- and Cerebrovascular Disease: From Molecular Mechanisms to Clinical Application. *J. Cardiovasc. Dev. Dis.* **2023**, *10*, 258. https://doi.org/10.3390/jcdd10060258

Received: 22 May 2023
Accepted: 12 June 2023
Published: 14 June 2023

Thrombosis has a pivotal role in the pathophysiology of acute cardiovascular events such as myocardial infarction and stroke [1]. Thus, an impressive effort has been done to better understand all those mechanisms able to promote intravascular thrombosis, in order to draw a better antithrombotic strategy to avoid future events. This Special Issue, by collecting the points of view of authoritative international research groups, gives an updated overview of the state-of-the-art as well as of the most promising future prospects for mechanisms of thrombosis and antithrombotic therapy.

The first article published comes from Prof Angiolillo's group [2]. It explores the role of antiplatelet monotherapy with P2Y12 inhibitors in patients undergoing percutaneous revascularization. By critically analyzing the available literature, the Researchers conclude that monotherapy with P2Y12 inhibitors is a current possibility that however needs to be better evaluated in well-designed clinical trials to carefully establish its onset after a period of DAPT. This paper underscores that antiplatelet monotherapy with a P2Y12 inhibitor has been already considered a reasonable antiplatelet strategy by the European and American guidelines in patients treated with PCI. Moreover, it gives complete information about the RCTs specifically designed to obtain new insights for P2Y12 inhibitor monotherapy even in patients with STEMI, and for long-term secondary prevention in patients with stable coronary artery disease.

Another historical actor on the stage of antiplatelet therapy is faced in the paper by Prof Cattaneo's group [3], who focuses on aspirin monotherapy, concluding that the enteric-coated formulation is less effective in terms of antiplatelet activity, especially in patients with more than 70 kg of weight. Therefore, by considering its more favorable pharmacological profile, plain aspirin should be the preferred formulation for cardiovascular prevention [3].

An important take-home message comes from the paper of Prof Davlouros' group [4]. Specifically, the Authors deal with a patient typology usually excluded by clinical trials, represented by cancer patients, who are at high risk of both ischemic and bleeding events. The high ischemic/bleeding risk of oncologic patients, due to the dysregulation of their hemostatic system by cancer, requires an appropriate duration and optimal antiplatelet therapy after PCI and/or acute coronary syndromes not treated with stent implantation. The use of new-generation DES may lead to shortened DAPT duration in all-comer patients, including patients with cancer. Current guidelines indicate that the optimal duration of DAPT should be 1–3 months, consisting of aspirin and clopidogrel, while TAT, if required, might only be administered for a short period of time (up to 1 week in the hospital), followed by a DOAC and a single oral antiplatelet agent (preferably clopidogrel). The advancements in other structural interventions, such as TAVR, PFO-ASD closure, and LAA occlusion, and non-cardiac diseases, such as PAD and CVA, may require DAPT, thus it is indisputable that a multidisciplinary approach is necessary for this to better balance thrombotic and bleeding risk [4].

Of course, antiplatelet therapy represents only one of the two faces of antithrombotic therapy since the coagulation pathway is involved in the pathophysiology of thrombosis too [5,6]. Thus, of extreme interest is the report by Prof. J. Badimon et al. that analyzes in detail the inhibition of factor XI/XIa as an attractive future option as antithrombotic therapy [7]. A better understanding of the contact pathway, especially of its significant role in thrombus stabilization and growth vs. in the initiation of clot formation, has opened up new targets for therapeutic intervention. FXI is one such promising target. FXI-directed strategies could offer similar protection against thrombotic events as DOACs, but with the advantage of lower bleeding risk. Growing strategies are now available to modulate FXI/FXIa, including ASOs, small molecules, antibodies, and aptamers. A wide variety of clinical scenarios may take advantage of this novel strategy. Several FXI-directed agents are currently undergoing clinical evaluations in phase II and phase III trials [7].

In line with this view on the future, the article proposed by Prof. G. Vilahur first comments on the key limitation of the currently used antithrombotic regimes in ischemic heart disease and ischemic stroke and then it looks at the emerging anticoagulant and antiplatelet agents in the pipeline with the potential to improve clinical outcomes. Some of these promising strategies are still under biological evaluation, others have reached the animal model test, and only few agents have been considered for randomized clinical trials [8].

Although many studies have clearly indicated that advantages of DOAC to prevent thrombotic events in specific clinical settings, some grey areas still exist when DOAC is compared to VKA [9]. This is the case of intracardiac thrombosis. On this important issue are focused the articles by Prof Pradhan [10] and Prof Chan [11]. Specifically, Pradhan et al. face the left ventricular thrombosis after myocardial infarction by critically analyzing the available literature including case reports, small trials, and meta-analysis, and supports the use of DOACs over VKA. Interestingly, an algorithm for the choice of agent and duration of the strategy has been also proposed [10]. On the other hand, the meta-analysis by Prof Chan's group on left atrial appendage thrombosis, which is main cause of cardioembolism and ischemic stroke, strengthens the role of DOACs also in this clinical context. Targeted clinical studies are warranted to better define its use in clinical practice to help clinicians in the choice of DOAC and duration of treatment [11].

Current guidelines recommend P2Y12 inhibitors plus aspirin as the gold standard of antiplatelet therapy in ACS and PCI-treated patients [12]. However, the onset of action for P2Y12 inhibitors is about one hour after administration [13]. Cangrelor, an injective P2Y12 inhibitor, by having an immediate effect on platelet aggregation, represents a new, extremely attractive therapeutic horizon in the use of parenteral P2Y12 inhibitors. Prof Andò and his research group analyzed the available data on the current use and potential of this new strategy [14]. The currently available oral P2Y12 have a relatively slow onset of action, so drug-naïve patients, and especially those with ACS, undergoing PCI lack of antiplatelet protection in the first hours after oral administration of antiplatelet therapy, thus the patients may be exposed to a greater thrombotic risk. Cangrelor might overcome this gap. Its effectiveness in drug-naïve patients undergoing PCI has been proved, both in the stable and the acute setting, by reducing early and 30-day ischemic outcomes, with particular emphasis on ischemia driven revascularization and early ST. Cangrelor appears to be a very safe drug with a low rate of bleeding and specifically of major events [14]. In this regard pharmacological research will open shortly a new debate about the optimal choice and timing to administer parenteral DAPT, since a new drug to be self-administered at home (selatogrel), or in the ambulance (zalunfiban) will be available in parallel to a drug to be administered in the hospital (cangrelor). Further RCTs are needed about the combination of parenteral and potent oral P2Y12 inhibitors in patients with ACS and about the optimal switching strategies.

Patients with peripheral artery disease have often been considered the Cinderella of patients with atherosclerosis in terms of thrombosis risk. Thus, the importance of antithrombotic therapy has been underestimated. The article by Prof. M. Bonaca et al. provides an overview of current evidence in different clinical settings in PAD and proposes an algorithm for antithrombotic therapy management in daily practice [15]. In patients undergoing revascularization, evidence supports more aggressive antithrombotic therapy, specifically, dual pathway inhibition after low extremity revascularization, irrespective of the type of intervention. The optimal management of patients undergoing revascularization for carotid and abdominal aortic disease remains to be better elucidated. The development of newer antithrombotic strategies, such as factor XIa inhibitors, may play an important role in this regard [15].

Finally, to complete the Special Issue, we find an original article from Prof. Xiao's research group in which the Authors provided further basic evidence on the role of metformin in stabilizing atherosclerotic plaque by binding MMP-9 and driving its degradation thus preserving the collagen content of plaque and improving atherosclerotic plaque stability [16]. Metformin remains a drug of choice for the treatment of type 2 diabetes with proven glucose-lowering effectiveness, safety, favorable effect on body weight, and low cost. Beyond these properties, an atherosclerotic stabilizing effect is here expanded and added to the known anti-inflammatory activity [16].

In summary, since the role of platelets and of the coagulation cascade in the pathophysiology of cardiovascular thrombosis have been extensively investigated, the research on antithrombotic options is extremely active and in progress. A more appropriate use of drugs already in use and the search for new safer drugs in terms of ischemic and bleeding risk might be considered a glimpse into the future to which this special issue contributes in an important way by acting as a hypothetical user manual.

Conflicts of Interest: The authors declare no conflict of interest.

References

1. Cimmino, G.; Di Serafino, L.; Cirillo, P. Pathophysiology and mechanisms of Acute Coronary Syndromes: Atherothrombosis, immune-inflammation, and beyond. *Expert Rev. Cardiovasc. Ther.* **2022**, *20*, 351–362. [CrossRef] [PubMed]
2. Zhou, X.; Angiolillo, D.J.; Ortega-Paz, L. P2Y(12) Inhibitor Monotherapy after Percutaneous Coronary Intervention. *J. Cardiovasc. Dev. Dis.* **2022**, *9*, 340. [CrossRef] [PubMed]
3. Clerici, B.; Cattaneo, M. Pharmacological Efficacy and Gastrointestinal Safety of Different Aspirin Formulations for Cardiovascular Prevention: A Narrative Review. *J. Cardiovasc. Dev. Dis.* **2023**, *10*, 137. [CrossRef] [PubMed]
4. Tsigkas, G.; Vakka, A.; Apostolos, A.; Bousoula, E.; Vythoulkas-Biotis, N.; Koufou, E.E.; Vasilagkos, G.; Tsiafoutis, I.; Hamilos, M.; Aminian, A.; et al. Dual Antiplatelet Therapy and Cancer; Balancing between Ischemic and Bleeding Risk: A Narrative Review. *J. Cardiovasc. Dev. Dis.* **2023**, *10*, 135. [CrossRef]
5. Cimmino, G.; Fischetti, S.; Golino, P. The Two Faces of Thrombosis: Coagulation Cascade and Platelet Aggregation. Are Platelets the Main Therapeutic Target? *J. Thromb. Circ. Open Access* **2017**, *3*, 1000117. [CrossRef]
6. Cimmino, G.; Cirillo, P. Tissue factor: Newer concepts in thrombosis and its role beyond thrombosis and hemostasis. *Cardiovasc. Diagn. Ther.* **2018**, *8*, 581–593. [CrossRef] [PubMed]
7. Badimon, J.J.; Escolar, G.; Zafar, M.U. Factor XI/XIa Inhibition: The Arsenal in Development for a New Therapeutic Target in Cardio- and Cerebrovascular Disease. *J. Cardiovasc. Dev. Dis.* **2022**, *9*, 437. [CrossRef] [PubMed]
8. Barriuso, I.; Worner, F.; Vilahur, G. Novel Antithrombotic Agents in Ischemic Cardiovascular Disease: Progress in the Search for the Optimal Treatment. *J. Cardiovasc. Dev. Dis.* **2022**, *9*, 397. [CrossRef] [PubMed]
9. Curcio, A.; Anselmino, M.; Di Biase, L.; Migliore, F.; Nigro, G.; Rapacciuolo, A.; Sergi, D.; Tomasi, L.; Pedrinelli, R.; Mercuro, G.; et al. The gray areas of oral anticoagulation for prevention of thromboembolic events in atrial fibrillation patients. *J. Cardiovasc. Med.* **2023**, *24*, e97–e105. [CrossRef] [PubMed]
10. Pradhan, A.; Bhandari, M.; Vishwakarma, P.; Salimei, C.; Iellamo, F.; Sethi, R.; Perrone, M.A. Anticoagulation for Left Ventricle Thrombus-Case Series and Literature Review for Use of Direct Oral Anticoagulants. *J. Cardiovasc. Dev. Dis.* **2023**, *10*, 41. [CrossRef] [PubMed]
11. Cheng, Y.Y.; Tan, S.; Hong, C.T.; Yang, C.C.; Chan, L. Left Atrial Appendage Thrombosis and Oral Anticoagulants: A Meta-Analysis of Risk and Treatment Response. *J. Cardiovasc. Dev. Dis.* **2022**, *9*, 351. [CrossRef] [PubMed]
12. Neumann, F.J.; Sousa-Uva, M.; Ahlsson, A.; Alfonso, F.; Banning, A.P.; Benedetto, U.; Byrne, R.A.; Collet, J.P.; Falk, V.; Head, S.J.; et al. 2018 ESC/EACTS Guidelines on myocardial revascularization. *Eur. Heart J.* **2019**, *40*, 87–165. [CrossRef] [PubMed]

13. Price, M.J.; Angiolillo, D.J.; Teirstein, P.S.; Lillie, E.; Manoukian, S.V.; Berger, P.B.; Tanguay, J.F.; Cannon, C.P.; Topol, E.J. Platelet reactivity and cardiovascular outcomes after percutaneous coronary intervention: A time-dependent analysis of the Gauging Responsiveness with a VerifyNow P2Y12 assay: Impact on Thrombosis and Safety (GRAVITAS) trial. *Circulation* **2011**, *124*, 1132–1137. [CrossRef] [PubMed]
14. Alagna, G.; Mazzone, P.; Contarini, M.; Ando, G. Dual Antiplatelet Therapy with Parenteral P2Y(12) Inhibitors: Rationale, Evidence, and Future Directions. *J. Cardiovasc. Dev. Dis.* **2023**, *10*, 163. [CrossRef] [PubMed]
15. Canonico, M.E.; Piccolo, R.; Avvedimento, M.; Leone, A.; Esposito, S.; Franzone, A.; Giugliano, G.; Gargiulo, G.; Hess, C.N.; Berkowitz, S.D.; et al. Antithrombotic Therapy in Peripheral Artery Disease: Current Evidence and Future Directions. *J. Cardiovasc. Dev. Dis.* **2023**, *10*, 164. [CrossRef] [PubMed]
16. Chen, X.; Wang, S.; Xu, W.; Zhao, M.; Zhang, Y.; Xiao, H. Metformin Directly Binds to MMP-9 to Improve Plaque Stability. *J. Cardiovasc. Dev. Dis.* **2023**, *10*, 54. [CrossRef] [PubMed]

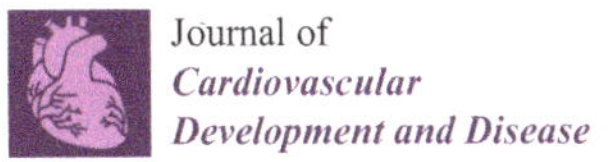
Journal of
Cardiovascular Development and Disease

MDPI

Review

$P2Y_{12}$ Inhibitor Monotherapy after Percutaneous Coronary Intervention

Xuan Zhou [1,2], Dominick J. Angiolillo [1,*] and Luis Ortega-Paz [1]

1 Division of Cardiology, University of Florida College of Medicine, Jacksonville, FL 32209, USA
2 Department of Internal Medicine, University of Alabama at Birmingham Montgomery, Montgomery, AL 36116, USA
* Correspondence: dominick.angiolillo@jax.ufl.edu; Tel.: +1-904-244-3378; Fax: +1-904-244-3102

Abstract: In patients with acute and chronic coronary artery disease undergoing percutaneous coronary intervention (PCI), dual antiplatelet therapy (DAPT) has been the cornerstone of pharmacotherapy for the past two decades. Although its antithrombotic benefit is well established, DAPT is associated with an increased risk of bleeding, which is independently associated with poor prognosis. The improvement of the safety profiles of drug-eluting stents has been critical in investigating and implementing shorter DAPT regimens. The introduction into clinical practice of newer generation oral $P2Y_{12}$ inhibitors such as prasugrel and ticagrelor, which provide more potent and predictable platelet inhibition, has questioned the paradigm of standard DAPT durations after coronary stenting. Over the last five years, several trials have assessed the safety and efficacy of $P2Y_{12}$ inhibitor monotherapy after a short course of DAPT in patients treated with PCI. Moreover, ongoing studies are testing the role of $P2Y_{12}$ inhibitor monotherapy immediately after PCI in selected patients. In this review, we provide up-to-date evidence on the efficacy and safety of $P2Y_{12}$ inhibitor monotherapy after a short period of DAPT compared to DAPT in patients undergoing PCI as well as outcomes associated with $P2Y_{12}$ inhibitor monotherapy compared to aspirin for long-term prevention.

Keywords: $P2Y_{12}$ inhibitor; monotherapy; percutaneous coronary intervention; dual antiplatelet therapy; high bleeding risk; high on-treatment platelet reactivity; randomized controlled trial

Citation: Zhou, X.; Angiolillo, D.J.; Ortega-Paz, L. $P2Y_{12}$ Inhibitor Monotherapy after Percutaneous Coronary Intervention. *J. Cardiovasc. Dev. Dis.* **2022**, *9*, 340. https://doi.org/10.3390/jcdd9100340

Academic Editor: Krzysztof J. Filipiak

Received: 11 September 2022
Accepted: 5 October 2022
Published: 6 October 2022

1. Introduction

Percutaneous coronary intervention (PCI) with stent implantation has emerged as the predominant revascularization strategy in patients with obstructive coronary artery disease (CAD) [1–3]. After PCI, antiplatelet therapy plays a pivotal role in preventing stent-related complications such as stent thrombosis and secondary prevention for non-stent-related ischemic events such as myocardial infarction (MI) and stroke [4–6]. The combination of aspirin and an oral $P2Y_{12}$ receptor inhibitor, known as dual antiplatelet therapy (DAPT), has become the guideline-recommended standard strategy after PCI based on data derived from more than 35 randomized clinical trials (RCTs) [1,2,7–10].

Clopidogrel is the most prescribed oral $P2Y_{12}$ inhibitor [11]. In particular, clopidogrel is the only guideline recommended $P2Y_{12}$ inhibitor after PCI in patients with chronic coronary syndromes (CCS) [1,2,7,8]. However, clopidogrel is a prodrug that requires hepatic cytochrome P450 2C19 (CYP2C19) metabolism to its active form, which leads to high variability in its pharmacodynamic (PD) effects [12,13]. Importantly, patients who persist with high platelet reactivity (HPR) while on clopidogrel are at increased risk of thrombotic events after PCI [14]. Indeed, patients with acute coronary syndromes (ACS) are at increased risk for HPR. Thus, the newer generation $P2Y_{12}$ inhibitors prasugrel and ticagrelor characterized by potent and predictable antiplatelet effects are preferred over clopidogrel as the standard of care in patients with ACS [1,2,9,15].

Even though the efficacy of DAPT is well established, it is also associated with an unavoidable increased risk of bleeding, which is associated with poor outcomes, including increased mortality [16]. Several investigations have led to defining the phenotype of patients more prone to bleeding, setting the foundation for introducing the high bleeding risk (HBR) concept [17]. In 2019, the Academic Research Consortium (ARC) formally defined HBR patients as those who are at risk of ≥4% of having type 3 or 5 bleeding according to the bleeding academic research consortium (BARC) or ≥1% of intracranial hemorrhage (ICH), both at 1 year [18]. Moreover, the ARC-HBR proposed a diagnostic criterion based on clinical and laboratory characteristics that has been classified into major and minor criteria, the presence of 1 major or 2 minor criteria are needed to fulfil the HBR definition.

Overall, these observations have prompted investigations evaluating "bleeding avoidance strategies" for patients undergoing PCI. The goal of these approaches is to minimize bleeding risk while preserving efficacy. Bleeding reduction strategies are directed to optimize the choice, duration, and modulation of DAPT (Figure 1). Amongst these, the strategy of discontinuation of aspirin after a short period of DAPT and maintaining $P2Y_{12}$ inhibitor monotherapy has been a subject of extensive investigation. This strategy was first investigated in patients requiring concomitant use of an oral anticoagulant agent. The details of this approach go beyond the scope of this manuscript and are described elsewhere [19,20]. In this manuscript, we provide an overview of $P2Y_{12}$ inhibitor monotherapy after a short course of DAPT in patients undergoing PCI without an indication of anticoagulation as well as the impact of $P2Y_{12}$ inhibitor monotherapy compared to aspirin for long term secondary prevention in patients with CCS.

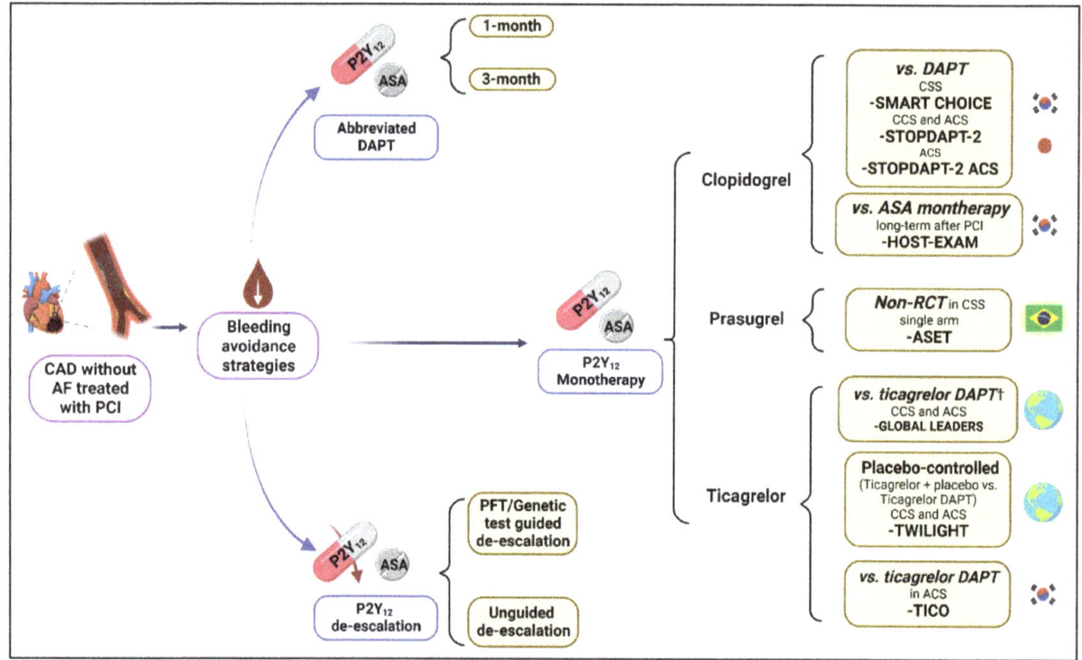

Figure 1. Selected bleeding avoidance strategies in patients without AF undergoing PCI. AF, atrial fibrillation; ACS, acute coronary syndrome; ASA, aspirin; CAD, coronary artery disease; CCS, chronic coronary syndrome; DAPT, dual antiplatelet therapy; PCI, percutaneous coronary intervention; PFT, platelet function test; RCT, randomized controlled trial.

2. Rationale for $P2Y_{12}$ Inhibitor Monotherapy

Platelet activation is a complex biological mechanism involving multiple activating factors such as thromboxane A_2 and adenosine diphosphate (ADP), which represent the targets of aspirin and $P2Y_{12}$ inhibitors, respectively [21]. Aspirin irreversibly blocks cyclooxygenase-1 (COX-1), the key enzyme in the arachidonic acid pathway of thromboxane A_2 generation. On the other hand, $P2Y_{12}$ inhibitors prevent ADP-mediated platelet activation by receptor blocking effect [22]. The exact mechanism can vary according to the type of drug. Clopidogrel and prasugrel (thienopyridines) require conversion to an active metabolite and mediate irreversible inhibition. Meanwhile, ticagrelor (non-thienopyridine) is a direct and reversible receptor antagonist [13]. Prasugrel and ticagrelor provide more potent and predictable platelet inhibition compared to clopidogrel [23,24]. These better PD profiles of prasugrel and ticagrelor compared to clopidogrel translate into lower ischemic/thrombotic events in pivotal RCTs, at the expense of increased bleeding events [25,26]. All these pivotal investigations have been performed on a background of aspirin therapy, under the notion that aspirin and $P2Y_{12}$ inhibitors (mainly demonstrated with clopidogrel) have synergetic effects on platelet inhibition, representing the foundation for the use of DAPT [27,28].

Although DAPT has remained the standardized therapy after PCI, the usage and duration of aspirin have been challenged based on three major arguments. First, the synergism between aspirin and $P2Y_{12}$ inhibitors was mainly established by early studies on aspirin with clopidogrel [28]. In the presence of potent $P2Y_{12}$ blockade, in vitro pharmacodynamic investigations have shown that aspirin does not provide much additional antiplatelet effect [29]. This was also confirmed in a series of ex vivo pharmacodynamic studies [30–32]. While withdrawal of aspirin indeed eliminates its specific inhibitory effects mediated by the COX-1 pathway, other platelet signaling pathways are still affected by potent $P2Y_{12}$ blockade [20,33]. Second, aspirin is associated with gastrointestinal (GI) adverse effects, from mild dyspepsia to ulceration and GI bleeding [34]. Systemically, aspirin irreversibly and non-selectively inhibits COX enzyme, leads to systemic prostaglandin depletion that compromises gastric mucosal barrier function and increases acid secretion [34]. Locally, aspirin may reduce surface hydrophobicity and destabilize the phospholipid barrier, which makes the mucosa susceptible to direct injury by gastric acid [35]. Although several approaches are used to mitigate aspirin gastric injury (i.e., consumption with food, proton pump inhibitors, and new aspirin formulations), the most effective way to reduce aspirin GI effects is by minimizing aspirin treatment duration [36]. Third, the introduction of newer drug-eluting stents has markedly decreased the rate of stent thrombosis, and the widespread usage of lipid-lowering therapies has further reduced the incidence of MI unrelated to the stent, which was assumed to be in part driven by the beneficial effects of DAPT [37].

3. Current Evidence of $P2Y_{12}$ Inhibitor Monotherapy

Over the last years, several large-scale RCTs have assessed the safety and efficacy of aspirin-free antiplatelet strategies after coronary stenting (Figure 2 and Table 1). Two main approaches have been assessed: (a) trials comparing $P2Y_{12}$ monotherapy versus conventional DAPT regimens after PCI and (b) trials comparing $P2Y_{12}$ inhibitors vs aspirin monotherapy for long-term secondary prevention.

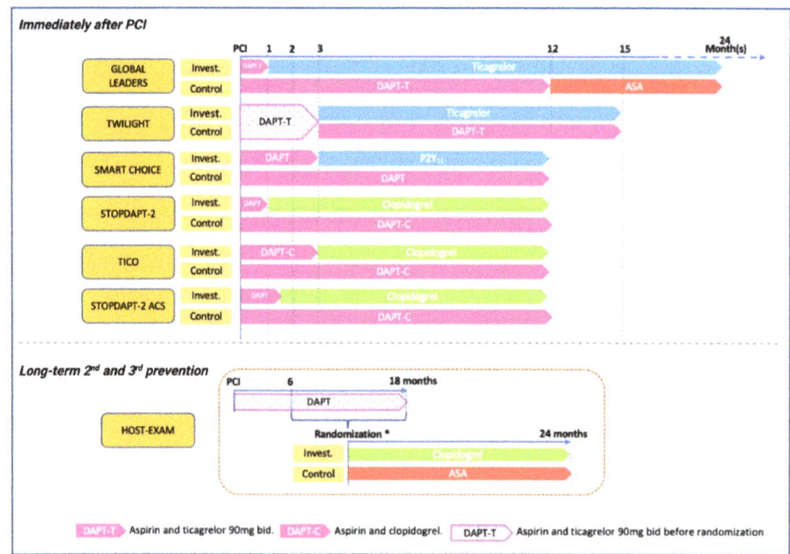

Figure 2. Randomized controlled trials of $P2Y_{12}$ inhibitor monotherapy in patients treated with PCI. ASA, aspirin; DAPT, dual antiplatelet therapy; DAPT-C, clopidogrel-based dual antiplatelet therapy; DAPT-T, ticagrelor-based dual antiplatelet therapy; Invest., investigational group; PCI, percutaneous coronary intervention. * In HOST-EXAM trial, event-free patients who maintained DAPT for 6–18 months after PCI were randomized.

Table 1. Randomized controlled trials for $P2Y_{12}$ inhibitor monotherapy in patients treated with PCI.

Studies	Experimental Group	Control Group *	Primary Outcome	Key Secondary Outcome
Immediately after PCI				
GLOBAL LEADERS 2018 (*n* = 15,968)	Ticagrelor-based DAPT for 1 month, then Ticagrelor monotherapy	ASA + clopidogrel (53%) ASA + ticagrelor (47%)	At 24 months, all-cause death, new Q-wave MI (RR, 0.87; 95% CI, [0.75–1.01]; $p = 0.073$)	BARC 3 or 5 bleeding (RR, 0.97; 95%CI, [0.78–1.20]; $p = 0.770$)
TWILIGHT 2019 (*n* = 7119)	Ticagrelor-based DAPT for 3 months, then Ticagrelor monotherapy	ASA + Ticagrelor	At 12 months, BARC 2–5 bleeding (HR, 0.56; 95% CI, [0.45–0.68]; $p < 0.001$)	BARC 3 or 5 bleeding (HR, 0.49; 95%CI, [0.33–0.74]; $p < 0.001$)
SMART CHOICE 2019 (*n* = 2993)	Clopidogrel (76.9%) Prasugrel (4.1%) Ticagrelor (19.0%) DAPT for 3 months, then monotherapy	ASA + clopidogrel (77.6%) ASA + Prasugrel (4.5%) ASA + ticagrelor (17.9%)	At 12 months, all-cause death, MI, stroke (difference, 0.4%; one-sided 95%CI, [−∞–1.3%]; $p = 0.007$ for non-inferiority)	BARC 2–5 bleeding (HR, 0.58; 95%CI [0.36–0.92]; $p = 0.020$)
STOPDAPT-2 2019 (*n* = 3045)	Clopidogrel based DAPT, then clopidogrel monotherapy	ASA + clopidogrel	At 12 months, CV death, MI, stroke, stent thrombosis, or TIMI major or minor bleeding (HR, 0.64; 95%CI, [0.42–0.98]; $p < 0.001$ for noninferiority; $p = 0.04$ for superiority)	TIMI major or minor bleeding (HR, 0.26; 95%CI, [0.11–0.64]; $p = 0.004$) -Ischemic endpoints (HR, 0.79; 95%CI, [0.49–1.29]; $p = 0.340$)
TICO (ACS) 2019 (*n* = 3056)	Ticagrelor-based DAPT, then ticagrelor monotherapy	ASA + ticagrelor	At 12 months, all-cause death, MI, stent thrombosis, stroke, target vessel revascularization and major bleeding (HR, 0.66; 95%CI, [0.48–0.92]; $p = 0.01$)	-TIMI major bleeding (HR, 0.56; 95%CI, [0.34–0.91]; $p = 0.02$) MACCE (HR, 0.69; 95%CI, [0.45–1.06]; $p = 0.09$)
STOPDAPT-2 ACS 2022 (*n* = 4169)	Clopidogrel-based DAPT, then Ticagrelor monotherapy	ASA + clopidogrel	At 12 months, CV death, MI, stroke, stent thrombosis, or TIMI major or minor bleeding (HR, 1.44; 95%CI, [0.80–1.62]; $p_{noninferiority} = 0.06$)	TIMI major or minor bleeding (HR, 0.46; 95%CI, [0.23–0.94]; $p = 0.03$) Significant increased risk of MI (HR, 1.91; 95%CI, [1.06–3.44]; $p = 0.03$)
Long-term 2nd and 3rd prevention				
HOST–EXAM 2020 (*n* = 5438)	Clopidogrel monotherapy, for 24 months	ASA monotherapy	At 24 months, all-cause death, non-fatal MI, stroke, readmission due to ACS, BARC 3–5 bleeding (HR, 0.73; 95%CI, [0.59–0.90]; $p = 0.003$)	BARC 3–5 bleeding (HR, 0.63; 95%CI, [0.41–0.97]; $p = 0.035$)

* Complete details about regimen duration are shown in Figure 1. ACS, acute coronary syndrome; ASA, aspirin; CAD, coronary artery disease; CCS, chronic coronary syndrome; CI, confidence interval; CV, cardiovascular; DAPT, dual antiplatelet therapy; HR, hazard ratio; PCI, percutaneous coronary intervention; PFT, platelet function test; RCT, randomized controlled trial; RR, rate ratio; TIMI, Thrombolysis in Myocardial Infarction.

4. P2Y$_{12}$ Monotherapy versus DAPT after PCI

4.1. Clopidogrel

SMART-CHOICE (Comparison Between P2Y$_{12}$ Antagonist Monotherapy vs Dual Antiplatelet Therapy in Patients Undergoing Implantation of Coronary Drug-Eluting Stents) was an open-label RCT comparing 3-month DAPT followed by P2Y$_{12}$ inhibitor monotherapy vs. standard 12-month DAPT after PCI in terms of major adverse cardiac and cerebrovascular events (MACCE) in a non-inferiority analysis [38]. A total of 2993 patients were enrolled. There were no restrictions on the type of P2Y$_{12}$ inhibitor or clinical presentation. The P2Y$_{12}$ inhibitor monotherapy was noninferior compared to DAPT in MACCE (Hazard ratio [HR], 1.19; 95% Confidence interval [CI], [$-\infty$%–1.3%]; $p_{\text{noninferiority}} = 0.007$). There were no significant differences in the primary endpoint components, but there was a significantly lower BARC 2–5 bleeding rate in the P2Y$_{12}$ inhibitor monotherapy than the DAPT group (HR, 0.58; 95%CI [0.36–0.92]; $p = 0.020$).

Two main post-hoc analyses have been reported. First, the clopidogrel–only cohort (80% of the total sample size), there were no significant differences between clopidogrel monotherapy versus clopidogrel–based DAPT in MACCE (HR, 1.02; 95%CI, [0.64–1.65]; $p = 0.100$) and BARC 2–5 bleeding (HR, 0.71; 95%CI, [0.42–1.21]; $p = 0.150$) [39]. Second, in the platelet reactivity sub-study ($n = 833$), 108 (13.0%) patients had HPR who had a significantly increased risk of MACCE compared to those without HPR (8.7% vs. 1.5%; HR, 3.03; 95%CI, [1.06–8.69]; $p = 0.038$) [40]. However, the treatment effect of clopidogrel monotherapy for the 12-month MACCE was not significantly different compared with DAPT in patients with HPR or without HPR (HR, 0.71; 95%CI, [0.18–2.73]; $p = 0.628$ and HR, 2.58; 95%CI, [0.68–9.77]; $p = 0.161$; $p_{\text{interaction}} = 0.170$). These results suggest that the main driver of adverse events was the HPR status rather than the allocated treatment, denoting the importance of optimizing platelet inhibition [41].

STOPDAPT-2 (Short and Optimal Duration of Dual Antiplatelet Therapy After Everolimus-Eluting Cobalt–Chromium Stent) was a prospective, open-labeled RCT comparing 1 month of DAPT (clopidogrel or prasugrel 3.75 mg od) followed by clopidogrel monotherapy versus 12 months DAPT with aspirin and clopidogrel in patients who underwent PCI [42]. A total of 3045 participants were recruited. The primary endpoint was a composite of ischemic (cardiovascular death, MI, stroke, or stent thrombosis) and bleeding endpoints (Thrombolysis in Myocardial Infarction [TIMI] major or minor bleeding) at 12 months. Clopidogrel monotherapy group met the prespecified criteria for noninferiority and superiority compared to the standard DAPT (HR, 0.64; 95%CI, [0.42–0.98]; $p < 0.001$ for noninferiority, $p = 0.04$ for superiority). There was no difference in the ischemic endpoints (HR, 0.79; 95%CI, [0.49–1.29]; $p = 0.340$), but there was a significant lower bleeding rate in the clopidogrel monotherapy than 12 months of DAPT (HR, 0.26; 95%CI, [0.11–0.64]; $p = 0.004$).

STOPDAPT-2 ACS (Short and Optimal Duration of Dual Antiplatelet Therapy-2 Study for the Patients With ACS) trial was a prospective, open-label RCT with the same design as the STOPDAPT-2, but including only patients with ACS, the ACS cohorts of both trials were combined (3008 newly enrolled and 1161 pooled form previous trial, in total 4169 patients were randomized) [43]. At the 1-year follow-up, 1–2 months DAPT (aspirin and clopidogrel) followed by clopidogrel monotherapy failed to meet the noninferior criteria compared to the 12-month DAPT (HR, 1.44; 95%CI, [0.80–1.62]; $p_{\text{noninferiority}} = 0.06$). The rate of major bleeding was significantly lower in the monotherapy group compared to the DAPT (HR, 0.46; 95%CI, [0.23–0.94]; $p = 0.03$). However, there was a significant increase in MI in the monotherapy group compared to the DAPT group (HR, 1.91; 95%CI, [1.06–3.44]; $p = 0.03$). The underlying reasons for which there was an increased risk of adverse events in the ACS cohort in patients treated with monotherapy compared to standard DAPT remains unclear but may be likely attributed to the presence of HPR among patients treated with clopidogrel only and no added antiplatelet effect given the withdrawal of aspirin.

STOPDAPT-2 Total Cohort the STOPDAPT investigators performed a prespecified pooled STOPDAPT-2 and STOPDAPT-2-ACS ($n = 5997$ in total), the rationale for this pooled analysis was that in both trials there had a lower-than-expected event rate that could affect the trials results [44]. The authors followed the same methodology and endpoints as in the main trials. One-month DAPT was noninferior but not superior to 12-month DAPT for the primary endpoint (HR, 0.94; 95%CI, [0.70–1.27]; $p_{noninferiority} = 0.001$ and $p_{superiority} = 0.68$). There was no significant risk-difference for the cardiovascular endpoint between groups (HR, 1.24; 95% CI, [0.88–1.75]; $p = 0.23$), but one-month DAPT was associated with a lower risk of the bleeding than 12-month DAPT (HR, 0.38 95%CI, [0.21–0.70]; $p = 0.002$). When the results were analyzed according to clinical presentation (ACS vs. CCS), one-month DAPT was associated with a lower risk for major bleeding than 12-month DAPT in ACS or CCS patients (HR, 0.46; 95%CI, [0.23–0.94]; $p = 0.03$. and HR, 0.26; 95%CI, [0.09–0.79]; $p = 0.02$; $p_{interaction} = 0.40$), but there was a numerical increase in cardiovascular events in ACS patients, but not in CCS patients (HR, 1.50; 95%CI, [0.99–2.27]; $p = 0.053$, and HR, 0.74; 95%CI, [0.38–1.45]; $p = 0.39$; $p_{interaction} = 0.08$).

4.2. *Prasugrel*

ASET (Acetyl Salicylic Elimination Trial) was a pilot, prospective, open-label, single-arm non-randomized study assessing the safety of prasugrel monotherapy in patients with CCS. All participants ($n = 201$) were on standard DAPT at the time of the index PCI, after successful PCI with platinum-chromium everolimus-eluting stent (Pt-EES), aspirin was discontinued and prasugrel was loaded and maintained for 3 months [45]. The primary ischemic endpoint was the composite of cardiac death, spontaneous target vessel MI, or definite stent thrombosis. The primary bleeding endpoint was major bleeding. There was only one event (cardiac death following intracranial bleeding). The compelling results of the ASET trial should be interpreted in the light of its small and very selected population and low lesion complexity.

4.3. *Ticagrelor*

GLOBAL LEADERS (A Clinical Study Comparing Two Forms of Antiplatelet Therapy After Stent Implantation) trial was a prospective, open-label RCT. Patients were randomized after successful PCI with a biolimus A9-eluting stent to either aspirin plus 90 mg ticagrelor twice daily for 1 month, followed by 23 months of ticagrelor monotherapy (90 mg, twice daily) or standard DAPT with clopidogrel (for patients with stable CAD) or ticagrelor (for patients with ACS) for 12 months, followed by aspirin monotherapy for another 12 months. A total of 15,968 patients were enrolled. The primary efficacy endpoint was all-cause death or non-fatal new Q-wave MI, and the primary safety endpoint was major bleeding, defined as BARC 3 or 5 bleeding. At 2 years, ticagrelor monotherapy was not superior to standard DAPT for reducing the primary efficacy (RR, 0.87; 95%CI, [0.75–1.01]; $p = 0.073$) or safety endpoints (RR, 0.97; 95%CI, [0.78–1.20]; $p = 0.770$). The adherence rate at two years was 77.6% in the experimental group and 93.1% in the control group, consistent with the premature ticagrelor discontinuation rate (25%) observed in other studies and mainly related to adverse events such as bleeding and dyspnea [46,47].

One of the main limitations of the GLOBAL LEADERS trial was the lack of independent event adjudication. Therefore, the prespecified GLASSY (GLOBAL LEADERS Adjudication Sub-Study) study was conducted following the same methodology as the main trial [48]. The study included approximately 47% of the main trial sample size enrolled in the top 20 enrolling sites. At 2 years, ticagrelor monotherapy was noninferior but not superior to standard 12 months DAPT for reducing the primary efficacy endpoint (RR, 0.85; 95%CI, [0.72–0.99]; $p_{noninferiority} < 0.001$ and $p_{superiority} = 0.046$ at alpha of 2.5%). There were no significant differences between groups in major bleeding regardless of the definition.

The prespecified [49–56] and selected post-hoc analyses [57–61] performed by the GLOBAL LEADERS investigators for exploring the effect size of the intervention on different subgroups are shown in Table S1.

TWILIGHT (Ticagrelor with Aspirin or Alone in High-Risk Patients after Coronary Intervention) was prospective, double-blind, placebo-controlled RCT that compared ticagrelor plus placebo vs. ticagrelor-based DAPT in event-free and high-risk PCI patients who completed 3 months of DAPT with aspirin and ticagrelor [62]. The primary endpoint was defined as clinically relevant bleeding (BARC 2, 3, or 5). The key secondary endpoint was the composite of all-cause death, nonfatal MI, or nonfatal stroke. A total of 7119 patient were randomized. At 1 year, the incidence of clinically relevant bleeding was significantly lower in the ticagrelor monotherapy group than in the ticagrelor-based DAPT group (HR, 0.56; 95%CI, [0.45–0.68]; $p < 0.001$). The secondary endpoint of BARC type 3 or 5 bleeding was also significantly less in the ticagrelor monotherapy group (HR, 0.49; 95%CI, [0.33–0.74]; $p < 0.001$). In the key secondary ischemic composite endpoint, ticagrelor monotherapy was non-inferior to ticagrelor-based DAPT group (HR, 0.99; 95%CI, [0.78–1.24]; $p_{noninferiority} < 0.001$).

The main results of the TWILIGHT trial have been shown to be consistent in several subgroup analyses such as age [63], gender [64], East Asian ethnicity [65], DM status [66], CKD status [67], prior MI [68], clinical presentation [69], stent used [70], and HBR status [71]. Overall, all indicate a reduced risk of clinically relevant bleeding and without a significant increase in ischemic events. A complete list of the prespecified and post-hoc analyses performed by the TWILIGHT investigators are shown in Table S2.

TICO (Ticagrelor Monotherapy After 3 Months in the Patients Treated With New Generation Sirolimus-eluting Stent for Acute Coronary Syndrome) trial was prospective, open-label RCT comparing ticagrelor monotherapy after 3 months of DAPT versus ticagrelor-based DAPT for 12 months in patients with ACS treated with PCI [72]. The primary outcome was a net adverse clinical event (NACE, composite of MACCE [composite of all-cause death, MI, stent thrombosis, stroke, or target vessel revascularization] and TIMI major bleeding). A total of 3056 patients were randomized. At 1 year, ticagrelor monotherapy significantly reduced NACE compared to ticagrelor-based DAPT (HR, 0.66; 95%CI, [0.48–0.92]; $p = 0.01$). There was significant reduction in major bleeding between two groups (HR, 0.56; 95%CI, [0.34–0.91]; $p = 0.02$), but not in MACCE (HR, 0.69; 95%CI, [0.45–1.06]; $p = 0.09$).

The main results of the TICO trial have been shown to be consistent in several subgroup analyses such as DM status [73], high-ischemic risk [74], ST-segment elevation myocardial infarction (STEMI) [75], and HBR status [76]. A complete list of the prespecified and post-hoc analyses performed by the TICO investigators are shown in Table S3.

4.4. Meta-Analysis

Several meta-analyses have been reported. However, the most comprehensive data reported are the individual patient data metanalysis by Valgimigli et al. [77]. In total, 24,096 patients from the GLASSY, SMART-CHOICE, STOPDAPT-2, TICO, and TWILIGHT trials were included. The primary efficacy endpoint was defined as a composite of all-cause death, MI, and stroke, and the key safety endpoint was major bleeding (BARC type 3 or 5). In the intention-treat analysis, $P2Y_{12}$ monotherapy was non-inferior but not superior to DAPT for the primary endpoint (HR, 0.93; 95%CI, [0.79–1.09]; $p = 0.005$ for noninferiority; $p = 0.380$). The bleeding risk was significantly lower with $P2Y_{12}$ inhibitor monotherapy than DAPT (HR, 0.49; 95%CI, [0.39–0.63]; $p < 0.001$). In the subgroup analysis, there was a significant interaction of sex in the effect size of $P2Y_{12}$ monotherapy and DAPT, there was a significant reduction in the primary endpoint in women but not in men (HR, 0.64; 95%CI, [0.46–0.89] and HR, 1.00; 95%CI, [0.83–1.19]; $p_{interaction} = 0.02$). The interaction was mainly driven by a reduction of cardiovascular mortality in women but not in men (HR, 0.31; 95%CI, [0.15–0.65] and HR, 0.86; 95%CI, [0.59–1.25]; $p_{interaction} = 0.02$). Furthermore, there was no significant interaction of the type of $P2Y_{12}$ inhibitor (clopidogrel vs. newer $P2Y_{12}$ inhibitor [mainly ticagrelor]) in the primary endpoint (HR, 0.94; 95%CI, [0.66–1.33] and HR, 0.89; 95%CI, [0.75–1.06]; $p_{interaction} = 0.16$) or major bleeding (HR, 0.60; 95%CI, [0.34–1.06] and HR, 0.47; 95%CI, [0.36–0.62]; $p_{interaction} = 0.41$).

5. $P2Y_{12}$ Inhibitor versus Aspirin Monotherapy for Long-Term Secondary Prevention

CAPRIE (A Randomized Blinded Trial of Clopidogrel Versus Aspirin in Patients at Risk of Ischaemic Events) trial was a prospective double-blind RCT reported in 1996 comparing clopidogrel monotherapy with aspirin (325 mg daily) monotherapy in patients with atherosclerotic vascular disease (defined as recent ischemic stroke, recent MI, or symptomatic peripheral arterial disease) [78]. A total of 19,185 patients were enrolled with a mean follow-up of 1.91 years. The primary endpoint was a composite of ischemic stroke, MI, or vascular death, which was significantly lower in the clopidogrel monotherapy group than the aspirin group (relative risk reduction, 8.7%; 95%CI, [0.3–16.5]; $p = 0.043$). Clopidogrel monotherapy had a significant lower rate of gastrointestinal hemorrhage events (patients ever reporting: 2.0% vs. 2.7%; $p < 0.05$ and severe gastrointestinal hemorrhage: 0.5% vs. 0.7%; $p < 0.05$). Moreover, clopidogrel monotherapy had a better upper GI tolerability than aspirin alone, with significant less indigestion/nausea/vomiting reported (patients ever reporting: 15.0% vs. 17.56%; $p < 0.05$) [78]. Despite the benefits of clopidogrel over aspirin, aspirin has remained the mainstay of therapy considering its reduced costs with clopidogrel being recommended over aspirin only in patients who could not tolerate or with hypersensitivity to aspirin. However, over two decades later with the availability of generic formulations of clopidogrel, there has been a re-appraisal for $P2Y_{12}$ inhibitor monotherapy for long-term secondary prevention.

HOST-EXAM (Harmonizing Optimal Strategy for Treatment of Coronary Artery Stenosis-Extended Antiplatelet Monotherapy) trial was a prospective, open-label RCT comparing clopidogrel monotherapy or aspirin monotherapy for 24 months in event-free patients who were on DAPT for 6–18 months after PCI ($n = 5530$) [79]. The primary endpoint was a composite of all-cause death, non-fatal MI, stroke, readmission due to ACS, and major bleeding (BARC 3–5). At 2 years, clopidogrel monotherapy significantly reduced the primary endpoint compared to aspirin monotherapy (HR, 0.73; 95%CI, [0.59–0.90]; $p = 0.003$), driven by both the ischemic composite endpoint (HR, 0.68; 95%CI, [0.52–0.87]; $p = 0.003$) and major bleeding (HR, 0.63; 95%CI, [0.41–0.97]; $p = 0.035$).

GLOBAL LEADERS investigators performed a post-hoc landmark analysis between the first and second year of follow-up in patients who were event free during the first year [80]. In particular, during this period, patients were on ticagrelor monotherapy and aspirin monotherapy. There was a lower rate of MI in the ticagrelor monotherapy compared to the aspirin monotherapy group (adjusted HR, 0.74; 95%CI, [0.58–0.96]; $p = 0.022$), but at the expense of a higher rate of major bleeding (adjusted HR, 1.89; 95%CI, [1.03–3.45]; $p = 0.005$).

Meta-Analysis

The $P2Y_{12}$ inhibitor or aspirin monotherapy as secondary prevention in patients with coronary artery disease: an individual patient data meta-analysis of randomized trials (PANTHER) trial assessed the role of long-term $P2Y_{12}$ monotherapy compared to aspirin monotherapy for the prevention of recurrent events in patients with CAD [81]. This analysis included 24,325 patients from seven RCTs. The primary endpoint was the composite of cardiovascular or vascular death, any non-fatal MI, and any non-fatal stroke. At a median of 557 days, $P2Y_{12}$ monotherapy was associated with a significant reduction in the primary endpoint compared to aspirin monotherapy (HR, 0.88; 95%CI, [0.79–0.97]; $p = 0.014$). The $P2Y_{12}$ monotherapy was associated with a significant reduction in MI (HR, 0.89; 95%CI, [0.81–0.98]; $p = 0.020$) and definite/probable stent thrombosis (HR, 0.46; 95%CI, [0.23–0.92]; $p = 0.028$) without a significant reduction in major bleedings (HR, 0.87; 95%CI, [0.70–1.09]; $p = 0.230$), and all cause-death (HR, 1.04; 95%CI, [0.91–1.20]; $p = 0.560$). Concerning the bleeding causes, $P2Y_{12}$ monotherapy was associated with a significant reduction in gastrointestinal bleeding (HR, 0.75; 95%CI, [0.57–0.97]; $p = 0.027$) and ICH (HR, 0.32; 95%CI, [0.14–0.75]; $p = 0.009$).

6. Guidelines on P2Y$_{12}$ Inhibitor Monotherapy

Several scientific societies have incorporated P2Y$_{12}$ monotherapy among their recommendations in patients treated with PCI. The 2020 European Society of Cardiology (ESC) guidelines for the management of non-ST-elevation acute coronary syndrome (NSTE-ACS) recommend stopping aspirin after 3–6 months should be considered, depending on the balance between the ischemic and bleeding risk [9]. The 2021 American College of Cardiology (ACC), American Heart Association (AHA), and Society for Cardiovascular Angiography and Interventions (SCAI) guidelines for coronary artery revascularization which were developed after the ESC guidelines and thus had more data available, state that in selected patients undergoing PCI, shorter duration DAPT (1–3 months) is reasonable, with subsequent transition to P2Y$_{12}$ inhibitor monotherapy to reduce the risk of bleeding events (Table 2) [2]. For long-term secondary prevention, clopidogrel is recommended in patients who cannot take aspirin due to intolerance or hypersensitivity [8].

Table 2. Clinical guidelines recommendations concerning P2Y$_{12}$ inhibitor monotherapy.

Cardiology Societies	Clinical Scenario	Recommendations	Level of Evidence *	Class of Recommendation *
ESC	NSTE-ACS [10] (2020)	After stent implantation in patients undergoing a strategy of DAPT, stopping aspirin after 3–6 months should be considered, depending on the balance between the ischemic and bleeding risk.	IIa	A
	Chronic coronary syndrome [9] (2019)	Clopidogrel 75 mg daily is recommended as an alternative to aspirin in patients with aspirin intolerance.	I	B
ACC/AHA/ SCAI	Coronary artery revascularization [2] (2021)	In selected patients undergoing PCI, shorter-duration DAPT (1–3 months) is reasonable, with subsequent transition to P2Y$_{12}$ inhibitor monotherapy to reduce the risk of bleeding events.	A	2a

* Details of the specific methodology of level of evidence and class of recommendation are provided in each guideline. ESC, European Society of cardiology; American College of Cardiology, American Heart Association, and Society for Cardiovascular Angiography and Interventions; NSTE-ACS, non-ST elevation acute coronary syndrome; DAPT, dual antiplatelet therapy; PCI, percutaneous coronary intervention.

7. Ongoing Studies of P2Y$_{12}$ Inhibitor Monotherapy

The role of P2Y$_{12}$ monotherapy in patients treated with PCI is currently a topic of extensive research with more than 10 ongoing RCTs (Table 3 and Figure 3). Overall, most of the ongoing trials are focused on ACS patients. In particular, ULTIMATE-DAPT is a placebo-controlled RCT that will recruit event-free patients after 1 month of DAPT and compare ticagrelor plus placebo or ticagrelor-based DAPT for 11 months. The MATE and CAGEFREE II trials are investigating a de-escalation strategy consisting of 1 month of DAPT, followed by 5 months of ticagrelor monotherapy, and finalized by 6 months of clopidogrel or aspirin monotherapy. Among HBR or ACS patients, STOPDAPT-3 will compare a short course if clopidogrel-based DAPT with standard clopidogrel DAPT duration. The BULK-STEMI will determine the efficacy of ticagrelor monotherapy after 3 months of ticagrelor-based DAPT in patients presenting with STEMI. Two studies, ASET-JAPAN and NEO-MINDSET, will also assess the role of prasugrel monotherapy, with peri-PCI aspirin only instead of short-term aspirin in other studies. Moreover, in the setting of prolonged antiplatelet therapy after a standard DAPT, SMART-CHOICE II, OPT-BIRISK, and SMART-CHOICE III trials will assess different long-term P2Y$_{12}$ monotherapy regimens vs. DAPT or ASA monotherapy.

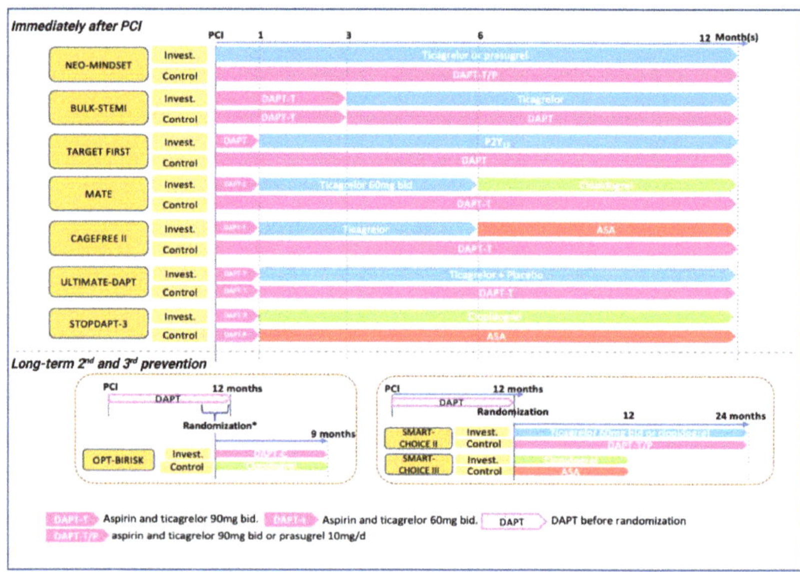

Figure 3. Ongoing randomized controlled trials of P2Y$_{12}$ inhibitor monotherapy in patients treated with PCI. ASA, aspirin; DAPT, dual antiplatelet therapy; DAPT-C, clopidogrel-based antiplatelet therapy; DAPT-T, ticagrelor-based dual antiplatelet therapy; DAPT-T/P, ticagrelor-based or prasugrel-based dual antiplatelet therapy; Invest., investigational group; PCI, percutaneous coronary intervention. * OPT-BIRISK trial is randomizing patients with high ischemic or bleeding risk who already finished 9–12 months of DAPT.

Table 3. Ongoing clinical trials for P2Y$_{12}$ inhibitor monotherapy in patients undergoing PCI.

Studies	Design	Population	Experimental Group	Control Group	Primary Outcome	Key Secondary Outcomes
RCTs immediately after PCI						
NEO-MINDSET (n = 3400) (NCT04360720)	Open-label RCT 12 months follow-up	ACS	Ticagrelor or prasugrel monotherapy	ASA + ticagrelor or prasugrel	Ischemic: all-cause death, cerebrovascular accident, MI or urgent target vessel revascularization Bleeding: BARC type 2, 3 or 5	Stent thrombosis BARC 1–5 bleeding Cost-effectiveness ratio
ULTIMATE-DAPT (n = 3486) (NCT03971500)	Placebo-controlled RCT 12 months follow-up	No MACCE or major bleeding within 30 days	Ticagrelor and placebo for 11 months	ASA + ticagrelor for 11 months	MACCE, clinical-relevant bleeding (BARC $\geq$ 2), target vessel failure	Net adverse clinical events
STOPDAPT-3 (n = 3110) (NCT04609111)	Open-label RCT 12 months follow-up	Patients with HBR or ACS	ASA + prasugrel for 1 month followed by clopidogrel monotherapy 11 months	ASA + prasugrel 1 month, ASA monotherapy 11 months	BARC 3 or 5 bleeding; cardiovascular composite (cardiovascular death, MI, ischemic stroke, definite stent thrombosis)	Target lesion/vessel failure and revascularization
BULK-STEMI (n = 1002) (NCT04570345)	Open-label RCT 12 months follow-up	STEMI	Ticagrelor monotherapy after 3 months of DAPT (ASA + ticagrelor)	ASA + P2Y$_{12}$ inhibitor after 3 months of DAPT (ASA + ticagrelor)	MACCE (all-cause death, MI, cerebrovascular event, stent thrombosis) and bleeding events (BARC 3 or 5)	
TARGET FIRST (n = 2246) (NCT04753749)	Open-label RCT 12 months follow-up	NSTEMI or STEMI with complete revascularization	P2Y$_{12}$ monotherapy after 1 month of DAPT	12 months of DAPT	All-cause death, non-fatal MI, stent thrombosis, stroke, or bleeding events (BARC 3 or 5)	

Table 3. *Cont.*

Studies	Design	Population	Experimental Group	Control Group	Primary Outcome	Key Secondary Outcomes
MATE (*n* = 2856) (NCT04937699)	Open-label RCT 12 months follow-up	ACS and high bleeding risk	ASA + ticagrelor (60 mg bid) for 1 month → ticagrelor monotherapy (60 mg bid) for 5 months → clopidogrel monotherapy for 6 months	ASA+ ticagrelor	All-cause death, non-fatal MI, stroke, BARC type 2, 3 or 5 bleeding	
CAGEFREE II (*n* = 1908) (NCT04971356)	Open-label RCT 12 months follow-up	ACS treated with drug-coated balloon	ASA + ticagrelor for 1 month → ticagrelor monotherapy for 5 months → ASA monotherapy for 6 months	ASA + ticagrelor	All-cause death, stroke, MI, revascularization, BARC 3 or 5 bleeding	Stent thrombosis rates
Non-randomized single-arm study						
PIONEER IV CHINA (*n* = 285) (NCT05015699)	Open-label single arm 12 months follow-up	PCI with HT supreme DES	Ticagrelor monotherapy after 1 month of DAPT	None	All-cause death, stroke, MI, coronary revascularization	
ASET–JAPAN (*n* = 400) (NCT05117866)	Open-label single arm 3 months follow-up for CCS, 12 months for ACS	NSTE–ACS and CCS	Prasugrel (loading: 20 mg; maintenance: 3.75 mg/d) 3 months in CCS and 12 months in NSTE–ACS	None	Ischemic: cardiac death, target-vessel MI, definite stent thrombosis Bleeding: BARC 3 or 5 bleeding	
Long-term 2nd and 3rd prevention						
OPT–BIRISK (*n* = 7700) (NCT03431142)	Open-label RCT 9 months follow-up	ACS patients received 9–12 months of DAPT with high ischemic or bleeding risk	Clopidogrel for 9 months	ASA + clopidogrel for 9 months	BARC type 2–5 bleeding	MACCE
SMART–CHOICE II (*n* = 1520) (NCT03119012)	Open-label RCT 36 months follow-up after index procedure	No major MACCE at 12 month after BRS implantation	Clopidogrel or ticagrelor (60 mg bid) monotherapy for 24 months	ASA + clopidogrel or ticagrelor (60 mg bid)	Death, MI, cerebrovascular events	BARC 2, 3, 5 bleeding Revascularization Stent thrombosis
SMART–CHOICE III (*n* = 5000) (NCT04418479)	Open-label RCT 12 months follow-up	Patient finished 12 months of DAPT with high risk of recurrent ischemic events	Clopidogrel monotherapy	ASA monotherapy	MACCE	BARC 3/5 bleeding

ACS, acute coronary syndrome; ASA, aspirin; BARC, Bleed Academic Research Consortium; BRS, Bioresorbable scaffold; DAPT, dual antiplatelet therapy; HBR, high bleeding risk; MACCE, major adverse cardiac and cerebrovascular events; MI, myocardial infarction; RCT, randomized controlled trial; STEMI, ST elevation myocardial infarction; NSTEMI, non-ST-elevation myocardial infarction. The dosages without specific notes are: aspirin, 81–100 mg daily; ticagrelor, 90 mg twice daily; prasugrel, 10 mg daily.

8. Gaps in Evidence

There are still several gaps in the knowledge that require further research. First, five out of seven trials studying $P2Y_{12}$ monotherapy enrolled exclusively East Asian populations, who have lower ischemic risk and a higher tendency of serious bleeding than Caucasians (i.e., East Asian Paradox), limiting extrapolation of many of the study findings to other ethnicities [82]. Second, as a potent $P2Y_{12}$ inhibitor, compared to ticagrelor, prasugrel has advantages including its once daily regimen and the less respiratory side effect, which greatly improves adherence. However, there are no dedicated RCTs of prasugrel monotherapy. Third, although HBR patients could benefit more from $P2Y_{12}$ monotherapy as a bleeding reduction strategy, there are no dedicated RCTs in HBR patients and the current evidence is derived from post-hoc analysis. Fourth, four out seven trials used clopidogrel as the main $P2Y_{12}$ inhibitor, platelet function testing or CYP2C19 genotyping to assess the probability of HPR was not performed in any of these trials and it is unclear if adverse events could be related to clopidogrel poor responders [41,83]. Ultimately, $P2Y_{12}$ monotherapy has been mainly compared with standard DAPT regimens and it is unknown how this strategy compares with other bleeding avoidance strategies, including short

DAPT with discontinuation of $P2Y_{12}$ inhibitor and maintaining aspirin or de-escalation DAPT approaches (e.g., switching from ticagrelor/prasugrel to clopidogrel or reducing the dose of ticagrelor/prasugrel) [84]. The current gaps in knowledge and ongoing trials are summarized in Table 4.

Table 4. Current gaps in the evidence and potential research opportunities in the $P2Y_{12}$ monotherapy.

Current Gaps	Ongoing Studies and Potential Research Opportunities
Population: • Most recent clopidogrel monotherapy trials exclusively recruited Asian population, known to have different thrombotic and hemorrhaging profiles, thus limiting their external validity in western populations	
Clinical presentation: • For ACS patients, data are controversial. In particular, the role of clopidogrel monotherapy. • STEMI-focused trials are still needed	• OPT–BIRISK, NEO–MINDSET, STOPDAPT-3, MATE, CAGEFREE II exclusively for ACS patients • BULK–STEMI, TARGET FIRST use STEMI as a major inclusion criterion
Specific conditions: • Studies on HBR patients are missing • Dedicated trials assessing treatment for patients with on-treatment HPR are missing. • Platelet function testing or CYP2C19 genotyping were not performed in clopidogrel trials	• STOPDAPT-3 and MATE study HBR as inclusion criteria • HPR-focused studies are warranted with deliciated platelet function test • CYP2C19 genotyping needs to be performed in future clopidogrel trials
Specific medications: • Data with prasugrel monotherapy is limited	• NEO–MINDSET, ASET–JAPAN will include prasugrel monotherapy
Comparison with other strategies: • It is unknown if $P2Y_{12}$ monotherapy provides a significant benefit compared to other bleeding avoidance strategies (i.e., de-escalation or abbreviated DAPT regimens)	• Dedicated RCTs are needed to compare clinical outcomes between patients treated with $P2Y_{12}$ monotherapy vs. other bleeding avoidance strategies

ACS, acute coronary syndrome; DAPT, dual antiplatelet therapy; HBR, high bleeding risk; HPR, high platelet reactivity; RCT, randomized controlled trial; STEMI, ST elevation myocardial infarction; NSTEMI.

9. Practical Implications

The $P2Y_{12}$ monotherapy is an emerging strategy to be considered among the available bleeding avoidance strategies in selected patients taking into consideration the following. First, the safety and efficacy of monotherapy outside of RCTs are very limited, underscoring that the eligible patients are those who meet the specific selection criteria of the RCTs [85]. It should be underscored that these trials are heterogeneous in terms of enrolled populations (Western countries vs. East Asian countries) which could impact the thrombotic and bleeding risk profiles of the studied populations. Furthermore, previous studies have shown that different bleeding avoidance strategies (i.e., abbreviated DAPT vs. de-escalation) are associated with different impact on clinical outcomes, suggesting that the selected strategy should be tailored according to patient characteristics and desired outcomes [84]. Moreover, procedural characteristics could also raise the concern about the outcomes in patients treated with complex PCI. Nevertheless, post-hoc analyses of these trials have not shown impaired outcomes among patients treated with complex PCI [86]. Second, the clinical presentation and the selected $P2Y_{12}$ inhibitor appear to impact outcomes. In particular, prasugrel, and ticagrelor are recommended over clopidogrel in patients with ACS. In the GLOBAL LEADERS, TWILIGHT, and TICO trials, patients with ACS treated with ticagrelor monotherapy reduced bleeding without affecting ischemic outcomes. However, in patients with ACS and clopidogrel monotherapy, the STOPDAPT-2 ACS trial showed reduced bleeding but increased ischemic events [43]. On the other hand, in CCS, clopidogrel appears to be a safe and effective drug, as shown in the SMART-CHOICE and STOPDAPT-2 trials [39,42]. Moreover, ticagrelor can also be an option in CCS with high ischemic risk as reported in the TWILIGHT trial [62]. Third, most of these trials were designed with run-in phases and randomized only event-free patients after a short course of DAPT (i.e., 1–3 months). Therefore, in daily clinical practice, the decision to drop aspirin and continue $P2Y_{12}$ inhibitor monotherapy should be made according to these protocols. Ultimately, $P2Y_{12}$ inhibitor monotherapy has been compared mainly with standard DAPT (i.e., guideline-recommended duration) up to one year after the index PCI or randomization. Therefore, the clinical benefit of $P2Y_{12}$ inhibitor monotherapy compared to other DAPT regimens and beyond the following 12–15 months of PCI is uncertain. Nevertheless, the only recent piece of information about $P2Y_{12}$ monotherapy for long-term 24 months in event-free patients who were on DAPT for 6–18 months after PCI) comes from the HOST-EXAM trial, which suggests that clopidogrel monotherapy is safe and effective strategy compared to aspirin monotherapy [79].

10. Conclusions

Although DAPT with aspirin and a $P2Y_{12}$ inhibitor is the standard care and guideline-recommended strategy in patients treated with PCI, recent pharmacodynamic studies have shown limited synergistic effects of aspirin in addition to potent oral $P2Y_{12}$ inhibitors and have challenged the need for DAPT to achieve optimal platelet inhibition. In fact, while DAPT is associated with a reduction in ischemic events, it also increases bleeding, the risk of which is proportional to the intensity and duration of DAPT. As thrombotic complications mostly occur early after PCI, while bleeding accrues over the time, bleeding reduction strategies have been developed so that enhanced antithrombotic effects are present in the early phases post-PCI end then reduced afterwards. To this extent, several RCTs have assessed the role of $P2Y_{12}$ inhibitor monotherapy compared to a standard DAPT regimen. Overall, $P2Y_{12}$ inhibitor monotherapy is safe and effective for reducing bleeding without compromising ischemic outcomes in event-free patients treated with PCI after a short course of DAPT. In particular, ticagrelor has shown optimal results in patients with ACS, whereas clopidogrel and ticagrelor have been safe and effective for preventing recurrent events in CCS. The $P2Y_{12}$ inhibitor monotherapy has already been incorporated in European and American guidelines as a reasonable antiplatelet strategy in patients treated with PCI. Over ten RCTs are ongoing to confirm previous findings and provide new insights $P2Y_{12}$ inhibitor monotherapy immediately after PCI, the role of prasugrel, and outcomes in patients with STEMI. Ultimately, ongoing research is warranted to define whether $P2Y_{12}$ inhibitor monotherapy should be preferred over aspirin for long-term secondary prevention in patients with CCS.

Supplementary Materials: The following supporting information can be downloaded at: https://www.mdpi.com/article/10.3390/jcdd9100340/s1, Table S1: Prespecified and selected post-hoc analyses of GLOBAL LEADERS trial; Table S2: Prespecified and selected post-hoc analyses of TWILIGHT trial; Table S3: Prespecified and selected post-hoc analyses of TICO trial.

Author Contributions: X.Z.: writing—original draft preparation, L.O.-P. and D.J.A.: writing—review and editing. All authors have read and agreed to the published version of the manuscript.

Funding: This research received no external funding.

Institutional Review Board Statement: Not applicable.

Informed Consent Statement: Not applicable.

Data Availability Statement: Not applicable.

Conflicts of Interest: Angiolillo declares that he has received consulting fees or honoraria from Abbott, Amgen, AstraZeneca, Bayer, Biosensors, Boehringer Ingelheim, Bristol-Myers Squibb, Chiesi, Daiichi-Sankyo, Eli Lilly, Haemonetics, Janssen, Merck, Novartis, PhaseBio, PLx Pharma, Pfizer, and

Sanofi; D.J.A. also declares that his institution has received research grants from Amgen, AstraZeneca, Bayer, Biosensors, CeloNova, CSL Behring, Daiichi-Sankyo, Eisai, Eli Lilly, Gilead, Idorsia, Janssen, Matsutani Chemical Industry Co., Merck, Novartis, and the Scott R. MacKenzie Foundation. Other authors have nothing to declare.

Abbreviations

ACC	American College of Cardiology
ACS	acute coronary syndrome
ADP	adenosine diphosphate
AHA	American Heart Association
ARC	Academic Research Consortium
BARC	bleeding academic research consortium
BRS	bioresorbable scaffold
CAD	coronary artery disease
CCS	chronic coronary syndrome
CI	confidence interval
COX-1	cyclooxygenase-1
CV	cardiovascular
CYP2C19	hepatic cytochrome P450 2C19
DAPT	dual antiplatelet therapy
DAPT-C	clopidogrel-based dual antiplatelet therapy
DAPT-T	ticagrelor-based dual antiplatelet therapy
DAPT-T/P	ticagrelor-based or prasugrel-based dual antiplatelet therapy
DM	diabetes mellitus
ESC	European Society of Cardiology
GI	gastrointestinal
HBR	high bleeding risk
HPR	high platelet reactivity
HR	hazard ratio
MACCE	major adverse cardiac and cerebrovascular events
MI	myocardial infarction
NACE	net adverse clinical event
NSTE-ACS	non-ST-elevation acute coronary artery syndrome
PCI	percutaneous coronary intervention
PFT	platelet function test
POCE	patient-oriented composite endpoints
Pt-EES	platinum-chromium everolimus-eluting stent
RCT	randomized controlled trial
SCAI	Society for Cardiovascular Angiography and Interventions
SIHD	stable ischemic heart disease
STEMI	ST elevation myocardial infarction
TIMI	Thrombolysis in Myocardial Infarction

References

1. Neumann, F.J.; Sousa-Uva, M.; Ahlsson, A.; Alfonso, F.; Banning, A.P.; Benedetto, U.; Byrne, R.A.; Collet, J.P.; Falk, V.; Head, S.J.; et al. 2018 ESC/EACTS Guidelines on myocardial revascularization. *Eur. Heart J.* **2019**, *40*, 87–165. [CrossRef] [PubMed]
2. Lawton, J.S.; Tamis-Holland, J.E.; Bangalore, S.; Bates, E.R.; Beckie, T.M.; Bischoff, J.M.; Bittl, J.A.; Cohen, M.G.; DiMaio, J.M.; Don, C.W.; et al. 2021 ACC/AHA/SCAI Guideline for Coronary Artery Revascularization: Executive Summary: A Report of the American College of Cardiology/American Heart Association Joint Committee on Clinical Practice Guidelines. *Circulation* **2022**, *145*, e4–e17. [CrossRef]
3. Alkhouli, M.; Alqahtani, F.; Kalra, A.; Gafoor, S.; Alhajji, M.; Alreshidan, M.; Holmes, D.R.; Holmes, D.R. Trends in Characteristics and Outcomes of Patients Undergoing Coronary Revascularization in the United States, 2003–2016. *JAMA Netw. Open* **2020**, *3*, e1921326. [CrossRef] [PubMed]
4. Mauri, L.; Kereiakes, D.J.; Yeh, R.W.; Driscoll-Shempp, P.; Cutlip, D.E.; Steg, P.G.; Normand, S.-L.T.; Braunwald, E.; Wiviott, S.D.; Cohen, D.J.; et al. Twelve or 30 Months of Dual Antiplatelet Therapy after Drug-Eluting Stents. *N. Engl. J. Med.* **2014**, *371*, 2155–2166. [CrossRef]

5. Brugaletta, S.; Gomez-Lara, J.; Ortega-Paz, L.; Jimenez-Diaz, V.; Jimenez, M.; Jiménez-Quevedo, P.; Diletti, R.; Mainar, V.; Campo, G.; Silvestro, A.; et al. 10-Year Follow-Up of Patients with Everolimus-Eluting Versus Bare-Metal Stents After ST-Segment Elevation Myocardial Infarction. *J. Am. Coll. Cardiol.* **2021**, *77*, 1165–1178. [CrossRef]
6. Coughlan, J.J.; Maeng, M.; Räber, L.; Brugaletta, S.; Aytekin, A.; Jensen, L.O.; Bär, S.; Ortega-Paz, L.; Laugwitz, K.-L.; Madsen, M.; et al. Ten-year patterns of stent thrombosis after percutaneous coronary intervention with new- versus early-generation drug-eluting stents: Insights from the DECADE cooperation. *Rev. Esp. Cardiol.* **2022**. [CrossRef] [PubMed]
7. Valgimigli, M.; Bueno, H.; Byrne, R.A.; Collet, J.-P.; Costa, F.; Jeppsson, A.; Jüni, P.; Kastrati, A.; Kolh, P.; Mauri, L.; et al. 2017 ESC focused update on dual antiplatelet therapy in coronary artery disease developed in collaboration with EACTS: The Task Force for dual antiplatelet therapy in coronary artery disease of the European Society of Cardiology (ESC) and of the European Association for Cardio-Thoracic Surgery (EACTS). *Eur. Heart J.* **2018**, *39*, 213–260.
8. Knuuti, J.; Revenco, V. 2019 ESC Guidelines for the diagnosis and management of chronic coronary syndromes. *Eur. Heart J.* **2020**, *41*, 407–477. [CrossRef]
9. Collet, J.-P.; Thiele, H.; Barbato, E.; Barthélémy, O.; Bauersachs, J.; Bhatt, D.L.; Dendale, P.; Dorobantu, M.; Edvardsen, T.; Folliguet, T.; et al. 2020 ESC Guidelines for the management of acute coronary syndromes in patients presenting without persistent ST-segment elevation. *Eur. Heart J.* **2021**, *42*, 1289–1367. [CrossRef] [PubMed]
10. Evidence Review Committee Members; Bittl, J.A.; Baber, U.; Bradley, S.M.; Wijeysundera, D.N. Duration of Dual Antiplatelet Therapy: A Systematic Review for the 2016 ACC/AHA Guideline Focused Update on Duration of Dual Antiplatelet Therapy in Patients with Coronary Artery Disease: A Report of the American College of Cardiology/American Heart Association Task Force on Clinical Practice Guidelines. *J. Am. Coll. Cardiol.* **2016**, *68*, 1116–1139.
11. Angiolillo, D.J.; Galli, M.; Collet, J.P.; Kastrati, A.; O'Donoghue, M.L. Antiplatelet therapy after percutaneous coronary intervention. *EuroIntervention* **2022**, *17*, e1371–e1396. [PubMed]
12. Angiolillo, D.J.; Fernandez-Ortiz, A.; Bernardo, E.; Alfonso, F.; Macaya, C.; Bass, T.A.; Costa, M.A. Variability in Individual Responsiveness to Clopidogrel: Clinical Implications, Management, and Future Perspectives. *J. Am. Coll. Cardiol.* **2007**, *49*, 1505–1516. [CrossRef] [PubMed]
13. Franchi, F.; Angiolillo, D.J. Novel antiplatelet agents in acute coronary syndrome. *Nat. Rev. Cardiol.* **2014**, *12*, 30–47. [CrossRef] [PubMed]
14. Sibbing, D.; Aradi, D.; Alexopoulos, D.; Berg, J.T.; Bhatt, D.L.; Bonello, L.; Collet, J.-P.; Cuisset, T.; Franchi, F.; Gross, L.; et al. Updated Expert Consensus Statement on Platelet Function and Genetic Testing for Guiding $P2Y_{12}$ Receptor Inhibitor Treatment in Percutaneous Coronary Intervention. *JACC: Cardiovasc. Interv.* **2019**, *12*, 1521–1537. [CrossRef]
15. Ibanez, B.; James, S.; Agewall, S.; Antunes, M.J.; Bucciarelli-Ducci, C.; Bueno, H.; Caforio, A.L.P.; Crea, F.; Goudevenos, J.A.; Halvorsen, S. 2017 ESC Guidelines for the management of acute myocardial infarction in patients presenting with ST-segment elevation: The Task Force for the management of acute myocardial infarction in patients presenting with ST-segment elevation of the European Society of Cardiology (ESC). *Eur. Heart J.* **2018**, *39*, 119–177.
16. Capodanno, D.; Bhatt, D.L.; Gibson, C.M.; James, S.; Kimura, T.; Mehran, R.; Rao, S.V.; Steg, P.G.; Urban, P.; Valgimigli, M.; et al. Bleeding avoidance strategies in percutaneous coronary intervention. *Nat. Rev. Cardiol.* **2021**, *19*, 117–132. [CrossRef]
17. Cao, D.; Chandiramani, R.; Chiarito, M.; E Claessen, B.; Mehran, R. Evolution of antithrombotic therapy in patients undergoing percutaneous coronary intervention: A 40-year journey. *Eur. Heart J.* **2020**, *42*, 339–351. [CrossRef]
18. Urban, P.; Mehran, R.; Colleran, R.; Angiolillo, D.J.; A Byrne, R.; Capodanno, D.; Cuisset, T.; Cutlip, D.; Eerdmans, P.; Eikelboom, J.; et al. Defining high bleeding risk in patients undergoing percutaneous coronary intervention: A consensus document from the Academic Research Consortium for High Bleeding Risk. *Eur. Heart J.* **2019**, *40*, 2632–2653. [CrossRef]
19. Galli, M.; Capodanno, D.; Andreotti, F.; Crea, F.; Angiolillo, D.J. Safety and efficacy of $P2Y_{12}$ inhibitor monotherapy in patients undergoing percutaneous coronary interventions. *Expert Opin. Drug Saf.* **2020**, *20*, 9–21. [CrossRef]
20. Capodanno, D.; Baber, U.; Bhatt, D.L.; Collet, J.-P.; Dangas, G.; Franchi, F.; Gibson, C.M.; Gwon, H.-C.; Kastrati, A.; Kimura, T.; et al. $P2Y_{12}$ inhibitor monotherapy in patients undergoing percutaneous coronary intervention. *Nat. Rev. Cardiol.* **2022**, 1–16. [CrossRef]
21. Angiolillo, D.J.; Ueno, M.; Goto, S. Basic Principles of Platelet Biology and Clinical Implications. *Circ. J.* **2010**, *74*, 597–607. [CrossRef] [PubMed]
22. Storey, R.F.; Newby, L.J.; Heptinstall, S. Effects of P2Y 1 and P2Y 12 receptor antagonists on platelet aggregation induced by different agonists in human whole blood. *Platelets* **2001**, *12*, 443–447. [CrossRef] [PubMed]
23. Ariotti, S.; van Leeuwen, M.; Brugaletta, S.; Leonardi, S.; Akkerhuis, K.M.; Rimoldi, S.F.; Janssens, G.N.; Ortega-Paz, L.; Gianni, U.; Berge, J.C.V.D.; et al. Effects of Ticagrelor, Prasugrel, or Clopidogrel at Steady State on Endothelial Function. *J. Am. Coll. Cardiol.* **2018**, *71*, 1289–1291. [CrossRef] [PubMed]
24. Ariotti, S.; Ortega-Paz, L.; van Leeuwen, M.; Brugaletta, S.; Leonardi, S.; Akkerhuis, K.M.; Rimoldi, S.F.; Janssens, G.; Gianni, U.; van den Berge, J.C. Effects of Ticagrelor, Prasugrel, or Clopidogrel on Endothelial Function and Other Vascular Biomarkers: A Randomized Crossover Study. *JACC Cardiovasc. Interv.* **2018**, *11*, 1576–1586. [CrossRef]
25. Wallentin, L.; Becker, R.C.; Budaj, A.; Cannon, C.P.; Emanuelsson, H.; Held, C.; Horrow, J.; Husted, S.; James, S.; Katus, H.; et al. Ticagrelor versus Clopidogrel in Patients with Acute Coronary Syndromes. *N. Engl. J. Med.* **2009**, *361*, 1045–1057. [CrossRef]

26. Wiviott, S.D.; Braunwald, E.; McCabe, C.H.; Montalescot, G.; Ruzyllo, W.; Gottlieb, S.; Neumann, F.-J.; Ardissino, D.; De Servi, S.; Murphy, S.A.; et al. Prasugrel versus Clopidogrel in Patients with Acute Coronary Syndromes. *N. Engl. J. Med.* **2007**, *357*, 2001–2015. [CrossRef]
27. Payne, D.; Hayes, P.; Jones, C.; Belham, P.; Naylor, A.; Goodall, A. Combined therapy with clopidogrel and aspirin significantly increases the bleeding time through a synergistic antiplatelet action. *J. Vasc. Surg.* **2002**, *35*, 1204–1209. [CrossRef]
28. Capodanno, D.; Angiolillo, D.J. When Less Becomes More: Insights on the Pharmacodynamic Effects of Aspirin Withdrawal in Patients with Potent Platelet P2Y $_{12}$ Inhibition Induced by Ticagrelor. *J. Am. Heart Assoc.* **2020**, *9*, e019432. [CrossRef]
29. Armstrong, P.C.J.; Leadbeater, P.D.; Chan, M.V.; Kirkby, N.S.; Jakubowski, J.A.; Mitchell, J.A.; Warner, T.D. In the presence of strong P2Y$_{12}$ receptor blockade, aspirin provides little additional inhibition of platelet aggregation. *J. Thromb. Haemost.* **2010**, *9*, 552–561. [CrossRef]
30. Johnson, T.W.; Baos, S.; Collett, L.; Hutchinson, J.L.; Nkau, M.; Molina, M.; Aungraheeta, R.; Reilly-Stitt, C.; Bowles, R.; Reeves, B.C.; et al. Pharmacodynamic Comparison of Ticagrelor Monotherapy Versus Ticagrelor and Aspirin in Patients After Percutaneous Coronary Intervention: The TEMPLATE (Ticagrelor Monotherapy and Platelet Reactivity) Randomized Controlled Trial. *J. Am. Heart Assoc.* **2020**, *9*, e016495. [CrossRef]
31. Baber, U.; Zafar, M.U.; Dangas, G.; Escolar, G.; Angiolillo, D.J.; Sharma, S.K.; Kini, A.S.; Sartori, S.; Joyce, L.; Vogel, B.; et al. Ticagrelor with or Without Aspirin After PCI: The TWILIGHT Platelet Substudy. *J. Am. Coll. Cardiol.* **2020**, *75*, 578–586. [CrossRef]
32. Hennigan, B.W.; Good, R.; Adamson, C.; Parker, W.A.; Martin, L.; Anderson, L.; Campbell, M.; Serruys, P.W.; Storey, R.F.; Oldroyd, K.G. Recovery of platelet reactivity following cessation of either aspirin or ticagrelor in patients treated with dual antiplatelet therapy following percutaneous coronary intervention: A GLOBAL LEADERS substudy. *Platelets* **2020**, *33*, 141–146. [CrossRef] [PubMed]
33. Franchi, F.; Rollini, F.; Faz, G.; Rivas, J.R.; Rivas, A.; Agarwal, M.; Briceno, M.; Wali, M.; Nawaz, A.; Silva, G.; et al. Pharmacodynamic Effects of Vorapaxar in Prior Myocardial Infarction Patients Treated With Potent Oral P2Y $_{12}$ Receptor Inhibitors With and Without Aspirin: Results of the VORA-PRATIC Study. *J. Am. Heart Assoc.* **2020**, *9*, e015865. [CrossRef] [PubMed]
34. Li, Z.; Wang, Z.; Shen, B.; Chen, C.; Ding, X.; Song, H. Effects of aspirin on the gastrointestinal tract: Pros vs. cons. *Oncol. Lett.* **2020**, *20*, 2567–2578. [CrossRef]
35. Lavie, C.J.; Howden, C.W.; Scheiman, J.; Tursi, J. Upper Gastrointestinal Toxicity Associated with Long-Term Aspirin Therapy: Consequences and Prevention. *Curr. Probl. Cardiol.* **2017**, *42*, 146–164. [CrossRef]
36. Angiolillo, D.J.; Prats, J.; Deliargyris, E.N.; Schneider, D.J.; Scheiman, J.; Kimmelstiel, C.; Steg, P.G.; Alberts, M.; Rosengart, T.; Mehran, R.; et al. Pharmacokinetic and Pharmacodynamic Profile of a Novel Phospholipid Aspirin Formulation. *Clin. Pharmacokinet.* **2022**, *61*, 465–479. [CrossRef]
37. Moon, J.Y.; Franchi, F.; Rollini, F.; Angiolillo, D.J. Evolution of Coronary Stent Technology and Implications for Duration of Dual Antiplatelet Therapy. *Prog. Cardiovasc. Dis.* **2018**, *60*, 478–490. [CrossRef] [PubMed]
38. Hahn, J.Y.; Song, Y.B.; Oh, J.H.; Chun, W.J.; Park, Y.H.; Jang, W.J.; Im, E.S.; Jeong, J.O.; Koh, Y.Y.; Yun, K.H. Effect of P2Y$_{12}$ Inhibitor Monotherapy vs Dual Antiplatelet Therapy on Cardiovascular Events in Patients Undergoing Percutaneous Coronary Intervention: The SMART-CHOICE Randomized Clinical Trial. *JAMA* **2019**, *321*, 2428–2437. [CrossRef]
39. Kim, J.; Jang, W.J.; Lee, W.S.; Choi, K.H.; Lee, J.M.; Park, T.K.; Yang, J.H.; Choi, J.-H.; Bin Song, Y.; Choi, S.-H.; et al. P2Y$_{12}$ inhibitor monotherapy after coronary stenting according to type of P2Y$_{12}$ inhibitor. *Heart* **2021**, *107*, 1077–1083. [CrossRef] [PubMed]
40. Lee, S.H.; Lee, S.Y.; Chun, W.J.; Song, Y.B.; Choi, S.H.; Jeong, J.O.; Oh, S.K.; Yun, K.H.; Koh, Y.Y.; Koh, Y.Y.; et al. Clopidogrel Monotherapy in Patients with and without On-Treatment High Platelet Reactivity: A SMART-CHOICE sub-study. *EuroIntervention* **2021**. [CrossRef] [PubMed]
41. Angiolillo, D.J.; Capodanno, D.; Danchin, N.; Simon, T.; Bergmeijer, T.O.; Ten Berg, J.M.; Sibbing, D.; Price, M.J. Derivation, Validation, and Prognostic Utility of a Prediction Rule for Nonresponse to Clopidogrel: The ABCD-GENE Score. *JACC Cardiovasc Interv.* **2020**, *13*, 606–617. [CrossRef] [PubMed]
42. Watanabe, H.; Domei, T.; Morimoto, T.; Natsuaki, M.; Shiomi, H.; Toyota, T.; Ohya, M.; Suwa, S.; Takagi, K.; Nanasato, M.; et al. Effect of 1-Month Dual Antiplatelet Therapy Followed by Clopidogrel vs 12-Month Dual Antiplatelet Therapy on Cardiovascular and Bleeding Events in Patients Receiving PCI: The STOPDAPT-2 Randomized. *Clin. Trial JAMA* **2019**, *321*, 2414–2427. [CrossRef] [PubMed]
43. Watanabe, H.; Morimoto, T.; Natsuaki, M.; Yamamoto, K.; Obayashi, Y.; Ogita, M.; Suwa, S.; Isawa, T.; Domei, T.; Yamaji, K.; et al. Comparison of Clopidogrel Monotherapy After 1 to 2 Months of Dual Antiplatelet Therapy With 12 Months of Dual Antiplatelet Therapy in Patients with Acute Coronary Syndrome: The STOPDAPT-2 ACS Randomized Clinical Trial. *JAMA Cardiol.* **2022**, *7*, 407–417. [CrossRef] [PubMed]
44. Obayashi, Y.; Watanabe, H.; Morimoto, T.; Yamamoto, K.; Natsuaki, M.; Domei, T.; Yamaji, K.; Suwa, S.; Isawa, T.; Watanabe, H.; et al. Clopidogrel Monotherapy After 1-Month Dual Antiplatelet Therapy in Percutaneous Coronary Intervention: From the STOPDAPT-2 Total Cohort. *Circ. Cardiovasc. Interv.* **2022**, *15*, e012004. [CrossRef]
45. Kogame, N.; Guimaraes, P.O.; Modolo, R.; De Martino, F.; Tinoco, J.; Ribeiro, E.E.; Kawashima, H.; Ono, M.; Hara, H.; Wang, R.; et al. Aspirin-Free Prasugrel Monotherapy Following Coronary Artery Stenting in Patients With Stable CAD: The ASET Pilot Study. *JACC Cardiovasc. Interv.* **2020**, *13*, 2251–2262. [CrossRef] [PubMed]

46. Arora, S.; Shemisa, K.; Vaduganathan, M.; Qamar, A.; Gupta, A.; Garg, S.K.; Kumbhani, D.J.; Mayo, H.; Khalili, H.; Pandey, A.; et al. Premature Ticagrelor Discontinuation in Secondary Prevention of Atherosclerotic CVD: JACC Review Topic of the Week. *J. Am. Coll. Cardiol.* **2019**, *73*, 2454–2464. [CrossRef]
47. Ortega-Paz, L.; Brugaletta, S.; Ariotti, S.; Akkerhuis, K.M.; Karagiannis, A.; Windecker, S.; Valgimigli, M.; On behalf of the HI-TECH Investigators. Adenosine and Ticagrelor Plasma Levels in Patients with and Without Ticagrelor-Related Dyspnea. *Circulation* **2018**, *138*, 646–648. [CrossRef]
48. GUSTO Investigators. An international randomized trial comparing four thrombolytic strategies for acute myocardial infarction. *N. Eng. J. Med.* **1993**, *329*, 673–682. [CrossRef]
49. Gao, C.; Takahashi, K.; Garg, S.; Hara, H.; Wang, R.; Kawashima, H.; Ono, M.; Montalescot, G.; Haude, M.; Slagboom, T.; et al. Regional variation in patients and outcomes in the GLOBAL LEADERS trial. *Int. J. Cardiol.* **2020**, *324*, 30–37. [CrossRef]
50. Chichareon, P.; Modolo, R.; Kerkmeijer, L.; Tomaniak, M.; Kogame, N.; Takahashi, K.; Chang, C.C.; Komiyama, H.; Moccetti, T.; Talwar, S.; et al. Association of Sex with Outcomes in Patients Undergoing Percutaneous Coronary Intervention: A Subgroup Analysis of the GLOBAL LEADERS Randomized Clinical Trial. *JAMA Cardiol.* **2020**, *5*, 21–29. [CrossRef] [PubMed]
51. Tomaniak, M.; Chichareon, P.; Modolo, R.; Takahashi, K.; Chang, C.C.; Kogame, N.; Spitzer, E.; Buszman, P.E.; van Geuns, R.-J.M.; Valkov, V.; et al. Ticagrelor monotherapy beyond one month after PCI in ACS or stable CAD in elderly patients: A pre-specified analysis of the GLOBAL LEADERS trial. *EuroIntervention* **2020**, *15*, e1605–e1614. [CrossRef] [PubMed]
52. Ono, M.; Chichareon, P.; Tomaniak, M.; Kawashima, H.; Takahashi, K.; Kogame, N.; Modolo, R.; Hara, H.; Gao, C.; Wang, R.; et al. The association of body mass index with long-term clinical outcomes after ticagrelor monotherapy following abbreviated dual antiplatelet therapy in patients undergoing percutaneous coronary intervention: A prespecified sub-analysis of the GLOBAL LEADERS Trial. *Clin. Res. Cardiol.* **2020**, *109*, 1125–1139. [CrossRef] [PubMed]
53. Chichareon, P.; Modolo, R.; Kogame, N.; Takahashi, K.; Chang, C.-C.; Tomaniak, M.; Botelho, R.; Eeckhout, E.; Hofma, S.; Trendafilova-Lazarova, D.; et al. Association of diabetes with outcomes in patients undergoing contemporary percutaneous coronary intervention: Pre-specified subgroup analysis from the randomized GLOBAL LEADERS study. *Atherosclerosis* **2020**, *295*, 45–53. [CrossRef] [PubMed]
54. Tomaniak, M.; Chichareon, P.; Klimczak-Tomaniak, D.; Takahashi, K.; Kogame, N.; Modolo, R.; Wang, R.; Ono, M.; Hara, H.; Gao, C.; et al. Impact of renal function on clinical outcomes after PCI in ACS and stable CAD patients treated with ticagrelor: A prespecified analysis of the GLOBAL LEADERS randomized clinical trial. *Clin. Res. Cardiol.* **2020**, *109*, 930–943. [CrossRef]
55. Kogame, N.; Chichareon, P.; De Wilder, K.; Takahashi, K.; Modolo, R.; Chang, C.C.; Tomaniak, M.; Komiyama, H.; Chieffo, A.; Colombo, A.; et al. Clinical relevance of ticagrelor monotherapy following 1-month dual antiplatelet therapy after bifurcation percutaneous coronary intervention: Insight from GLOBAL LEADERS trial. *Catheter. Cardiovasc. Interv.* **2019**, *96*, 100–111. [CrossRef] [PubMed]
56. Gao, C.; Buszman, P.; Buszman, P.; Chichareon, P.; Modolo, R.; Garg, S.; Takahashi, K.; Kawashima, H.; Wang, R.; Chang, C.C.; et al. Influence of Bleeding Risk on Outcomes of Radial and Femoral Access for Percutaneous Coronary Intervention: An Analysis from the GLOBAL LEADERS Trial. *Can. J. Cardiol.* **2021**, *37*, 122–130. [CrossRef]
57. Franzone, A.; McFadden, E.P.; Leonardi, S.; Piccolo, R.; Vranckx, P.; Serruys, P.W.; Hamm, C.; Steg, P.G.; Heg, D.; Branca, M.; et al. Ticagrelor alone or conventional dual antiplatelet therapy in patients with stable or acute coronary syndromes. *EuroIntervention* **2020**, *16*, 627–633. [CrossRef]
58. Serruys, P.W.; Tomaniak, M.; Chichareon, P.; Modolo, R.; Kogame, N.; Takahashi, K.; Chang, C.C.; Spitzer, E.; Walsh, S.J.; Adlam, D.; et al. Patient-oriented composite endpoints and net adverse clinical events with ticagrelor monotherapy following percutaneous coronary intervention: Insights from the randomised GLOBAL LEADERS trial. *EuroIntervention* **2019**, *15*, e1090–e1098. [CrossRef]
59. Takahashi, K.; Chichareon, P.; Modolo, R.; Kogame, N.; Chang, C.C.; Tomaniak, M.; Moschovitis, A.; Curzen, N.; Haude, M.; Jung, W.; et al. Impact of ticagrelor monotherapy on two-year clinical outcomes in patients with long stenting: A post hoc analysis of the GLOBAL LEADERS trial. *EuroIntervention* **2020**, *16*, 634–644. [CrossRef]
60. Chichareon, P.; Onuma, Y.; van Klaveren, D.; Modolo, R.; Kogame, N.; Takahashi, K.; Chang, C.C.; Tomaniak, M.; Asano, T.; Katagiri, Y.; et al. Validation of the updated logistic clinical SYNTAX score for all-cause mortality in the GLOBAL LEADERS trial. *EuroIntervention* **2019**, *15*, e539–e546. [CrossRef] [PubMed]
61. Serruys, P.W.; Takahashi, K.; Chichareon, P.; Kogame, N.; Tomaniak, M.; Modolo, R.; Chang, C.C.; Komiyama, H.; Soliman, O.; Wykrzykowska, J.; et al. Impact of long-term ticagrelor monotherapy following 1-month dual antiplatelet therapy in patients who underwent complex percutaneous coronary intervention: Insights from the Global Leaders trial. *Eur. Heart J.* **2019**, *40*, 2595–2604. [CrossRef] [PubMed]
62. Mehran, R.; Baber, U.; Sharma, S.K.; Cohen, D.J.; Angiolillo, D.J.; Briguori, C.; Cha, J.Y.; Collier, T.; Dangas, G.; Dudek, D.; et al. Ticagrelor with or without Aspirin in High-Risk Patients after PCI. *N. Engl. J. Med.* **2019**, *381*, 2032–2042. [CrossRef]
63. Angiolillo, D.J.; Cao, D.; Baber, U.; Sartori, S.; Zhang, Z.; Dangas, G.; Mehta, S.; Briguori, C.; Cohen, D.J.; Collier, T.; et al. Impact of Age on the Safety and Efficacy of Ticagrelor Monotherapy in Patients Undergoing PCI. *JACC: Cardiovasc. Interv.* **2021**, *14*, 1434–1446. [CrossRef] [PubMed]

64. Vogel, B.; Baber, U.; Cohen, D.J.; Sartori, S.; Sharma, S.K.; Angiolillo, D.J.; Farhan, S.; Goel, R.; Zhang, Z.; Briguori, C.; et al. Sex Differences Among Patients with High Risk Receiving Ticagrelor with or Without Aspirin After Percutaneous Coronary Intervention: A Subgroup Analysis of the TWILIGHT Randomized Clinical Trial. *JAMA Cardiol.* **2021**, *6*, 1032–1041. [CrossRef]
65. Han, Y.; Claessen, B.E.; Chen, S.L.; Chunguang, Q.; Zhou, Y.; Xu, Y.; Hailong, L.; Chen, J.; Qiang, W.; Zhang, R.; et al. Ticagrelor with or Without Aspirin in Chinese Patients Undergoing Percutaneous Coronary Intervention: A TWILIGHT China Substudy. *Circ. Cardiovasc. Interv.* **2022**, *15*, e009495. [CrossRef]
66. Angiolillo, D.J.; Baber, U.; Sartori, S.; Briguori, C.; Dangas, G.; Cohen, D.J.; Mehta, S.R.; Gibson, C.M.; Chandiramani, R.; Huber, K.; et al. Ticagrelor with or Without Aspirin in High-Risk Patients with Diabetes Mellitus Undergoing Percutaneous Coronary Intervention. *J. Am. Coll. Cardiol.* **2020**, *75*, 2403–2413. [CrossRef]
67. Stefanini, G.G.; Briguori, C.; Cao, D.; Baber, U.; Sartori, S.; Zhang, Z.; Dangas, G.; Angiolillo, D.J.; Mehta, S.; Cohen, D.J.; et al. Ticagrelor monotherapy in patients with chronic kidney disease undergoing percutaneous coronary intervention: TWILIGHT-CKD. *Eur. Heart J.* **2021**, *42*, 4683–4693. [CrossRef] [PubMed]
68. Chiarito, M.; Baber, U.; Cao, D.; Sharma, S.K.; Dangas, G.; Angiolillo, D.J.; Briguori, C.; Cohen, D.J.; Dudek, D.; Džavík, V.; et al. Ticagrelor Monotherapy After PCI in High-Risk Patients With Prior MI: A Prespecified TWILIGHT Substudy. *JACC Cardiovasc. Interv.* **2022**, *15*, 282–293. [CrossRef] [PubMed]
69. Baber, U.; Dangas, G.; Angiolillo, D.J.; Cohen, D.J.; Sharma, S.K.; Nicolas, J.; Briguori, C.; Cha, J.Y.; Collier, T.; Dudek, D.; et al. Ticagrelor alone vs. ticagrelor plus aspirin following percutaneous coronary intervention in patients with non-ST-segment elevation acute coronary syndromes: TWILIGHT-ACS. *Eur. Heart J.* **2020**, *41*, 3533–3545. [CrossRef]
70. Dangas, G.; Baber, U.; Sharma, S.; Giustino, G.; Sartori, S.; Nicolas, J.; Goel, R.; Mehta, S.; Cohen, D.; Angiolillo, D.A.; et al. Safety and efficacy of ticagrelor monotherapy according to drug-eluting stent type: The TWILIGHT-STENT study. *EuroIntervention* **2022**, *17*, 1330–1339. [CrossRef] [PubMed]
71. Escaned, J.; Cao, D.; Baber, U.; Nicolas, J.; Sartori, S.; Zhang, Z.; Dangas, G.; Angiolillo, D.J.; Briguori, C.; Cohen, D.J.; et al. Ticagrelor monotherapy in patients at high bleeding risk undergoing percutaneous coronary intervention: TWILIGHT-HBR. *Eur. Heart J.* **2021**, *42*, 4624–4634. [CrossRef]
72. Kim, B.K.; Hong, S.J.; Cho, Y.H.; Yun, K.H.; Kim, Y.H.; Suh, Y.; Cho, J.Y.; Her, A.Y.; Cho, S.; Jeon, D.W.; et al. Effect of Ticagrelor Monotherapy vs Ticagrelor with Aspirin on Major Bleeding and Cardiovascular Events in Patients With Acute Coronary Syndrome: The TICO Randomized Clinical Trial. *JAMA* **2020**, *323*, 2407–2416. [CrossRef]
73. Yun, K.H.; Cho, J.Y.; Lee, S.Y.; Rhee, S.J.; Kim, B.K.; Hong, M.K.; Jang, Y.; Oh, S.K. The TICO Investigators Ischemic and Bleeding Events of Ticagrelor Monotherapy in Korean Patients with and Without Diabetes Mellitus: Insights From the TICO Trial. *Front. Pharmacol.* **2021**, *11*, 620906. [CrossRef] [PubMed]
74. Lee, S.-J.; Lee, Y.-J.; Kim, B.-K.; Hong, S.-J.; Ahn, C.-M.; Kim, J.-S.; Ko, Y.-G.; Choi, D.; Hong, M.-K.; Jang, Y. Ticagrelor Monotherapy Versus Ticagrelor with Aspirin in Acute Coronary Syndrome Patients with a High Risk of Ischemic Events. *Circ. Cardiovasc. Interv.* **2021**, *14*, e010812. [CrossRef]
75. Lee, S.-J.; Cho, J.Y.; Kim, B.-K.; Yun, K.H.; Suh, Y.; Cho, Y.-H.; Kim, Y.H.; Her, A.-Y.; Cho, S.; Jeon, D.W.; et al. Ticagrelor Monotherapy Versus Ticagrelor with Aspirin in Patients With ST-Segment Elevation Myocardial Infarction. *JACC Cardiovasc. Interv.* **2021**, *14*, 431–440. [CrossRef] [PubMed]
76. Lee, Y.-J.; Suh, Y.; Kim, J.-S.; Cho, Y.-H.; Yun, K.H.; Kim, Y.H.; Cho, J.Y.; Her, A.-Y.; Cho, S.; Jeon, D.W.; et al. Ticagrelor Monotherapy After 3-Month Dual Antiplatelet Therapy in Acute Coronary Syndrome by High Bleeding Risk: The Subanalysis From the TICO Trial. *Korean Circ. J.* **2022**, *52*, 324–337. [CrossRef]
77. Valgimigli, M.; Gragnano, F.; Branca, M.; Franzone, A.; Baber, U.; Jang, Y.; Kimura, T.; Hahn, J.-Y.; Zhao, Q.; Windecker, S.; et al. $P2Y_{12}$ inhibitor monotherapy or dual antiplatelet therapy after coronary revascularisation: Individual patient level meta-analysis of randomised controlled trials. *BMJ* **2021**, *373*, n1332. [CrossRef]
78. CAPRIE Steering Committee. A randomised, blinded, trial of clopidogrel versus aspirin in patients at risk of ischaemic events (CAPRIE). CAPRIE Steering Committee. *Lancet* **1996**, *348*, 1329–1339. [CrossRef]
79. Koo, B.-K.; Kang, J.; Park, K.W.; Rhee, T.-M.; Yang, H.-M.; Won, K.-B.; Rha, S.-W.; Bae, J.-W.; Lee, N.H.; Hur, S.-H.; et al. Aspirin versus clopidogrel for chronic maintenance monotherapy after percutaneous coronary intervention (HOST-EXAM): An investigator-initiated, prospective, randomised, open-label, multicentre trial. *Lancet* **2021**, *397*, 2487–2496. [CrossRef]
80. Ono, M.; Hara, H.; Kawashima, H.; Gao, C.; Wang, R.; Wykrzykowska, J.J.; Piek, J.J.; Garg, S.; Hamm, C.; Steg, P.G.; et al. Ticagrelor monotherapy versus aspirin monotherapy at 12 months after percutaneous coronary intervention: A landmark analysis of the GLOBAL LEADERS trial. *EuroIntervention* **2022**, *18*, e377–e388. [CrossRef] [PubMed]
81. Valgimigli, M. Patients with coronary artery disease should receive $P2Y_{12}$ inhibitor instead of aspirin. *Vol ESC Congress* **2022**, *2022*.
82. Kim, H.K.; Tantry, U.S.; Smith, S.C., Jr.; Jeong, M.H.; Park, S.-J.; Kim, M.H.; Lim, D.-S.; Shin, E.-S.; Park, D.-W.; Huo, Y.; et al. The East Asian Paradox: An Updated Position Statement on the Challenges to the Current Antithrombotic Strategy in Patients with Cardiovascular Disease. *Thromb. Haemost.* **2020**, *121*, 422–432. [CrossRef] [PubMed]
83. Galli, M.; Ortega-Paz, L.; Franchi, F.; Rollini, F.; Angiolillo, D.J. Precision medicine in interventional cardiology: Implications for antiplatelet therapy in patients undergoing percutaneous coronary intervention. *Pharmacogenomics* **2022**, *23*, 723–737. [CrossRef] [PubMed]

84. Laudani, C.; Greco, A.; Occhipinti, G.; Ingala, S.; Calderone, D.; Scalia, L.; Agnello, F.; Legnazzi, M.; Mauro, M.S.; Rochira, C.; et al. Short Duration of DAPT Versus De-Escalation After Percutaneous Coronary Intervention for Acute Coronary Syndromes. *JACC Cardiovasc. Interv.* **2022**, *15*, 268–277. [CrossRef] [PubMed]
85. Otsuka, T.; Ueki, Y.; Kavaliauskaite, R.; Zanchin, T.; Bär, S.; Stortecky, S.; Pilgrim, T.; Valgimigli, M.; Meier, B.; Heg, D.; et al. Single antiplatelet therapy with use of prasugrel in patients undergoing percutaneous coronary intervention. Catheterization and cardiovascular interventions. *Off. J. Soc. Card. Angiogr. Interv.* **2021**, *98*, E213–E221.
86. Ortega-Paz, L.; Angiolillo, D.J. Optimal antiplatelet therapy in patients at high bleeding risk undergoing complex percutaneous coronary intervention. *Eur. Heart J.* **2022**, *43*, 3115–3117. [CrossRef]

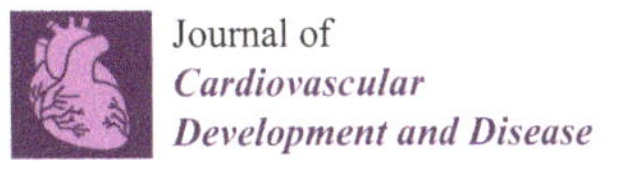
Journal of
Cardiovascular Development and Disease

MDPI

Review

Pharmacological Efficacy and Gastrointestinal Safety of Different Aspirin Formulations for Cardiovascular Prevention: A Narrative Review

Bianca Clerici [1,*] and Marco Cattaneo [2]

[1] Divisione di Medicina Generale II, ASST Santi Paolo e Carlo, Dipartimento di Scienze della Salute, Università degli Studi di Milano, Via Di Rudinì 8, 20142 Milano, Italy
[2] Fondazione Arianna Anticoagulazione, Via Paolo Fabbri 1/3, 40138 Bologna, Italy
* Correspondence: bianca.clerici@unimi.it

Abstract: Aspirin inhibits platelet function by irreversibly inhibiting the synthesis of thromboxane A2 (TxA2). Aspirin, at low doses, is widely used for cardiovascular prevention. Gastrointestinal discomfort, mucosal erosions/ulcerations and bleeding are frequent complications of chronic treatment. To reduce these adverse effects, different formulations of aspirin have been developed, including enteric-coated (EC) aspirin, the most widely used aspirin formulation. However, EC aspirin is less effective than plain aspirin in inhibiting TxA2 production, especially in subjects with high body weight. The inadequate pharmacological efficacy of EC aspirin is mirrored by lower protection from cardiovascular events in subjects weighing >70 kg. Endoscopic studies showed that EC aspirin causes fewer erosions of the gastric mucosa compared to plain aspirin (which is absorbed in the stomach) but causes mucosal erosions in the small intestine, where it is absorbed. Several studies demonstrated that EC aspirin does not reduce the incidence of clinically relevant gastrointestinal ulceration and bleeding. Similar results were found for buffered aspirin. Although interesting, the results of experiments on the phospholipid-aspirin complex PL2200 are still preliminary. Considering its favorable pharmacological profile, plain aspirin should be the preferred formulation to be used for cardiovascular prevention.

Keywords: aspirin; coronary artery disease; cerebrovascular disease; diabetes mellitus; essential thrombocythemia; platelet function; thromboxane; gastrointestinal bleeding; enteric-coated aspirin; cardiovascular prevention

Citation: Clerici, B.; Cattaneo, M. Pharmacological Efficacy and Gastrointestinal Safety of Different Aspirin Formulations for Cardiovascular Prevention: A Narrative Review. *J. Cardiovasc. Dev. Dis.* **2023**, *10*, 137. https://doi.org/10.3390/jcdd10040137

Academic Editor: Krzysztof J. Filipiak

Received: 23 February 2023
Revised: 18 March 2023
Accepted: 20 March 2023
Published: 23 March 2023

1. Background

Acetylsalicylic acid, the active principle of aspirin, irreversibly inhibits the activity of platelet cyclo-oxygenase-1 (COX-1), thereby inhibiting the platelet production of the pro-aggregatory and vasoconstrictor molecule thromboxane A2 (TxA2) [1,2]. Due to its inhibitory effect on platelet function, aspirin is widely used as an antithrombotic drug for the treatment of acute coronary syndromes and cerebrovascular accidents and for their secondary prevention; its role in the primary prevention of these disorders is less well established [3]. A common complication of chronic treatment with aspirin is the increased risk of gastrointestinal (GI) discomfort, mucosal erosions/ulcerations and bleeding, which are frequently observed despite the fact that prevention of thrombosis can be obtained by administering low-dose aspirin (75–100 mg o.d.) [2]. A placebo-controlled study showed that the incidence of bleeding peptic ulcers in subjects on cardiovascular prophylaxis with low-dose aspirin was 40–80% higher than in placebo-treated subjects [4], while a Danish cohort study of 27,694 individuals showed that the standardized incidence rate ratio of upper GI bleeding (UGIB) was 2.6 among users of low-dose aspirin [5]. A meta-analysis of 24 randomized clinical trials (RCTs) on the risk of GI hemorrhage with long term use (at least 1 year) of aspirin as an antiplatelet agent compared to placebo or no treatment showed that the pooled odds ratio for GI bleeding in 65,987 participants was 1.68 (95% CI, 1.51–1.88) [6]. As GI bleeding in survivors of myocardial infarction is independently associated with increased risk of death [adjusted hazard ratio 2.54 (95% CI, 1.66–3.89)] [7] its prevention is of outmost importance. Moreover, it is important to emphasize that chronic use of aspirin is associated not only with gastric complications but also with a variety of lesions in the small bowel, including multiple petechiae, loss of villi, erosions, and round, irregular, or punched-out ulcers [8]. With the aim of decreasing GI toxicity, different formulations of aspirin have been developed, including enteric-coated (EC) aspirin (tablets coated with cellulose, silicon, or other inactive ingredients) [9], buffered aspirin (tablets added with buffering agents) [10], and, more recently, PL2200 (a modified-release lipid-based aspirin) [11]. Among these formulations, EC aspirin has been thoroughly studied in terms of pharmacokinetics (PK) and pharmacodynamics (PD) and is the most widely used formulation for the prevention of arterial thrombotic events. Coating aspirin tablets prevents aspirin absorption in the stomach, thus hypothetically decreasing its GI toxicity, which was mostly attributed to the local effects of the drug. However, clear evidence that EC aspirin is safer than non-EC aspirin (which we will refer to as "plain aspirin" in the rest of the manuscript) in terms of incidence of gastric discomfort and bleeding is lacking. In addition to its dubious advantages in terms of GI safety, it must be emphasized that many reports indicate that EC aspirin is inefficiently absorbed by the intestine in some subjects and, consequently, is unable to inhibit platelet function adequately.

Herein, we will review the PK, pharmacological and clinical efficacy, and GI safety of EC aspirin as well as, when available, other formulations, compared to plain aspirin.

2. Pharmacokinetics of Different Aspirin Formulations

Plain aspirin is absorbed in the stomach, where the low pH favors its absorption and protects the active principle from inactivation. EC aspirin, on the other hand, reaches the small intestine, where the higher pH favors drug deacetylation rather than its absorption [12]. A lower bioavailability of EC aspirin compared with plain aspirin can thus be expected. Aspirin is rapidly hydrolyzed to its metabolite salicylic acid by intestinal, plasma, and hepatic esterases [13], and has therefore a systemic bioavailability of only approximately 50% [14], with a C_{max} and an $AUC_{0-24\,h}$ that are much lower than those of salicylic acid [15]. After oral administration of 100 mg tablets to healthy subjects, T_{max} is about 0.5 h for plain aspirin [16–18] and about 4–5 h for EC aspirin [15,17,18], while C_{max} and AUC are slightly lower with EC aspirin [17,18]. After its absorption, aspirin acetylates platelet COX-1 in the pre-systemic circulation [14], as demonstrated by the fact that inhibition of TxB2 (the stable metabolite of TxA2) production [14] and the appearance of acetylated COX-1 in platelets [15] are detectable before the active principle is measurable in the systemic circulation. Maximal inhibition of TxB2 production in healthy subjects was observed 1–1.5 h after oral dosing with 100 mg plain aspirin [18] and 6–8 h after oral dosing with 100 mg EC aspirin [15,18].

At high doses, buffered aspirin [19,20] and PL2200 [21] displayed PK and PD bioequivalence with plain aspirin, while the bioequivalence of low doses (81–100 mg), which are commonly used in cardiovascular disease (CVD) prophylaxis, has not yet been assessed.

The PK and PD properties of plain aspirin, EC aspirin, buffered aspirin, and PL2200 are summarised in Table 1.

Table 1. Pharmacokinetics and pharmacodynamics of plain aspirin, enteric-coated aspirin, buffered aspirin, and PL2200 after oral administration to healthy subjects.

	Plain Aspirin (100 mg Tablets)	EC Aspirin (100 mg Tablets)	Buffered Aspirin (325 mg Tablets)	PL2200 (325 mg Tablets)
Preparations	Uncoated tablets	Tablets coated with inactive ingredients	Aspirin associated with buffering agents *	Complex of aspirin and lipidic excipients
Site of absorption	Stomach	Small intestine	Stomach	Duodenum
Time to maximal plasma concentration of aspirin	0.5 h	4 h	0.4 h	1 h
Time to maximal inhibition of thromboxane B2 production	1–1.5 h	6–8 h	1 h	2 h

* Calcium carbonate, magnesium oxide, magnesium carbonate; abbreviations: EC, enteric coated.

3. Pharmacological and Clinical Efficacy of Different Aspirin Formulations

At the beginning of the 21st century, several studies reported a high prevalence of poor pharmacological response to aspirin in treated patients, which was often referred to as "aspirin resistance" [22]. However, a careful analysis of the published studies revealed major flaws in the evaluation of the pharmacological response to aspirin, which was studied using inappropriate and unspecific tests of platelet function [22]. In fact, most attempts to evaluate the efficacy of aspirin using in vivo and in vitro platelet function tests, such as the bleeding time, platelet aggregation assays, and the PFA-100 system, failed to provide consistent data that may be used when discussing the matter of aspirin resistance because of the poor specificity, accuracy, reproducibility, and standardization of the aforementioned tests [22]. The most accurate method to study aspirin resistance is to measure the degree of inhibition of TxA2 formation after drug administration by dosing its stable analogue TxB2 in serum under controlled conditions [23]. The inhibition of at least 95% of serum TxB2 formation has long been considered necessary to prevent thromboxane-dependent platelet activation [24]. Some studies that accurately addressed the issue of aspirin response by measuring serum TxB2 showed inadequate pharmacological inhibition almost exclusively in subjects treated with EC aspirin, as summarized in the following paragraphs.

3.1. *Studies of Healthy Subjects or Patients on Chronic Treatment for Stable Coronary Artery Disease*

In the year 2005, Maree et al. measured serum TxB2 levels in 131 stable coronary artery disease (CAD) patients with a median age of 63 years on chronic low-dose (75 mg o.d.) EC aspirin treatment [25]. In this study population, a suboptimal inhibition of TxB2 formation was found in as many as 44% of the patients. In the same patients, the effects of EC aspirin on platelet aggregation were also studied. Although platelet aggregation tests are less accurate and precise than TxB2 measurement to test the pharmacologic efficacy of aspirin, the authors used arachidonic acid (AA), instead of other platelet agonists as in other studies, which is the specific platelet agonist triggering the COX1/TxA2 pathway of platelet aggregation. As expected, inadequate inhibition of AA-induced platelet aggregation was observed more frequently among patients with high serum TxB2 levels. The in vitro addition of aspirin to patients' platelet-rich plasma (PRP) samples abolished the residual AA-induced platelet aggregation, thus implying that insufficient bioavailability of aspirin after oral EC aspirin administration was responsible for the inadequate pharmacological response that had been observed in these patients. A very interesting finding of this study was that predictors of poor response to EC aspirin included young age and high body weight. In the following year, the same group of investigators showed that equivalent doses of EC aspirin are less effective than plain aspirin in inhibiting serum TxB2 formation in 71 healthy subjects aged 20 to 50 years [12]. However, in this study, poor pharmacological response to EC aspirin was observed more frequently among subjects with high body weight. The inverse relationship between pharmacological response to EC aspirin and body weight was again confirmed by a study of 148 CAD patients on chronic treatment with 75 mg o.d. EC aspirin for at least three months [26]. Finally, very high percentages of poor responders, defined as <95% inhibition of TxA2 production, were observed among healthy subjects 4 h (39/146, 29%) or 8 h (14/199, 7%) after ingestion of 100 mg EC aspirin, versus none among 40 healthy subjects after ingestion of plain aspirin [27]. Even within the class of EC aspirin, there is variability in the ability to inhibit platelet production of TxA2, as shown by Cox et al., who compared two EC aspirin preparations with plain aspirin: both EC preparations were less effective than plain aspirin in inhibiting TxA2 production, but there was no bioequivalence between the two EC preparations [28].

3.2. *Studies of Patients Affected by Diseases Associated with Particularly High Cardiovascular Risk*

3.2.1. Patients with Diabetes Mellitus

In a randomized, single-blinded, triple-crossover study [29], 40 obese diabetic patients not requiring insulin received three different 325 mg aspirin preparations: plain aspirin, PL2200, and EC aspirin for three days. Aspirin poor responsiveness, defined as <99% inhibition of TxB2 formation in serum at any time during the first 72 h of the study, occurred in a higher proportion of patients receiving the EC preparation (52.8%), compared with plain aspirin (15.8%) or PL2200 (8.1%). Therefore, some degree of aspirin hyporesponsiveness in diabetic patients was observed independently of the aspirin formulation used, although it was much higher in patients treated with EC aspirin. However, it must be noted that the chosen criterion to define aspirin responsiveness in this study was extremely strict (>99% inhibition of TxB2 production), which likely accounts for the high prevalence of "poor responders" also in patients treated with plain aspirin. PK studies, confirming the results of previous reports [16–18], showed that T_{max} was significantly lower, while C_{max} and AUC were significantly higher for plain aspirin and PL2200 compared with EC aspirin, suggesting that the observed poor responsiveness to EC aspirin was due to reduced absorption and bioavailability of aspirin. A small study of 42 patients with acute stroke reported that the prevalence of poor pharmacological response to EC aspirin compared to plain aspirin was higher in diabetic patients [30]. In conclusion, plain aspirin should be the preferred formulation for use in diabetic patients.

3.2.2. Patients with Essential Thrombocythemia

Patients with the myeloproliferative neoplasm Essential Thrombocythemia (ET) are at heightened risk for cardiovascular events and, consequently, are prophylactically treated with low-dose aspirin, in analogy with patients with another myeloproliferative neoplasm, Polycythemia Vera, unless their platelet count is >1000 $\times$ 10^9/L, which is associated with high bleeding risk [31]. Several studies reported that these patients may be poor responders to aspirin because the 24 h serum levels of TxB2 were higher than in normal subjects [32]. However, these studies actually tested the recovery of platelet ability to synthesize TxB2 after aspirin administration, rather than the pharmacological response to the drug [31]. In a more recent cross-over study, we showed that poor responsiveness to aspirin is attributable to the use of EC aspirin in these patients [18]. Indeed, our study showed that, in a high proportion of ET patients, serum TxB2 levels are not decreased by 100 mg o.d. EC aspirin, whereas they are adequately suppressed in the same patients by 100 mg o.d. plain aspirin. This difference was attributable to impaired and variable absorption of EC aspirin, with consequent higher T_{max} and lower C_{max} and AUC compared with those of healthy subjects treated with 100 mg o.d. EC aspirin. In contrast, all PK parameters in ET patients were comparable to those of healthy subjects after the oral administration of plain aspirin. In partial agreement with previous reports, we found that the 24 h post-dose serum TxB2 levels were higher in ET patients than in healthy controls, independent of the aspirin formulation used. This difference was attributable to the increased entry of newly formed non-acetylated platelets in the circulation, caused by increased platelet production (which characterizes the disease). Twice daily administration of 100 mg plain aspirin corrected this abnormality in ET patients, suggesting that ET patients with high platelet counts (>400–450 $\times$ 10^9/L) might benefit from 12-h administration of the plain aspirin [18,31].

3.3. Studies with Clinical End Points

It is impossible to provide accurate and solid information on the differences between plain aspirin and EC aspirin in preventing cardiovascular events because no direct comparisons between the two formulations have been made in high-quality, large RCTs. However, some indirect evidence exists that EC aspirin could be less effective than plain aspirin.

Rothwell et al., reviewed seven RCTs of low-dose aspirin (75–100 mg o.d.) in the primary prevention of vascular events, which collected data on body weight, height, and individual subject data on baseline characteristics [33]. The most relevant finding of the study was that the ability of 75–100 mg aspirin to reduce cardiovascular events decreased with increasing body weight of the treated subjects: vascular events were reduced by aspirin in subjects weighing 50–69 kg (hazard ratio 0.75 [95% CI 0.65–0.85]) but not in those weighing 70 kg or more (0.95 [0.86–1.04]; 1.09 [0.93–1.29]). The inverse relation between body weight and the efficacy of aspirin was confirmed by the observation that also the increased risk of major bleeding on low-dose aspirin versus control was lost in participants weighing 90 kg or more. Findings were similar in men and women, in people with diabetes, in trials of aspirin in secondary prevention, and in relation to height. Aspirin-mediated reductions in long-term risk of colorectal cancer were also weight-dependent.

Among the seven trials on low-dose aspirin in primary prevention included in Rothwell's analysis, four employed EC aspirin [34–37] and one used a delayed-release formulation [38]. The body weight dependence of the effect of low-dose aspirin on cardiovascular events was observed for all formulations, but the loss of effect in participants weighing 70 kg or more was much more evident for EC or delayed-release aspirin [33]. This finding is in perfect agreement with the demonstrations that a poor pharmacological response to EC aspirin is observed more frequently among subjects with high body weight [12,25,26]. Therefore, it is plausible to hypothesize that, given the large prevalence of adult subjects weighing >70 kg who need cardiovascular protection by aspirin, a higher efficacy of aspirin would have been observed if plain aspirin, instead of EC aspirin, had been used for primary (and secondary) prophylaxis of cardiovascular events.

4. Gastrointestinal Injury and Bleeding with Different Aspirin Formulations

As already mentioned, aspirin formulations alternative to plain aspirin were developed with the aim of decreasing GI discomfort, mucosal erosions/ulcerations and bleeding that are associated with chronic treatment with plain aspirin. The effective safety advantage of these formulations (EC aspirin in most instances) over plain aspirin was tested in some studies.

4.1. Endoscopic Studies in Asymptomatic Healthy Subjects

Some studies tested the effects of the acute administration (5–7 days) of plain aspirin compared with EC aspirin on the prevalence of gastric mucosal erosion and submucosal hemorrhage in healthy asymptomatic subjects who underwent endoscopic examination at the end of treatment (in some studies, endoscopy had also been performed at the beginning of the study, to have a baseline picture of the status of the volunteers). Both formulations of aspirin were given (in a cross-over design for some studies) at doses ranging from 100 mg daily [39], up to 300–325 mg daily [40–44] or even 2.4–3.9 g [40,41,45]. All studies demonstrated that treatment with EC aspirin was associated with a lower prevalence of mucosal injuries, especially when very high doses of aspirin were used, which are commonly administered for the management of inflammatory states rather than for cardiovascular prevention. In none of the studies had episodes of GI bleeding or ulceration been detected. No differences in the frequency of lesions of the gastric mucosa were observed after the oral administration of plain aspirin and buffered aspirin [44].

After a 7-day course of 325 mg aspirin was administered to subjects at high risk of GI complications, endoscopic studies showed that PL2200 was associated with fewer gastric mucosal lesions than plain aspirin [16]. The comparative effects of PL2200 and plain aspirin at low doses and in longer-term studies are necessary to define more accurately the safety profile of PL2200 compared to plain aspirin.

4.2. Studies of Upper Gastrointestinal Bleeding or Ulceration in Patients on Chronic Treatment with Aspirin

In the year 1996, a multicenter case-control study by Kelly et al., aimed at assessing aspirin use in the week preceding the acute event or the day of the interview in incident cases of upper GI bleeding (UGIB) and matched controls derived from population census lists [47]. This study investigated the use of plain aspirin, EC aspirin, and buffered aspirin. Data analysis showed that the relative risks (RR) of UGIB for plain, EC, and buffered aspirin preparations at average daily doses of 325 mg or less were 2.6 (95% CI, 1.7–4.0), 2.7 (95% CI, 1.4–5.3) and 3.1 (95% CI, 1.3–7.6), respectively; there were insufficient data to compare the RR of UGIB for doses greater than 325 mg of plain aspirin with those of EC aspirin. The authors concluded that, given the similar RR of major UGIB (both gastric and duodenal) in subjects taking different 325 mg or less aspirin preparations, the systemic effects of the active principle might outweigh the differences in local toxicity, showing no clear benefit in the use of EC preparations. Results mirroring those of Kelly's study were provided by a population-based case-control study on the risk of upper GI complications (UGIC, bleeding and perforation) associated with the administration of 75–300 mg/day of aspirin [48]. This study used data from the UK-based General Practice Research Database; unlike Kelly's study, no direct contact was made with patients and controls to better define aspirin exposure, which was solely estimated according to database information. Moreover, only 13% of cases and 7% of controls were exposed to aspirin. Despite these limitations, the RR of UGIC was 2.3 (95% CI, 1.6–3.2) for EC aspirin and 1.9 (95% CI, 1.6–2.3) for plain aspirin, and the results did not change when only patients without antecedents of upper GI disorder were included in the analysis and after adjustment for the use of antiulcer drugs. A Danish population-based cohort study showed similar risks of UGIB in users of low-dose plain aspirin and EC aspirin (standardized incidence rate ratio, 2.6; 95% CI, 1.8–3.5 for plain aspirin vs. 2.6; 95% CI, 2.2–3.0 for EC aspirin) [5]. Only one case-control study on the risk of peptic ulcer bleeding in prophylactic (300 mg daily or less) aspirin users suggested that EC preparations might be safer than other preparations, although no aspirin preparation seemed to be free of the risk of peptic ulcer complications [49].

Garcìa Rodrìguez et al., reviewed the aforementioned four studies and two studies on buffered aspirin published between 1990 and 2001 [50]. The authors calculated a summary RR of serious UGIC (bleeding, perforation, or other serious upper GI events resulting in hospitalization or a visit to a specialist) of 2.6 (95% CI; 2.3, 2.9) for plain aspirin, 5.3 (95% CI; 3.0, 9.2) for buffered aspirin, and 2.4 (95% CI; 1.9, 2.9) for EC aspirin. They therefore concluded that aspirin formulation has little or no effect on the prevention of serious UGIC and hypothesized a likely greater impact of the systemic rather than topical effects of the drug, as suggested by the similar RR of duodenal and gastric lesions. Therefore, the lower incidence of gastric mucosal lesions in endoscopic studies might be explained by the topical effects of the drug, whereas the systemic effects might be predominant in the pathogenesis of UGIC.

The hypothesis about differences between local and systemic toxicity of aspirin is corroborated by evidence from additional studies with somewhat different designs. Some studies showed that the frequency of small bowel mucosal lesions detected by capsule endoscopy was higher in patients taking EC aspirin (which is absorbed in the small intestine) than in those taking non-EC aspirin formulations [51–53]. Moreover, although an endoscopic study showed that buffered aspirin formulations reduced the frequency of gastric mucosal erosion compared to plain aspirin [54], the use of buffered aspirin failed to decrease the incidence of peptic ulcer [55].

To summarize, the only source of evidence regarding the decreased GI toxicity of EC aspirin is represented by endoscopic studies, which showed fewer gastric mucosal lesions. However, lesions of the small bowel mucosa appeared to be more frequent with EC aspirin than with non-EC aspirin formulations. These data suggest that GI mucosal lesions are caused by topical effects of aspirin in the region of its absorption. Most importantly, event-driven studies of GI hemorrhage failed to provide data supporting the clinical benefit of EC aspirin or buffered aspirin, therefore suggesting that the systemic effects of the drug, which are unchanged by enteric coating, are to blame for the occurrence of clinically relevant GI complications and bleeding.

5. Use of Proton Pump Inhibitors during Chronic Aspirin Treatment

The European Society of Cardiology recommends the use of proton pump inhibitors (PPIs) in patients on chronic aspirin treatment who are at high risk of GI bleeding [56]. PPIs are effective in reducing upper GI clinical events in patients receiving aspirin in the context of dual antiplatelet therapy [57]. The risk, however, is only reduced, not abolished: randomization to PPI therapy reduced 180-day Kaplan-Meier estimates of the primary GI endpoint in low-dose aspirin recipients to 1.2% from 3.1% [57]. These results are in keeping with a previous literature review focused on PPIs effectiveness in patients taking aspirin as single antiplatelet therapy [58] and with the results of a Swedish cohort study [59], which highlighted that compliance to continuous use of PPIs was pivotal, as intermittent use was associated with increased risk of adverse GI outcomes and of aspirin discontinuation. As an alternative to PPIs, histamine H2 receptor antagonists (H2RAs) can be used, although they have been proven less effective than PPIs in the prevention of GI complications in patients on low dose aspirin alone [60] or in combination with anti-P2Y12 drugs [61].

6. Conclusions

The absorption of EC aspirin is delayed and erratic, resulting in less effective inhibition of the platelet production of TxA2, thus providing less effective inhibition of platelet function, especially in subjects with high body weight. Such inferiority in pharmacological efficacy seems to have a clinical impact, as shown by a meta-analysis of RCTs, predominantly on primary cardiovascular prevention, which revealed that lack of protection by low-dose aspirin in subjects weighing >70 kg was particularly evident in subjects treated with EC aspirin. On the other hand, there is no evidence that EC aspirin protects from clinically relevant GI bleeding and ulceration. Differences in the incidence of asymptomatic lesions of the GI mucosa detected by endoscopy reflect the effects of the drug on the site of its absorption: more lesions of the gastric mucosa can be observed after plain aspirin administration, while more lesions of the small bowel are observed after EC aspirin ingestion (the main differences between plain aspirin and EC aspirin are summarized in Figure 1).

The use of PPIs is recommended for patients on chronic aspirin with risk factors for GI bleeding, which include a history of peptic ulcer disease or gastrointestinal bleeding, older age, concomitant use of NSAIDs, concomitant use of anticoagulants or other platelet aggregation inhibitors, and the presence of severe co-morbidities [62]. Coformulations of aspirin and PPIs could be considered for patients for whom polypharmacy and poor compliance are a reason for concern. H2RAs can be considered as alternatives when PPIs are unavailable or contraindicated.

Considering its more favorable pharmacological profile, plain aspirin should be the preferred formulation for cardiovascular prevention. The improvement in P2Y12 inhibition obtained with the newer antiplatelet drugs prasugrel and ticagrelor, which have a more efficient PK than clopidogrel [63] could be replicated for COX-1 inhibition by using an older antiplatelet drug with a more efficient PK than EC aspirin, which is still the most widely used aspirin formulation in the setting of cardiovascular prevention.

Figure 1. Pharmacological profile, clinical efficacy, and safety of enteric coated (EC) aspirin compared to plain aspirin. The height of the histograms shown in the figure is not reflecting real data and should be interpreted as illustrative of the average values obtained in several studies with EC-aspirin relative to plain aspirin (higher, equal, lower). Pharmacokinetic parameters (usually measured in serum): Tmax = time to peak drug concentration; Cmax = peak drug concentration; AUC = Area Under the Curve (integral of drug concentration as a function of time). TxB2 = thromboxane B2 (a stable metabolite of thromboxane A2).

Author Contributions: Conceptualization, M.C.; Methodology, M.C.; Software, B.C.; Validation, B.C. and M.C.; Writing—Original Draft Preparation, B.C.; Writing—Review & Editing, M.C.; Supervision, M.C. All authors have read and agreed to the published version of the manuscript.

Funding: This research received no external funding.

Institutional Review Board Statement: Not applicable.

Informed Consent Statement: Not applicable.

Data Availability Statement: Not applicable.

Conflicts of Interest: The authors declare no conflict of interest.

References

1. Hamberg, M.; Svensson, J.; Samuelsson, B. Thromboxanes: A new group of biologically active compounds derived from prostaglandin endoperoxides. *Proc. Natl. Acad. Sci. USA* **1975**, *72*, 2994–2998. [CrossRef] [PubMed]
2. Patrono, C.; García Rodríguez, L.A.; Landolfi, R.; Baigent, C. Low-dose aspirin for the prevention of atherothrombosis. *N. Engl. J. Med.* **2005**, *353*, 2373–2383. [CrossRef]
3. Zheng, S.L.; Roddick, A.J. Association of Aspirin Use for Primary Prevention with Cardiovascular Events and Bleeding Events: A Systematic Review and Meta-analysis [published correction appears in JAMA. *JAMA* **2019**, *321*, 2245, Erratum in *JAMA* **2019**, *321*, 277–287. [CrossRef]
4. Steering Committee of the Physicians Health Study Research Group. Final report of the aspirin component of the outgoing physicians health study. *N. Engl. J. Med.* **1989**, *321*, 129–135. [CrossRef] [PubMed]
5. Sørensen, H.T.; Mellemkjaer, L.; Blot, W.J.; Nielsen, G.L.; Steffensen, F.H.; McLaughlin, J.K.; Olsen, J.H. Risk of upper gastrointestinal bleeding associated with use of low-dose aspirin. *Am. J. Gastroenterol.* **2000**, *95*, 2218–2224. [CrossRef]
6. Derry, S.; Loke, Y.K. Risk of gastrointestinal haemorrhage with long term use of aspirin: Meta-analysis. *BMJ* **2000**, *321*, 1183–1187. [CrossRef]
7. Moukarbel, G.V.; Signorovitch, J.E.; Pfeffer, M.A.; McMurray, J.J.; White, H.D.; Maggioni, A.P.; Velazquez, E.J.; Califf, R.M.; Scheiman, J.M.; Solomon, S.D. Gastrointestinal bleeding in high risk survivors of myocardial infarction: The VALIANT Trial. *Eur. Heart J.* **2009**, *30*, 2226–2232. [CrossRef]

8. Endo, H.; Sakai, E.; Kato, T.; Umezawa, S.; Higurashi, T.; Ohkubo, H.; Nakajima, A. Small bowel injury in low-dose aspirin users. *J. Gastroenterol.* **2015**, *50*, 378–386. [CrossRef]
9. Banoob, D.W.; McCloskey, W.W.; Webster, W. Risk of gastric injury with enteric- versus nonenteric-coated aspirin. *Ann. Pharmacother.* **2002**, *36*, 163–166. [CrossRef]
10. Paul, W.D.; Dryer, R.L.; Routh, J.I. Effect of buffering agents on absorption of acetylsalicylic acid. *J. Amer Pharm. Assoc.* **1950**, *39*, 21–24. [CrossRef]
11. Angiolillo, D.J.; Prats, J.; Deliargyris, E.N.; Schneider, D.J.; Scheiman, J.; Kimmelstiel, C.; Steg, P.G.; Alberts, M.; Rosengart, T.; Mehran, R.; et al. Pharmacokinetic and Pharmacodynamic Profile of a Novel Phospholipid Aspirin Formulation. *Clin. Pharmacokinet.* **2022**, *61*, 465–479. [CrossRef] [PubMed]
12. Cox, D.; Maree, A.O.; Dooley, M.; Conroy, R.; Byrne, M.F.; Fitzgerald, D.J. Effect of enteric coating on antiplatelet activity of low-dose aspirin in healthy volunteers. *Stroke* **2006**, *37*, 2153–2158. [CrossRef] [PubMed]
13. Williams, F.M. Clinical significance of esterases in man. *Clin. Pharmacokinet.* **1985**, *10*, 392–403. [CrossRef] [PubMed]
14. Pedersen, A.K.; FitzGerald, G.A. Dose-related kinetics of aspirin. Presystemic acetylation of platelet cyclooxygenase. *N. Engl. J. Med.* **1984**, *311*, 1206–1211. [CrossRef] [PubMed]
15. Patrignani, P.; Tacconelli, S.; Piazuelo, E.; Di Francesco, L.; Dovizio, M.; Sostres, C.; Marcantoni, E.; Guillem-Llobat, P.; Del Boccio, P.; Zucchelli, M.; et al. Reappraisal of the clinical pharmacology of low-dose aspirin by comparing novel direct and traditional indirect biomarkers of drug action. *J. Thromb. Haemost.* **2014**, *12*, 1320–1330. [CrossRef]
16. Nagelschmitz, J.; Blunck, M.; Kraetzschmar, J.; Ludwig, M.; Wensing, G.; Hohlfeld, T. Pharmacokinetics and pharmacodynamics of acetylsalicylic acid after intravenous and oral administration to healthy volunteers. *Clin. Pharmacol.* **2014**, *6*, 51–59. [CrossRef]
17. Bochner, F.; Somogyi, A.A.; Wilson, K.M. Bioinequivalence of four 100 mg oral aspirin formulations in healthy volunteers. *Clin. Pharmacokinet.* **1991**, *21*, 394–399. [CrossRef]
18. Scavone, M.; Rizzo, J.; Femia, E.A.; Podda, G.M.; Bossi, E.; Caberlon, S.; Paroni, R.; Cattaneo, M. Patients with Essential Thrombocythemia may be Poor Responders to Enteric-Coated Aspirin, but not to Plain Aspirin. *Thromb. Haemost.* **2020**, *120*, 1442–1453. [CrossRef]
19. Batterman, R.C. Comparison of buffered and unbuffered acetylsalicylic acid. *N. Engl. J. Med.* **1958**, *258*, 213–219. [CrossRef] [PubMed]
20. Feldman, M.; Cryer, B. Aspirin absorption rates and platelet inhibition times with 325-mg buffered aspirin tablets (chewed or swallowed intact) and with buffered aspirin solution. *Am. J. Cardiol.* **1999**, *84*, 404–409. [CrossRef]
21. Angiolillo, D.J.; Bhatt, D.L.; Lanza, F.; Cryer, B.; Dong, J.F.; Jeske, W.; Zimmerman, R.R.; von Chong, E.; Prats, J.; Deliargyris, E.N.; et al. Pharmacokinetic/pharmacodynamic assessment of a novel, pharmaceutical lipid-aspirin complex: Results of a randomized, crossover, bioequivalence study. *J. Thromb. Thrombolysis* **2019**, *48*, 554–562. [CrossRef] [PubMed]
22. Cattaneo, M. Aspirin and clopidogrel: Efficacy, safety, and the issue of drug resistance. *Arterioscler. Thromb. Vasc. Biol.* **2004**, *24*, 1980–1987. [CrossRef] [PubMed]
23. Patrono, C.; Ciabattoni, G.; Pinca, E.; Pugliese, F.; Castrucci, G.; De Salvo, A.; Satta, M.A.; Peskar, B.A. Low dose aspirin and inhibition of thromboxane B2 production in healthy subjects. *Thromb. Res.* **1980**, *17*, 317–327. [CrossRef] [PubMed]
24. Reilly, I.A.; FitzGerald, G.A. Inhibition of thromboxane formation in vivo and ex vivo: Implications for therapy with platelet inhibitory drugs. *Blood* **1987**, *69*, 180–186. [CrossRef]
25. Maree, A.O.; Curtin, R.J.; Dooley, M.; Conroy, R.M.; Crean, P.; Cox, D.; Fitzgerald, D.J. Platelet response to low-dose enteric-coated aspirin in patients with stable cardiovascular disease. *J. Am. Coll. Cardiol.* **2005**, *46*, 1258–1263. [CrossRef]
26. Peace, A.; McCall, M.; Tedesco, T.; Kenny, D.; Conroy, R.M.; Foley, D.; Cox, D. The role of weight and enteric coating on aspirin response in cardiovascular patients. *J. Thromb. Haemost.* **2010**, *8*, 2323–2325. [CrossRef]
27. Grosser, T.; Fries, S.; Lawson, J.A.; Kapoor, S.C.; Grant, G.R.; FitzGerald, G.A. Drug resistance and pseudoresistance: An unintended consequence of enteric coating aspirin. *Circulation* **2013**, *127*, 377–385. [CrossRef]
28. Cox, D.; Fitzgerald, D.J. Lack of Bioequivalence Among Low-dose, Enteric-coated Aspirin Preparations. *Clin. Pharmacol. Ther.* **2018**, *103*, 1047–1051. [CrossRef]
29. Bhatt, D.L.; Grosser, T.; Dong, J.F.; Logan, D.; Jeske, W.; Angiolillo, D.J.; Frelinger, A.L., 3rd; Lei, L.; Liang, J.; Moore, J.E.; et al. Enteric Coating and Aspirin Nonresponsiveness in Patients with Type 2 Diabetes Mellitus. *J. Am. Coll. Cardiol.* **2017**, *69*, 603–612. [CrossRef]
30. Elshafei, M.N.; Imam, Y.; Alsaud, A.E.; Chandra, P.; Parray, A.; Abdelmoneim, M.S.; Obeidat, K.; Saeid, R.; Ali, M.; Ayadathil, R.; et al. The impact of enteric coating of aspirin on aspirin responsiveness in patients with suspected or newly diagnosed ischemic stroke: Prospective cohort study: Results from the (ECASIS) study. *Eur. J. Clin. Pharmacol.* **2022**, *78*, 1801–1811. [CrossRef]
31. Cattaneo, M. Aspirin in essential thrombocythemia: To whom? What formulation? What regimen? *Haematologica*, 2023; *Epub ahead of print*. [CrossRef] [PubMed]
32. Pascale, S.; Petrucci, G.; Dragani, A.; Habib, A.; Zaccardi, F.; Pagliaccia, F.; Pocaterra, D.; Ragazzoni, E.; Rolandi, G.; Rocca, B.; et al. Aspirin-insensitive thromboxane biosynthesis in essential thrombocythemia is explained by accelerated renewal of the drug target. *Blood* **2012**, *119*, 3595–3603. [CrossRef] [PubMed]

33. Rothwell, P.M.; Cook, N.R.; Gaziano, J.M.; Price, J.F.; Belch, J.F.F.; Roncaglioni, M.C.; Morimoto, T.; Mehta, Z. Effects of aspirin on risks of vascular events and cancer according to bodyweight and dose: Analysis of individual patient data from randomised trials. *Lancet* **2018**, *392*, 387–399. [CrossRef]
34. De Gaetano, G.; Collaborative Group of the Primary Prevention Project. Low-dose aspirin and vitamin E in people at cardiovascular risk: A randomised trial in general practice. Collaborative Group of the Primary Prevention Project. *Lancet* **2001**, *357*, 89–95. [CrossRef]
35. Belch, J.; MacCuish, A.; Campbell, I.; Cobbe, S.; Taylor, R.; Prescott, R.; Lee, R.; Bancroft, J.; MacEwan, S.; Shepherd, J.; et al. The prevention of progression of arterial disease and diabetes (POPADAD) trial: Factorial randomised placebo controlled trial of aspirin and antioxidants in patients with diabetes and asymptomatic peripheral arterial disease. *BMJ* **2008**, *337*, a1840. [CrossRef] [PubMed]
36. Ogawa, H.; Nakayama, M.; Morimoto, T.; Uemura, S.; Kanauchi, M.; Doi, N.; Jinnouchi, H.; Sugiyama, S.; Saito, Y.; Japanese Primary Prevention of Atherosclerosis with Aspirin for Diabetes (JPAD) Trial Investigators. Low-dose aspirin for primary prevention of atherosclerotic events in patients with type 2 diabetes: A randomized controlled trial. *JAMA* **2008**, *300*, 2134–2141, Erratum in *JAMA* **2009**, *301*, 1882; Erratum in *JAMA* **2012**, *308*, 1861. [CrossRef]
37. Fowkes, F.G.; Price, J.F.; Stewart, M.C.; Butcher, I.; Leng, G.C.; Pell, A.C.; Sandercock, P.A.; Fox, K.A.; Lowe, G.D.; Murray, G.D.; et al. Aspirin for prevention of cardiovascular events in a general population screened for a low ankle brachial index: A randomized controlled trial. *JAMA* **2010**, *303*, 841–848. [CrossRef]
38. Thrombosis prevention trial: Randomised trial of low-intensity oral anticoagulation with warfarin and low-dose aspirin in the primary prevention of ischaemic heart disease in men at increased risk. The Medical Research Council's General Practice Research Framework. *Lancet* **1998**, *351*, 233–241.
39. Dammann, H.G.; Burkhardt, F.; Wolf, N. Enteric coating of aspirin significantly decreases gastroduodenal mucosal lesions. *Aliment. Pharmacol. Ther.* **1999**, *13*, 1109–1114. [CrossRef]
40. Lanza, F.L.; Royer, G.L., Jr.; Nelson, R.S. Endoscopic evaluation of the effects of aspirin, buffered aspirin, and enteric-coated aspirin on gastric and duodenal mucosa. *N. Engl. J. Med.* **1980**, *303*, 136–138. [CrossRef]
41. Hawthorne, A.B.; Mahida, Y.R.; Cole, A.T.; Hawkey, C.J. Aspirin-induced gastric mucosal damage: Prevention by enteric-coating and relation to prostaglandin synthesis. *Br. J. Clin. Pharmacol.* **1991**, *32*, 77–83. [CrossRef] [PubMed]
42. Cole, A.T.; Hudson, N.; Liew, L.C.; Murray, F.E.; Hawkey, C.J.; Heptinstall, S. Protection of human gastric mucosa against aspirin—Enteric coating or dose reduction? *Aliment. Pharmacol. Ther.* **1999**, *13*, 187–193. [CrossRef] [PubMed]
43. Blondon, H.; Barbier, J.P.; Mahé, I.; Deverly, A.; Kolsky, H.; Bergmann, J.F. Gastroduodenal tolerability of medium dose enteric-coated aspirin: A placebo controlled endoscopic study of a new enteric-coated formulation versus regular formulation in healthy volunteers. *Fundam. Clin. Pharmacol.* **2000**, *14*, 155–157. [CrossRef] [PubMed]
44. Petroski, D. Endoscopic comparison of three aspirin preparations and placebo. *Clin. Ther.* **1993**, *15*, 314–320.
45. Hoftiezer, J.W.; Silvoso, G.R.; Burks, M.; Ivey, K.J. Comparison of the effects of regular and enteric-coated aspirin on gastroduodenal mucosa of man. *Lancet* **1980**, *2*, 609–612. [CrossRef]
46. Cryer, B.; Bhatt, D.L.; Lanza, F.L.; Dong, J.F.; Lichtenberger, L.M.; Marathi, U.K. Low-dose aspirin-induced ulceration is attenuated by aspirin-phosphatidylcholine: A randomized clinical trial. *Am. J. Gastroenterol.* **2011**, *106*, 272–277. [CrossRef]
47. Kelly, J.P.; Kaufman, D.W.; Jurgelon, J.M.; Sheehan, J.; Koff, R.S.; Shapiro, S. Risk of aspirin-associated major upper-gastrointestinal bleeding with enteric-coated or buffered product. *Lancet* **1996**, *348*, 1413–1416. [CrossRef]
48. De Abajo, F.J.; García Rodríguez, L.A. Risk of upper gastrointestinal bleeding and perforation associated with low-dose aspirin as plain and enteric-coated formulations. *BMC Clin. Pharmacol.* **2001**, *1*, 1. [CrossRef]
49. Weil, J.; Colin-Jones, D.; Langman, M.; Lawson, D.; Logan, R.; Murphy, M.; Rawlins, M.; Vessey, M.; Wainwright, P. Prophylactic aspirin and risk of peptic ulcer bleeding. *BMJ* **1995**, *310*, 827–830. [CrossRef]
50. García Rodríguez, L.A.; Hernández-Díaz, S.; De Abajo, F.J. Association between aspirin and upper gastrointestinal complications: Systematic review of epidemiologic studies. *Br. J. Clin. Pharmacol.* **2001**, *52*, 563–571. [CrossRef]
51. Endo, H.; Hosono, K.; Inamori, M.; Nozaki, Y.; Yoneda, K.; Fujita, K.; Takahashi, H.; Yoneda, M.; Abe, Y.; Kirikoshi, H.; et al. Characteristics of small bowel injury in symptomatic chronic low-dose aspirin users: The experience of two medical centers in capsule endoscopy. *J. Gastroenterol.* **2009**, *44*, 544–549. [CrossRef]
52. Endo, H.; Sakai, E.; Higurashi, T.; Yamada, E.; Ohkubo, H.; Iida, H.; Koide, T.; Yoneda, M.; Abe, Y.; Inamori, M.; et al. Differences in the severity of small bowel mucosal injury based on the type of aspirin as evaluated by capsule endoscopy. *Dig. Liver Dis.* **2012**, *44*, 833–838. [CrossRef] [PubMed]
53. Kedir, H.M.; Sisay, E.A.; Abiye, A.A. Enteric-Coated Aspirin and the Risk of Gastrointestinal Side Effects: A Systematic Review. *Int. J. Gen. Med.* **2021**, *14*, 4757–4763. [CrossRef] [PubMed]
54. Murray, F.E.; Hudson, N.; Atherton, J.C.; Cole, A.T.; Scheck, F.; Hawkey, C.J. Comparison of effects of calcium carbasalate and aspirin on gastroduodenal mucosal damage in human volunteers. *Gut* **1996**, *38*, 11–14. [CrossRef] [PubMed]
55. van Oijen, M.G.; Dieleman, J.P.; Laheij, R.J.; Sturkenboom, M.C.; Jansen, J.B.; Verheugt, F.W. Peptic ulcerations are related to systemic rather than local effects of low-dose aspirin. *Clin. Gastroenterol. Hepatol.* **2008**, *6*, 309–313. [CrossRef]
56. Visseren, F.L.J.; Mach, F.; Smulders, Y.M.; Carballo, D.; Koskinas, K.C.; Bäck, M.; Benetos, A.; Biffi, A.; Boavida, J.M.; Biffi, A.; et al. 2021 ESC Guidelines on cardiovascular disease prevention in clinical practice. *Eur. Heart J.* **2021**, *42*, 3227–3337, Erratum in *Eur. Heart J.* **2022**, *43*, 4468. [CrossRef]

57. Vaduganathan, M.; Bhatt, D.L.; Cryer, B.L.; Liu, Y.; Hsie, W.H.; Doros, G.; Cohen, M.; Lanas, A.; Schnitzer, T.J.; Shook, T.L.; et al. Proton-Pump Inhibitors Reduce Gastrointestinal Events Regardless of Aspirin Dose in Patients Requiring Dual Antiplatelet Therapy. *J. Am. Coll. Cardiol.* **2016**, *67*, 1661–1671. [CrossRef]
58. Mo, C.; Sun, G.; Lu, M.L.; Zhang, L.; Wang, Y.Z.; Sun, X.; Yang, S. Proton pump inhibitors in prevention of low-dose aspirin-associated upper gastrointestinal injuries. *World J. Gastroenterol.* **2015**, *21*, 5382–5392. [CrossRef]
59. Hedberg, J.; Sundström, J.; Thuresson, M.; Aarskog, P.; Oldgren, J.; Bodegard, J. Low-dose acetylsalicylic acid and gastrointestinal ulcers or bleeding—A cohort study of the effects of proton pump inhibitor use patterns. *J. Intern. Med.* **2013**, *274*, 371–380. [CrossRef]
60. Mo, C.; Sun, G.; Wang, Y.Z.; Lu, M.L.; Yang, Y.S. PPI versus Histamine H2 Receptor Antagonists for Prevention of Upper Gastrointestinal Injury Associated with Low-Dose Aspirin: Systematic Review and Meta-analysis. *PLoS ONE* **2015**, *10*, e0131558. [CrossRef]
61. Almufleh, A.; Ramirez, F.D.; So, D.; Le May, M.; Chong, A.Y.; Torabi, N.; Hibbert, B. H2 Receptor Antagonists versus Proton Pump Inhibitors in Patients on Dual Antiplatelet Therapy for Coronary Artery Disease: A Systematic Review. *Cardiology* **2018**, *140*, 115–123. [CrossRef] [PubMed]
62. Valkhoff, V.E.; Sturkenboom, M.C.; Kuipers, E.J. Risk factors for gastrointestinal bleeding associated with low-dose aspirin. *Best. Pract. Res. Clin. Gastroenterol.* **2012**, *26*, 125–140. [CrossRef] [PubMed]
63. Cattaneo, M. New P2Y(12) inhibitors. *Circulation* **2010**, *121*, 171–179. [CrossRef] [PubMed]

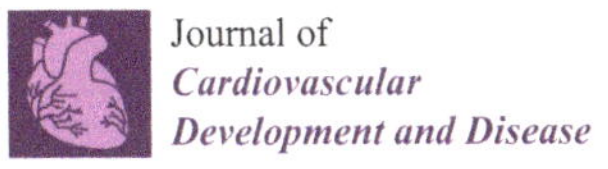
Journal of
Cardiovascular Development and Disease

MDPI

Review

Dual Antiplatelet Therapy and Cancer; Balancing between Ischemic and Bleeding Risk: A Narrative Review

Grigorios Tsigkas [1,*], Angeliki Vakka [1], Anastasios Apostolos [1,2], Eleni Bousoula [3], Nikolaos Vythoulkas-Biotis [1], Eleni-Evangelia Koufou [1], Georgios Vasilagkos [1], Ioannis Tsiafoutis [4], Michalis Hamilos [5], Adel Aminian [6] and Periklis Davlouros [1]

1 Department of Cardiology, University Hospital of Patras, 265 04 Patras, Greece; aggvakka@outlook.com (A.V.); anastasisapostolos@gmail.com (A.A.); vbnikos@gmail.com (N.V.-B.); elenikoufou17@gmail.com (E.-E.K.); giorgosvasilagkos@gmail.com (G.V.); pdav@med.upatras.gr (P.D.)
2 First Department of Cardiology, Hippocration General Hospital, National and Kapodistrian University of Athens, 157 72 Athens, Greece
3 Department of Cardiology, Tzaneio General Hospital, 185 36 Piraeus, Greece; ebousoula@gmail.com
4 First Department of Cardiology, Red Cross Hospital, 115 26 Athens, Greece; tsiafoutisg@yahoo.com
5 Department of Cardiology, Heraklion University Hospital, 715 00 Heraklion, Crete, Greece; mich.hamilos@gmail.com
6 Department of Cardiology, Centre Hospitalier Universitaire de Charleroi, 6042 Charleroi, Belgium; adaminian@hotmail.com
* Correspondence: grigoriostsigkas@gmail.com

Abstract: Cardiovascular (CV) events in patients with cancer can be caused by concomitant CV risk factors, cancer itself, and anticancer therapy. Since malignancy can dysregulate the hemostatic system, predisposing cancer patients to both thrombosis and hemorrhage, the administration of dual antiplatelet therapy (DAPT) to patients with cancer who suffer from acute coronary syndrome (ACS) or undergo percutaneous coronary intervention (PCI) is a clinical challenge to cardiologists. Apart from PCI and ACS, other structural interventions, such as TAVR, PFO-ASD closure, and LAA occlusion, and non-cardiac diseases, such as PAD and CVAs, may require DAPT. The aim of the present review is to review the current literature on the optimal antiplatelet therapy and duration of DAPT for oncologic patients, in order to reduce both the ischemic and bleeding risk in this high-risk population.

Keywords: cancer; acute coronary syndrome (ACS); percutaneous coronary intervention (PCI); dual antiplatelet therapy (DAPT); triple antithrombotic therapy (TAT); atrial fibrillation (AF); cardiotoxicity

Citation: Tsigkas, G.; Vakka, A.; Apostolos, A.; Bousoula, E.; Vythoulkas-Biotis, N.; Koufou, E.-E.; Vasilagkos, G.; Tsiafoutis, I.; Hamilos, M.; Aminian, A.; et al. Dual Antiplatelet Therapy and Cancer; Balancing between Ischemic and Bleeding Risk: A Narrative Review. *J. Cardiovasc. Dev. Dis.* **2023**, *10*, 135. https://doi.org/10.3390/jcdd10040135

Academic Editors: Giovanni Cimmino and Plinio Cirillo

Received: 21 January 2023
Revised: 15 March 2023
Accepted: 19 March 2023
Published: 23 March 2023

1. Introduction

Patients with cancer show a high prevalence of coronary artery disease (CAD) [1]. These diseases share common predisposing factors, such as obesity, diet, sedentary lifestyle, smoking, alcohol, and chronic inflammation [2]. Moreover, cancer itself increases the risk of cardiovascular disease by invading the cardiovascular system directly, releasing metabolites and cytokines, and leading to neurohormonal activation [3,4]. At the same time, anticancer therapies promote inflammation, vasospasm, endothelial dysfunction, plaque formation, and dysregulation of the hemostatic system [5,6].

Patients with cancer who have undergone percutaneous coronary intervention (PCI) and/or suffered an acute coronary syndrome (ACS) may need to discontinue dual antiplatelet therapy (DAPT) due to re-initiation of anticancer therapy, surgery, or biopsies [7]. In addition, since cancer can cause disorders in the hemostatic system, leading to both thrombosis and hemorrhage [6,8], deciding the optimal duration of DAPT in patients with cancer is challenging.

The aim of this study is to review the current literature on the optimal antiplatelet therapy and duration of DAPT for oncologic patients, in order to reduce both the ischemic and bleeding risk in this high-risk population.

2. Materials and Methods

A literature review was performed by searching the PubMed database for studies published in the English language up to January 2023. The following key words and their abbreviations were used: "cancer" OR "malignancy" OR "anticancer therapy" AND "dual antiplatelet therapy" OR "acute coronary syndrome" OR "percutaneous coronary intervention" OR "coagulation" OR "cardiotoxicity" OR "transcatheter aortic valve replacement" OR "patent foramen ovale—atrial septal defect closure" OR "left atrial appendage occlusion". Clinical guidelines, meta-analyses, systematic reviews, retrospective and prospective studies, narrative reviews, and case reports were included. Non-English-language articles and articles with unavailable full text were excluded from further analysis. The articles were considered eligible regarding their clinical relevance to the optimal agents and duration of DAPT in patients with cancer when DAPT is needed.

3. Biological and Clinical Aspects of Coagulation in Patients with Cancer

Malignancy may dysregulate hemostatic mechanisms, predisposing cancer patients to both thrombosis and hemorrhage [6]. Approximately 15% of patients with acute coronary syndrome (ACS) have concomitant cancer [7], including lung, prostate, gastric, pancreatic, and breast cancer [9]. The risk of venous thromboembolism (VTE) is fourfold to sevenfold higher in patients with active cancer [10], while approximately 10% of patients with solid cancer experience bleeding, and this incidence is even higher in patients with hematologic malignancies [6,11]. Moreover, according to the study of Guo et al., cancer patients who undergo PCI have a higher risk of thrombotic and ischemic events as well as bleeding after the procedure [12].

Thromboembolic events, which include arterial and venous thrombosis, thrombotic microangiopathy, non-bacterial thrombotic endocarditis, and veno-occlusive disease, can lead to ACS and ischemic stroke. Regarding bleeding, a fatal or a major bleeding event or an ongoing low-degree emission may happen, and it can be manifested either as a localized injury due to tumor invasion or as generalized bleeding predisposition [6,13]

These thromboembolic and bleeding manifestations, which are caused by the dysregulation of the hemostatic system provoked by the cancer, have been associated with both clinical and biological risk factors.

The clinical risk factors can be divided in three groups, regarding patient characteristics, cancer characteristics, and treatment characteristics [6,14–16], as shown in Figure 1. Concerning cancer characteristics, the incidence of VTE is higher in patients with pancreatic, gastric, and lung cancer; hematologic malignancies; and metastatic disease [17]. Malignancies that often cause bleeding include head and neck, lung, gastrointestinal, colorectal, and gynecologic malignancies; acute myelogenous leukemia; chronic lymphocytic leukemia; non-Hodgkin lymphoma; multiple myeloma; Waldenström's macroglobulinemia; and monoclonal gammopathy of unknown significance (MGUS) [18–20]. Patients with active lung cancer or colon cancer treated with PCI are more likely to have a 90-day readmission for acute myocardial infarction (AMI) after PCI, while patients with active colon cancer or metastatic cancer are more likely to have a 90-day readmission for bleeding after PCI [21].

As for biological factors, cancer cells can activate the hemostatic system by expressing and releasing molecules. Specifically, tumor cells activate the coagulation cascade by releasing procoagulant tissue factor, inflammatory cytokines, and microparticles, while they also activate endothelial cells, leukocytes, and platelets by expressing procoagulant proteins and releasing soluble factors [6,8].

Apart from the anticancer therapies, other causes of bleeding in oncologic patients are decreased synthesis of coagulation factors, vitamin K deficiency, excessive fibrinolysis, medication—such as anticoagulation and non-steroidal anti-inflammatory drugs—disseminated intravascular coagulation syndrome (DIC), acquired hemophilia, and acquired von Willebrand Disease [13,18,19].

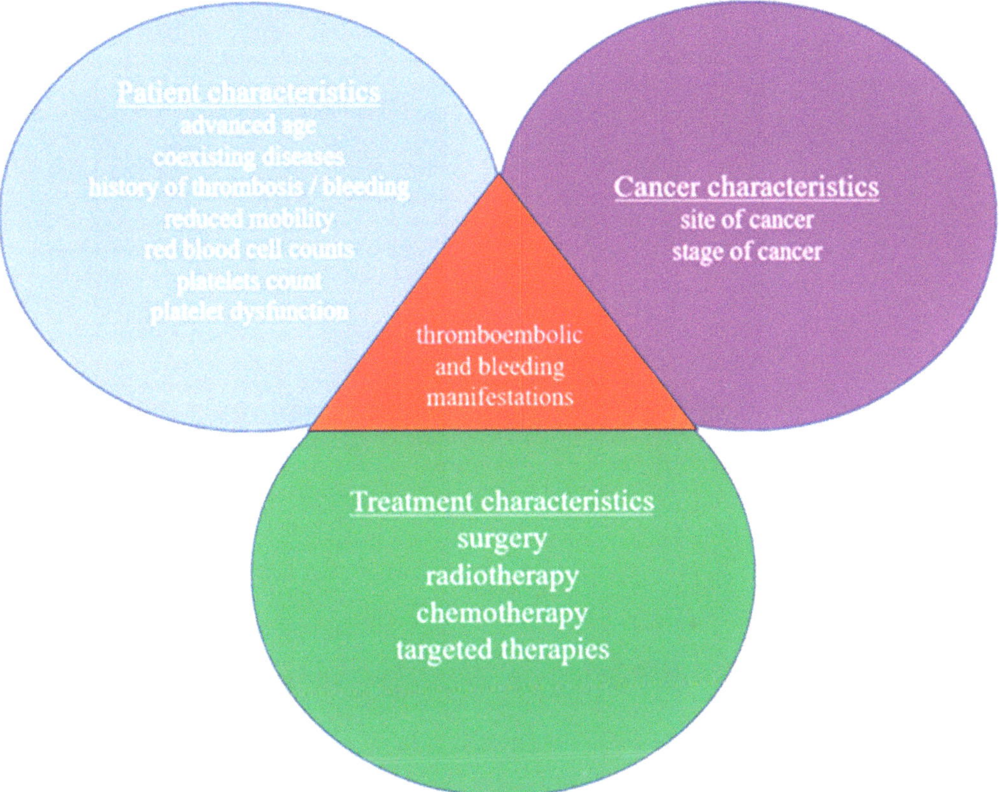

Figure 1. Clinical risk factors for both thromboembolic and bleeding manifestations in patients with cancer.

4. Cardiotoxicity Caused by Anticancer Treatment

Damage to the cardiovascular system can be caused by radiation treatment, chemotherapeutic agents, immunotherapies, and targeted therapies (Table 1). The mechanism by which anticancer treatment causes harm varies depending on the agent used [7].

Radiation treatment can cause endothelial injury, accelerated atherosclerosis, plaque rupture, and platelet aggregation since radiation produces free radicals, leading to oxidative stress, DNA damage, and inflammation [20,22].

Additionally, 5-fluorouracil is a chemotherapeutic drug most frequently used in breast, pancreatic, gastric, and colorectal cancer. A systematic review showed that patients treated with 5-fluorouracil developed chest pain as a result of myocardial infarction with ST elevation, most commonly during the first 2 days after administration [23]. The underlying mechanism is considered to be endothelium damage, which promotes inflammation, vasospasm, and plaque formation [24].

Another well-known chemotherapeutic agent, which is used mostly in patients with ovarian, testicular, or small cell lung cancer, is cisplatin. Long-term cardiac events may be related to LDL–HDL imbalance and endovascular damage caused by lipid peroxidation that causes platelet aggregation and thus thrombosis. In acute setting, cisplatin administration has been associated with vasospastic angina [25,26].

Bevacizumab is an anti-VEGF agent used as a first-line therapy for colorectal, lung, breast, and renal cancer. Ischemic heart disease is developed in one per 100 patients treated with bevacizumab and is developed due to endothelium dysfunction [27], while it has also be linked with hemorrhage and arterial thromboembolism [28].

Men with prostate cancer undergoing androgen deprivation therapy and women with breast cancer treated with aromatase inhibitors present an increased risk of cardiovascular events [29,30].

Tyrosine kinase inhibitors (TKIs), such as sorafenib and sunitinib, are both related with hypertension and ACS due to coronary vasospasm because TKIs decrease the vasodilator nitric oxide and increase the vasoconstrictor endothelin-1 [31]. Patients treated with nilotinib and ponatinib develop acute coronary occlusion and myocardial infarction due to progression of atherosclerosis [32].

Immune checkpoint inhibitors (monoclonal antibodies that block the immune brakers or regulators), such as ipilimumab (cytotoxic T lymphocyte-associated antigen-4 inhibitor), nivolumab (programmed death-1 inhibitor), and atezolizumab (programmed death-ligand 1 inhibitor), are related to major cardiovascular events such as myocardial infarction due to the acceleration of atherosclerosis and plaque rupture [33].

The use of the immunomodulatory drugs lenalidomide and pormalidomide in patients with multiple myeloma is associated with increased risk of ACS, but the underlying mechanism needs further investigation [34].

Ibrutinib, which reduces mortality in several B-cell malignancies and chronic lymphocytic leukemia, is associated with atrial fibrillation and increased bleeding risk [35].

Table 1. Agents associated with cardiovascular dysfunction.

Treatment	Incidence	Mechanism
Radiation [22]	Depends on the prescribed dose and the cardiac radiation exposure	Endothelial injury, acceleration of CAD, ACS
5-Fluorouracil [23,24]	2–18%	ACS, vasospasm
Cisplatin [25,26]	0.2–12%	ACS, acute thrombosis, acceleration of CAD
Bevacizumab [28,36]	0.52–1.7%	ACS, acute thrombosis
Leuprolide (GNRH agonist) [37]	2.6–5.6%	Angina, ACS, acceleration of CAD
Anastrozole (aromatase inhibitor) [29]	2%	ACS
Tyrosine kinase inhibitors:		
Sorafenib [31]	1%	Acute thrombosis
Sunitinib [31]	5–8%	Acute thrombosis, acceleration of CAD
Nilotinib [20,32]	8–12%	ACS, acceleration of CAD, AF
Ponatinib [32]	2%	ACS, acceleration of CAD
Ibrutinib [35]	8.8%	Bleeding diathesis, AF

ACS, acute coronary syndrome; CAD, coronary artery disease; AF, atrial fibrillation.

5. DAPT in Patients with Cancer Undergoing Elective PCI

Since cancer patients undergoing elective PCI have an increased ischemic and bleeding risk, the appropriate antiplatelet therapy remains a challenge. Clopidogrel is the main P2Y12 inhibitor used in these patients since prasugrel and ticagrelor have been associated with more bleeding events, and there are no data in the literature regarding their safety in cancer patients [38].

New technologies entering our quiver, such as new generation drug-eluted stents (DESs), have led to the possibility of shortening the DAPT duration to a minimum of 1 month [39]. After the first month, there is the possibility of extending DAPT up to 3–6 months depending on the patient's ischemic and bleeding risk, the type and stage of cancer, the need for surgery, and the current cancer treatment [7,20,40]. In their study of 75 patients undergoing PCI with drug-eluting stents (DESs), Balanescu et al. reported that the discontinuation of DAPT at 6 months after DES implantation did not increase the incidence of in-stent thrombosis and restenosis [41]. The shortening of DAPT was feasible and safe using newer generation DESs; thus, cancer therapies with high bleeding risk can be administered more quickly, resulting in potential survival benefits. The authors suggested that cancer therapies can be safely started again at <6 months and as early as 2 weeks after PCI with DESs. However, a retrospective, observational study comparing the outcomes of using bare metal stents (BMSs) or DESs in cancer patients with CAD did not show any significant difference between the number of revascularizations nor the all-cause mortality between cancer patients with CAD treated with BMSs versus DESs during a follow-up period of 34.1 months [42].

Finally, an alternative strategy for these high-risk patients to allow early DAPT discontinuation could be the evaluation of the coverage of the stent's struts with optical coherence tomography (OCT) [43,44]. In the PROTECT-OCT study, cancer patients who had a recent DES placement (1–12 months) and had to discontinue DAPT prematurely were evaluated using coronary angiograms and OCT [45]. Patients with satisfactory characteristics, such as appropriate stent strut coverage, expansion, apposition, and absence of in-stent restenosis or intraluminal masses, were considered low risk, and DAPT was discontinued, while the remaining patients were considered high risk and stopped DAPT after bridging with low-molecular-weight heparin. In a total of 40 patients, no cardiovascular event occurred in the low-risk group, and only one myocardial infarction occurred in the high-risk group, suggesting that the use of OCT could be useful in the management of this group of patients [45]. Nevertheless, further studies with more patients are required to exact more reliable conclusions.

However, the decision between optimal medical therapy and invasive therapy should be individualized, taking into consideration the cancer prognosis, type of cancer, cancer treatment, and patients' ischemic and bleeding risks, and it should be made after an extensive discussion between the various specialties involved [46].

6. DAPT in Patients with ACS and Cancer

Available data concerning ACS management among cancer patients are limited, making the clinical decision a challenge. Generally, treatment should be personalized according to the ACS subtype, the stage and type of cancer, and the patient prognosis [47], and cancer therapy should be temporarily interrupted, especially if a causal relationship is suspected [48].

The management of patients with cancer presenting with an ACS often requires a multidisciplinary and individualized approach [48]. An invasive strategy should be preferred in ST elevation myocardial infarction (STEMI) patients, as well as in NSTEMI patients who are unstable or are considered high risk. The use of third-generation DESs is indicated because of their lower risk of thrombosis and the need for a shorter duration of DAPT. On the contrary, in clinically stable NSTEMI patients, a conservative non-invasive strategy could be adopted, especially in the case of poor life expectancy and/or of a high risk of bleeding, such as patients with metastases, coagulopathies, or thrombocytopenia [49].

DAPT required after PCI poses a great concern in cancer patients, limiting the use of an invasive strategy. DAPT consisting of aspirin and clopidogrel is recommended in these patients, especially in cancer patients with a recent diagnosis (<1 year) or other coexisting bleeding risk factors [10,17]. On the contrary, newer P2Y12 antagonists, such as ticagrelor and prasugrel, should be avoided due to their high bleeding risk and the lack of data on this patient subset. However, ticagrelor or prasugrel may be used under strict surveillance of the bleeding risk in specific patients with previous stent thrombosis during treatment with clopidogrel [48]. According to the 2022 ESC Guidelines on cardio-oncology, the duration of DAPT should be as short as possible, with 1–3 months being proposed as the optimal duration [49–51]. If urgent surgery is necessary, interruption of clopidogrel is recommended, as in non-cancer patients [48]. We suggest that a 6-month DAPT may be considered in specific patients with cancer and ACS who are of high ischemic risk, according to risk criteria for extended treatment with a second antithrombotic agent in "2020 ESC Guidelines for the Management of Acute Coronary Syndromes in Patients Presenting without Persistent ST-Segment Elevation" [52] under careful monitoring (Figure 2).

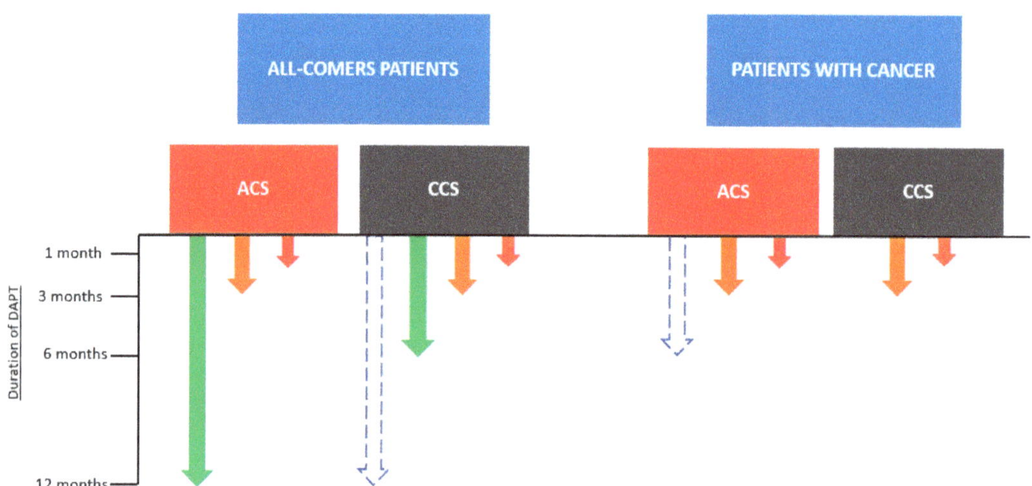

Figure 2. Strategies regarding the duration of DAPT after elective PCI and ACS in all-comers and cancer patients.

In Figure 2, for all-comer patients, green indicates a low bleeding risk according to "2019 ESC Guidelines for the Diagnosis and Management of Chronic Coronary Syndromes" [53] and "2020 ESC Guidelines for the Management of Acute Coronary Syndromes in Patients Presenting without Persistent ST-Segment Elevation" [52]. Orange indicates a high bleeding risk according to "2019 ESC Guidelines for the Diagnosis and Management of Chronic Coronary Syndromes" [53] and "2020 ESC Guidelines for the Management of Acute Coronary Syndromes in Patients Presenting without Persistent ST-Segment Elevation" [52]. Red indicates a very high bleeding risk according to "2019 ESC Guidelines for the Diagnosis and Management of Chronic Coronary Syndromes" [53] and "2020 ESC Guidelines for the Management of Acute Coronary Syndromes in Patients Presenting without Persistent ST-Segment Elevation" [52]. The dashed blue arrow indicates that in patients with high thrombotic risk and CCS (as described in "2020 ESC Guidelines for the Management of Acute Coronary Syndromes in Patients Presenting without Persistent ST-Segment Elevation" [52]), 12-month DAPT could be considered.

In Figure 2, for patients with cancer, orange indicates active malignancy (excluding non-melanoma skin cancer) within the past 12 months without any other bleeding risk factors. Red indicates active malignancy (excluding non-melanoma skin cancer) within the past 12 months plus at least one major or two minor criteria for high bleeding risk according to the Academic Research Consortium for High Bleeding Risk at the time of percutaneous coronary intervention [52], or according to criteria in "2022 ESC Guidelines on Cardio-Oncology Developed in Collaboration with the European Hematology Association (EHA), the European Society for Therapeutic Radiology and Oncology (ESTRO) and the International Cardio-Oncology Society (IC-OS)" [20], namely: a high risk of gastrointestinal or genitourinary bleeding, significant drug–drug interactions, severe renal dysfunction (creatinine clearance < 30 mL/min), significant liver disease (alanine aminotransferase/aspartate aminotransferase > 2 × ULN), or significant thrombocytopenia (platelet count < 50,000/μL). The dashed blue arrow indicates that in patients with active malignancy and ACS who are of high ischemic risk (as described in "2020 ESC Guidelines for the Management of Acute Coronary Syndromes in Patients Presenting without Persistent ST-Segment Elevation" [52]), 6-month DAPT could be considered. ACS stands for acute coronary syndrome, and CCS stands for chronic coronary syndrome.

Furthermore, the platelet count should be taken into consideration when DAPT is administered in patients with cancer. Aspirin is allowed if the platelet count is >10,000/μL, while DAPT initiation (with aspirin and clopidogrel) is allowed if the platelet count is >30,000/µL [7,54]. Ticagrelor, prasugrel, and glycoprotein IIb/IIIa inhibitors should be used with more caution in cancer patients and should be avoided in patients with a platelet count <50,000/µL/ [55]. In addition, if the platelet count is <20,000/µL, prophylactic platelet transfusion may be considered [56]. Taking into account the platelet count and the need for urgent surgery or chemotherapy, Radmilovic et al. also suggested a protocol [7] regarding the management of antiplatelet therapy in cancer patients taking into account the platelet count (Figure 3).

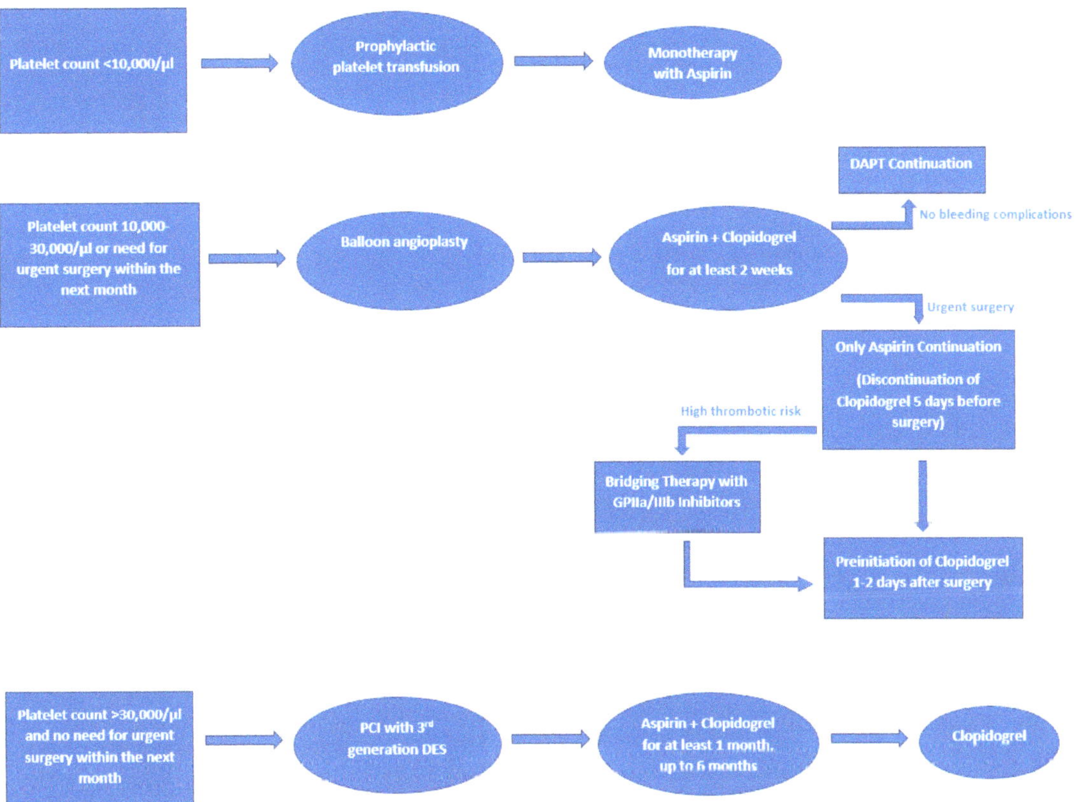

Figure 3. Modified algorithm of Radmilovic et al. for the management of antiplatelet therapy in cancer patients. DES = drug-eluting stent, DAPT = dual antiplatelet therapy, PCI = percutaneous coronary intervention.

Bleeding in patients with ACS increases mortality and requires clinical decisions on the continuation of DAPT. Bleeding could lead to anemia, which is an independent risk factor for ACS [57]. The severity of bleeding should be taken into consideration when deciding the continuation or discontinuation of antiplatelet therapy. In this way, DAPT can be maintained in cases of minor bleeding (such as hematomas). On the contrary, DAPT should be stopped in cases of severe bleeding (such as need for hospitalization or >2g/dL decrease in hemoglobin levels), and monotherapy with clopidogrel should be considered thereafter. In the case of life-threatening bleeding, all antiplatelet agents should be discontinued [7].

In many cancer patients, comorbidities such as atrial fibrillation (AF), venous thromboembolism, and valvular heart disease often coexist. However, triple antithrombotic therapy (TAT), which would be indicated in the absence of cancer, is not advised because of the significantly higher risk of bleeding [49]. Thus, the administration of a novel oral anticoagulant (NOAC) and a single oral antiplatelet agent (preferably clopidogrel) is preferred after a short period of triple antithrombotic therapy (TAT) (up to 1 week in the hospital) [48].

It should be noted that several scores (such as PARIS and DAPT) used to assess the bleeding risk regarding the duration of antiplatelet therapy after PCI have not been validated in patients with malignancy [49], while the PRECISE-DAPT score did not perform well for predicting bleeding in oncologic patients [58]. In addition, cancer has not been included in the most common risk scores, such as CHA2DS2-VASc and HAS-BLED, making the clinical decision of balancing the higher ischemic and bleeding risk even more difficult [49]. A risk assessment model (RAM) for VTE that is applicable to patients with specific types of solid tumor after the initiation of anticancer therapy is the COMPASS–CAT RAM [59].

Although platelet function testing (PFT) is not performed on a routine basis in patients with ACS or stable CAD, it may be a useful tool for guiding antiplatelet treatment escalation in patients with high platelet reactivity (HPR) on clopidogrel and screening for HPR on clopidogrel when DAPT de-escalation is necessary in complex cases [60]. Patients with advanced cancer display platelet hyperreactivity [61], with a higher number of platelets stably adhering to von Willebrand factor (VWF) and greater platelet surface coverage compared with those patients with early-stage cancer [62]. On the contrary, in patients with acute myeloid leukemia (AML) and thrombocytopenia, reduced platelet aggregation and platelet activation predict bleeding better than platelet count alone [63]. Thus, since platelet function is altered in patients with cancer, PFT may be useful in adjusting DAPT in oncologic patients, while taking into consideration their ischemic and bleeding risk.

7. Antiplatelet Therapy in Patients with Cancer Undergoing CABG

As complex PCI has become a reasonable and safe choice in daily practice in the majority of catheterization laboratories, the overall utilization of coronary artery bypass grafting (CABG) has decreased over time (250,677 in 2003 vs. 134,534 in 2015), while the proportion of those with comorbid cancer undergoing CABG has increased (7% vs. 12.6%, $p < 0.001$) [64].

Although most cancer patients with CAD are treated conservatively or with PCI, given the prevalence of complex coronary disease and the potential challenges of prolonged anticoagulation therapy in the presence of cancer, CABG may be sometimes the best option for these patients, mainly for those without active cancer or those with >1 year life expectancy [20]. Moreover, according to Guha et al., the presence of breast, lung, prostate, and colon cancer and lymphoma does not appear to be associated with increased in-hospital mortality in cancer versus non-cancer patients with CABG. However, there is a higher bleeding risk in CABG patients with breast and prostate cancer compared with non-cancer patients with CABG [64]. Even in non-cancer patients undergoing CABG, guidelines and clinical practice are not uniform and specific regarding DAPT therapy, especially in the setting of chronic coronary syndrome (CCS). According to the latest guidelines of the European Association for Cardio-Thoracic Surgery [65] and the American Heart Association [66], there is limited evidence regarding DAPT after CABG in CCS. Therefore, based on extensive evidence, aspirin is strongly recommended (class IA recommendation) for all patients after CABG by both American and European guidelines, whereas the use of DAPT in CABG patients without a separate indication (e.g., ACS) is graded as class IIb, meaning they only provide some benefit [67,68]. Thus, cancer patients undergoing CABG for CCS should be treated with a single antiplatelet agent.

Many patients with malignancies suffer from atrial fibrillation (AF) and/or VTE for which they should take anticoagulants. For this population undergoing CABG, a short course of combined antithrombotic therapy with an antiplatelet and an NOAC is recommended, followed by monotherapy with an NOAC lifelong [65].

8. Antithrombotic Therapy in Patients with AF and Cancer Undergoing PCI or Suffering from ACS

Different cancer types are associated with different bleeding risk profiles. Active cancers (especially hematologic malignancies and gastrointestinal cancers) and existing metastases increase the bleeding risk [69]. Moreover, cancer patients have a higher rate of bleeding after PCI compared with non-cancer patients [12]. Taking this into consideration, along with the fact that cancer patients with AF are already under anticoagulation therapy, it is preferred to keep DAPT as short as possible in cancer patients with AF after stent implantation or ACS.

According to the 2022 ESC Guidelines on Cardio-Oncology, when both anticoagulation and antiplatelet therapy are needed, TAT can be administered for a short period of time (up to 1 week in the hospital), and then an NOAC and single oral antiplatelet agent (preferably clopidogrel) is the default strategy [20]. The combination of NOACs plus a P2Y12 inhibitor was associated with less bleeding without a significant difference in major adverse cardiovascular events (MACE), compared with the use of vitamin K antagonists (VKA) plus DAPT [70–73]. A combination of VKA plus DAPT should be avoided due to dramatically increasing bleeding complications [72].

In patients treated with oral anticoagulants (OACs) undergoing PCI, bleeding complications occur mostly in the first period of treatment, and this risk remains elevated over time [74,75]. Therefore, in patients with additional risk factors for bleeding, the duration of aspirin therapy should not exceed the peri-PCI period, namely, during inpatient stay, until the time of discharge [74].

Clopidogrel is the most studied P2Y12 inhibitor in (≈88%) patients enrolled in trials of AF patients treated with an NOAC undergoing PCI [76–80]. Prasugrel should not be used concomitantly with an OAC, while ticagrelor may be a good alternative to clopidogrel for specific cases of cancer patients [74]. Considering that cancer patients treated with an OAC are at a high risk of bleeding, the duration of the P2Y12 inhibitor administration should be as short as possible after PCI or ACS, and then patients should continue with the OAC at the appropriate dose. Whether the P2Y12 inhibitor would be discontinued after 1, 3, or 6 months or in between probably depends on the specific profile of the patient and is up to the discretion of the treating physicians [74].

9. Antiplatelet Therapy in Patients with Cancer Undergoing Cardiac Structural Interventions

Apart from PCI, there are also other structural interventions, such transcatheter aortic valve replacement (TAVR), patent foramen ovale (PFO), or atrial septal defect (ASD) closure and left atrial appendage (LAA) occlusion, which include device implantation and require the appropriate antiplatelet therapy as a prevention measure for thrombosis.

9.1. TAVR

Nowadays, TAVR has gained significant ground in the management of aortic stenosis. While it was applied mainly in very high-risk patients, recent data support that TAVR is a feasible and safe option even for low-risk patients [81,82]. Moreover, TAVR is indicated for oncologic patients with active cancer or cancer in remission, as the existing literature supports that TAVR should be preferred when compared with medical treatment or surgical replacement [83–85].

The optimal antiplatelet post-TAVR therapy remains under investigation [86]. Capodanno and colleagues suggested that single-antiplatelet treatment (SAPT) should be administered in post-TAVR patients without any indication for DAPT. Aspirin should be preferred, whereas clopidogrel could be the alternative option [87]. The only indication in which DAPT should be chosen is for patients who need DAPT for another reason, such as coronary stenting during the last 3 months. In these patients, an individualized approach should be followed, and DAPT could be administered for no more than 6 months; then, it should be replaced with SAPT [88]. On the contrary, the recent OCEAN-TAVI Registry showed that the nonantithrombotic strategy after TAVR does not increase the risk of net adverse clinical events and reduces the bleeding risk in patients who do not need anticoagulation therapy [89], leading to the conclusion that a nonantithrombotic approach after TAVR may be feasible in specific oncologic patients with high bleeding risk. However, future studies are required to establish the necessity, duration, and agents of antiplatelet therapy after TAVR in patients with cancer.

9.2. PFO-ASD Closure

Thanks to the progress of interventional cardiology, transcatheter PFO and ASD occlusion have been established as feasible and safe procedures. After the procedure and until complete endothelization of the device, DAPT is required mainly for thrombosis prevention and secondarily for nickel release inhibition. Endothelization is estimated to be completed in 3–6 months, so the duration of DAPT should be adapted respectively [90–92]. Regarding the duration of DAPT in this field, the existing literature lacks large-scale, randomized trials to provide adequate data; current practice is established on based consensus statements and empirical approaches. Recently, Pristipino et al. published the first European position on the management of patients with PFO [93]. Based on current studies, the experts advised using DAPT for 1–6 months (strength: conditional, evidence level: A), which should be followed by SAPT with aspirin for at least 5 years (strength: conditional, evidence level: C) [94,95]. Interatrial shunt closure in the setting of active cancer remains poorly investigated. Taking into consideration the thrombogenicity of malignancies, percutaneous PFO closure could theoretically be beneficial acting as a protective shield against thrombus formation and embolization to cerebral circulation. Further studies are required to evaluate the benefit/harm ratio in patients with active malignancy undergoing PFO closure as a secondary prevention strategy [96].

9.3. LAA Occlusion

Fatal strokes are the main mortality cause in patients with AF, while the emboli are created in the LAA. Studies have reported that surgical LAA occlusion has been associated with reduced incidence of both fatal and non-fatal strokes. Notably, cancer patients who need anticoagulation therapy due to AF are commonly candidates for this specific intervention due to their high bleeding and ischemic risk [97]. However, no common line exists regarding the most appropriate antiplatelet therapy for either patients with cancer or generally people undergoing LAA closure. Chen et al. supported that either short-term DAPT for 6 weeks or SAPT should be preferred due to the hemorrhagic risk in people undergoing LAA closure [98]. A newer study showed that SAPT or even no therapy does not increase the ischemic risk [99], while a recent, non-randomized study found that SAPT instead of DAPT after LAA occlusion was associated with a reduction of bleeding complications, with no significant increase in the risk of thrombotic events [100]. Therefore, a tailored approach should be followed, pending for large-scale, suitably designed clinical trials.

10. Antiplatelet Therapy for Non-Cardiac Diseases in Patients with Cancer

Antiplatelet therapy also plays a pivotal role in non-cardiac diseases, such as peripheral artery disease (PAD) and cerebrovascular accidents (CVAs).

10.1. PAD

PAD often coexists with CAD, hypertension, dyslipidemia, and diabetes mellitus. Nowadays, the management of PAD includes percutaneous stent implantation regardless of the location of the lesions. Thanks to the newer drug-eluting stents, carotid artery stenting (CAS) has become a safe and feasible approach with comparable results to surgery. According to the recent ESC Guidelines about PAD [101], stent implantation in the carotid artery should be followed by DAPT (aspirin + clopidogrel) for 1 month (class IA recommendation). After this time frame, DAPT should be replaced with SAPT, with either aspirin or clopidogrel. A similar approach should be followed in patients with lower-extremities artery disease. After percutaneous revascularization, DAPT administration for 30 days is required prior to switching to SAPT (aspirin or clopidogrel). However, the existing literature for lower-extremities artery disease lacks large-scale, randomized studies, so the strength of evidence is limited (class IIa C). To date, special recommendations for cancer patients with concomitant PAD are not available. Thus, application of the guidelines relevant to the general population in this subpopulation should be considered. Nevertheless, a personalized approach based on the ischemic and bleeding risk of each patient should be followed.

10.2. CVAs

CVAs remain one of the major causes of mortality and disability globally. The progress of imaging techniques, reperfusion therapy, and improved medical treatment during the last decades has significantly increased the life expectancy of these patients [102]. Although the optimal antithrombotic treatment is important to minimize the incidence of ischemic CVAs, the optimal regimen remains under investigation.

A recent guideline by the European Stroke Organization (ESO) strongly advises the administration of DAPT (aspirin + clopidogrel) for 21 days in patients with a non-cardioembolic minor ischemic stroke or high-risk transient ischemic attack (TIA) during the last 24 h. Moreover, the experts recommend that DAPT (aspirin + ticagrelor) for 30 days may be beneficial in patients with non-cardioembolic mild-to-moderate ischemic stroke or high-risk TIA in the last 24 h [103].

According to a recent meta-analysis of randomized trials, short-term (for up to 3 months) DAPT seems to reduce the risk of recurrent stroke at the expense of a higher risk of major bleeding, compared with aspirin, in patients with high-risk TIA or mild to moderate ischemic strokes [104].

Cancer patients are at higher risk of suffering from acute CVAs and fatal strokes. In particular, patients with prostate, breast, and colorectum malignancies are more prone to fatal strokes [105]. Recently, Bang and colleagues [106] proposed that cancer-related strokes could be an emerging subtype of ischemic stroke, with unique underlying pathophysiological mechanisms. However, the existing literature and current evidence cannot adequately support the precise and tailored antithrombotic management of these patients.

11. Conclusions

Since oncologic patients are at high risk for both ischemic and bleeding events due to the dysregulation of their hemostatic system by cancer, the appropriate duration and the optimal agents of antiplatelet therapy after undergoing PCI and/or suffering from an ACS remain a challenge. The use of new technologies, such as DESs and OCT, may lead to shortened DAPT duration in all-comer patients, including patients with cancer. The optimal duration of DAPT is considered to be 1–3 months, consisting of aspirin and clopidogrel, while TAT can only be administered for a short period of time (up to 1 week in the hospital), followed by an NOAC and a single oral antiplatelet agent (preferably clopidogrel). Other structural interventions, such as TAVR, PFO-ASD closure, and LAA occlusion, and non-cardiac diseases, such as PAD and CVA, may require DAPT. Although further studies are needed in order to establish the optimal duration and agents of DAPT, it is indisputable that a personalized and multidisciplinary approach is necessary to increase the life expectancy and quality of life of patients with cancer and CVD, along with finding the balance between thrombotic and bleeding risk.

Author Contributions: Conceptualization, G.T. and A.V.; methodology, G.T.; writing—original draft preparation, G.T., A.V., A.A. (Anastasios Apostolos), E.B., N.V.-B., G.V., E.-E.K., I.T., M.H., A.A. (Adel Aminian) and P.D.; writing—review and editing, G.T., A.V., A.A. (Anastasios Apostolos), E.B., N.V.-B., I.T., M.H., A.A. (Adel Aminian) and P.D.; supervision, G.T. and P.D. All authors have read and agreed to the published version of the manuscript.

Funding: This research received no external funding.

Institutional Review Board Statement: Not applicable.

Informed Consent Statement: Not applicable.

Data Availability Statement: Not applicable.

Conflicts of Interest: The authors declare no conflict of interest.

References

1. Marenzi, G.; Cosentino, N.; Cardinale, D. Ischaemic and bleeding risk in cancer patients undergoing PCI: Another brick in the wall. *Eur. Heart J.* **2021**, *42*, 1035–1037. [CrossRef] [PubMed]
2. Koene, R.J.; Prizment, A.E.; Blaes, A.; Konety, S.H. Shared Risk Factors in Cardiovascular Disease and Cancer. *Circulation* **2016**, *133*, 1104–1114. [CrossRef] [PubMed]
3. Anker, M.S.; Hadzibegovic, S.; Lena, A.; Belenkov, Y.; Bergler-Klein, J.; de Boer, R.A.; Farmakis, D.; von Haehling, S.; Iakobishvili, Z.; Maack, C.; et al. Recent advances in cardio-oncology: A report from the "Heart Failure Association 2019 and World Congress on Acute Heart Failure 2019". *ESC Heart Fail.* **2019**, *6*, 1140–1148. [CrossRef] [PubMed]
4. de Boer, R.A.; Meijers, W.C.; van der Meer, P.; van Veldhuisen, D.J. Cancer and heart disease: Associations and relations. *Eur. J. Heart Fail.* **2019**, *21*, 1515–1525. [CrossRef] [PubMed]
5. Singh, S.; Singh, K. Atherosclerosis, Ischemia, and Anticancer Drugs. *Heart Views* **2021**, *22*, 127.
6. Falanga, A.; Marchetti, M.; Vignoli, A. Coagulation and cancer: Biological and clinical aspects. *J. Thromb. Haemost.* **2013**, *11*, 223–233. [CrossRef]
7. Radmilovic, J.; Di Vilio, A.; D'andrea, A.; Pastore, F.; Forni, A.; Desiderio, A.; Ragni, M.; Quaranta, G.; Cimmino, G.; Russo, V.; et al. The Pharmacological Approach to Oncologic Patients with Acute Coronary Syndrome. *J. Clin. Med.* **2020**, *9*, 3926. [CrossRef]
8. Falanga, A.; Russo, L.; Milesi, V. The coagulopathy of cancer. *Curr. Opin. Hematol.* **2014**, *21*, 423–429. [CrossRef]
9. Rohrmann, S.; Witassek, F.; Erne, P.; Rickli, H.; Radovanovic, D. Treatment of patients with myocardial infarction depends on history of cancer. *Eur. Heart J. Acute Cardiovasc. Care* **2018**, *7*, 639–645. [CrossRef]
10. Lee, L.H.; Nagarajan, C.; Tan, C.W.; Ng, H.J. Epidemiology of Cancer-Associated Thrombosis in Asia: A Systematic Review. *Front. Cardiovasc. Med.* **2021**, *8*, 435. [CrossRef]
11. Johnstone, C.; Rich, S.E. Bleeding in cancer patients and its treatment: A review. *Ann. Palliat. Med.* **2018**, *7*, 265–273. [CrossRef] [PubMed]
12. Guo, W.; Fan, X.; Lewis, B.R.; Johnson, M.P.; Rihal, C.S.; Lerman, A.; Herrmann, J. Cancer Patients Have a Higher Risk of Thrombotic and Ischemic Events After Percutaneous Coronary Intervention. *JACC. Cardiovasc. Interv.* **2021**, *14*, 1094–1105. [CrossRef] [PubMed]
13. Kamphuisen, P.W.; Beyer-Westendorf, J. Bleeding complications during anticoagulant treatment in patients with cancer. *Thromb. Res.* **2014**, *133* (Suppl. S2), S49–S55. [CrossRef] [PubMed]
14. Grover, S.P.; Hisada, Y.M.; Kasthuri, R.S.; Reeves, B.N.; MacKman, N. Cancer Therapy-Associated Thrombosis. *Arterioscler. Thromb. Vasc. Biol.* **2021**, *41*, 1291–1305. [CrossRef]
15. Wilts, I.T.; Bleker, S.M.; Van Es, N.; Büller, H.R.; Di Nisio, M.; Kamphuisen, P.W. Safety of anticoagulant treatment in cancer patients. *Expert Opin. Drug Saf.* **2015**, *14*, 1227–1236. [CrossRef]
16. Watson, N.; Al-Samkari, H. Thrombotic and bleeding risk of angiogenesis inhibitors in patients with and without malignancy. *J. Thromb. Haemost.* **2021**, *19*, 1852–1863. [CrossRef]
17. Wun, T.; White, R.H. Epidemiology of cancer-related venous thromboembolism. *Best Pract. Res. Clin. Haematol.* **2009**, *22*, 9–23. [CrossRef]
18. Mantha, S. Bleeding Disorders Associated with Cancer. *Cancer Treat. Res.* **2019**, *179*, 191–203.
19. Prommer, E. Management of bleeding in the terminally ill patient. *Hematology* **2005**, *10*, 167–175. [CrossRef]
20. Lyon, A.R.; López-Fernández, T.; Couch, L.S.; Asteggiano, R.; Aznar, M.C.; Bergler-Klein, J.; Boriani, G.; Cardinale, D.; Cordoba, R.; Cosyns, B.; et al. 2022 ESC Guidelines on cardio-oncology developed in collaboration with the European Hematology Association (EHA), the European Society for Therapeutic Radiology and Oncology (ESTRO) and the International Cardio-Oncology Society (IC-OS). *Eur. Heart J.* **2022**, *43*, 4229–4361. [CrossRef]
21. Kwok, C.S.; Wong, C.W.; Kontopantelis, E.; Barac, A.; Brown, S.A.; Velagapudi, P.; Hilliard, A.A.; Bharadwaj, A.S.; Chadi Alraies, M.; Mohamed, M.; et al. Percutaneous coronary intervention in patients with cancer and readmissions within 90 days for acute myocardial infarction and bleeding in the USA. *Eur. Heart J.* **2021**, *42*, 1019–1034. [CrossRef] [PubMed]
22. Belzile-Dugas, E.; Eisenberg, M.J. Radiation-Induced Cardiovascular Disease: Review of an Underrecognized Pathology. *J. Am. Heart Assoc. Cardiovasc. Cerebrovasc. Dis.* **2021**, *10*, 21686. [CrossRef] [PubMed]
23. Das, S.K.; Das, A.K.; William, M. 5-Fluorouracil-induced acute coronary syndrome. *Med. J. Aust.* **2019**, *211*, 255–257.e1. [CrossRef] [PubMed]
24. Polk, A.; Vaage-Nilsen, M.; Vistisen, K.; Nielsen, D.L. Cardiotoxicity in cancer patients treated with 5-fluorouracil or capecitabine: A systematic review of incidence, manifestations and predisposing factors. *Cancer Treat. Rev.* **2013**, *39*, 974–984. [CrossRef] [PubMed]

25. Hanchate, L.P.; Sharma, S.R.; Madyalkar, S. Cisplatin Induced Acute Myocardial Infarction and Dyslipidemia. *J. Clin. Diagn. Res.* **2017**, *11*, OD05–OD07. [CrossRef] [PubMed]
26. Karabay, K.O.; Yildiz, O.; Aytekin, V. Multiple coronary thrombi with cisplatin. *J. Invasive Cardiol.* **2014**, *26*, E18–E20.
27. Chen, X.L.; Lei, Y.H.; Liu, C.F.; Yang, Q.F.; Zuo, P.Y.; Liu, C.Y.; Chen, C.Z.; Liu, Y.W. Angiogenesis inhibitor bevacizumab increases the risk of ischemic heart disease associated with chemotherapy: A meta-analysis. *PLoS ONE* **2013**, *8*, e66721. [CrossRef]
28. Kounis, N.G.; Soufras, G.D.; Tsigkas, G.; Hahalis, G. Adverse cardiac events to monoclonal antibodies used for cancer therapy. *Oncoimmunology* **2014**, *3*, e27987. [CrossRef]
29. Amir, E.; Seruga, B.; Niraula, S.; Carlsson, L.; Ocaña, A. Toxicity of adjuvant endocrine therapy in postmenopausal breast cancer patients: A systematic review and meta-analysis. *J. Natl. Cancer Inst.* **2011**, *103*, 1299–1309. [CrossRef]
30. Ziehr, D.R.; Chen, M.H.; Zhang, D.; Braccioforte, M.H.; Moran, B.J.; Mahal, B.A.; Hyatt, A.S.; Basaria, S.S.; Beard, C.J.; Beckman, J.A.; et al. Association of androgen-deprivation therapy with excess cardiac-specific mortality in men with prostate cancer. *BJU Int.* **2015**, *116*, 358–365. [CrossRef]
31. Naib, T.; Steingart, R.M.; Chen, C.L. Sorafenib-associated multivessel coronary artery vasospasm. *Herz* **2011**, *36*, 348–351. [CrossRef] [PubMed]
32. Costa, I.B.S.D.S.; Andrade, F.T.D.A.; Carter, D.; Seleme, V.B.; Costa, M.S.; Campos, C.M.; Hajjar, L.A. Challenges and Management of Acute Coronary Syndrome in Cancer Patients. *Front. Cardiovasc. Med.* **2021**, *8*, 590016. [CrossRef] [PubMed]
33. Drobni, Z.D.; Alvi, R.M.; Taron, J.; Zafar, A.; Murphy, S.P.; Rambarat, P.K.; Mosarla, R.C.; Lee, C.; Zlotoff, D.A.; Raghu, V.K.; et al. Association Between Immune Checkpoint Inhibitors With Cardiovascular Events and Atherosclerotic Plaque. *Circulation* **2020**, *142*, 2299–2311. [CrossRef] [PubMed]
34. Cornell, R.F.; Ky, B.; Weiss, B.M.; Dahm, C.N.; Gupta, D.K.; Du, L.; Carver, J.R.; Cohen, A.D.; Engelhardt, B.G.; Garfall, A.L.; et al. Prospective Study of Cardiac Events During Proteasome Inhibitor Therapy for Relapsed Multiple Myeloma. *J. Clin. Oncol.* **2019**, *37*, 1946–1955. [CrossRef]
35. Abdel-Qadir, H.; Sabrie, N.; Leong, D.; Pang, A.; Austin, P.C.; Prica, A.; Nanthakumar, K.; Calvillo-Argüelles, O.; Lee, D.S.; Thavendiranathan, P. Cardiovascular Risk Associated With Ibrutinib Use in Chronic Lymphocytic Leukemia: A Population-Based Cohort Study. *J. Clin. Oncol.* **2021**, *39*, 3453–3462. [CrossRef]
36. Kapelakis, I.; Toutouzas, K.; Drakopoulou, M.; Michelongona, A.; Zagouri, F.; Mpamias, A.; Pliatsika, P.; Dimopoulos, M.-A.; Stefanadis, C.; Tousoulis, D. Bevacizumab increases the incidence of cardiovascular events in patients with metastatic breast or colorectal cancer. *Hell. J. Cardiol.* **2016**, *58*, 215–219. [CrossRef]
37. Lopes, R.D.; Higano, C.S.; Slovin, S.F.; Nelson, A.J.; Bigelow, R.; Sørensen, P.S.; Melloni, C.; Goodman, S.G.; Evans, C.P.; Nilsson, J.; et al. Cardiovascular Safety of Degarelix Versus Leuprolide in Patients With Prostate Cancer: The Primary Results of the PRONOUNCE Randomized Trial. *Circulation* **2021**, *144*, 1295–1307. [CrossRef]
38. Banasiak, W.; Zymlinski, R.; Undas, A. Optimal management of cancer patients with acute coronary syndrome. *Polish Arch. Intern. Med.* **2018**, *128*, 244–253. [CrossRef]
39. Apostolos, A.; Trigka, A.; Chlorogiannis, D.; Vasilagkos, G.; Chamakioti, M.; Spyropoulou, P.; Karamasis, G.; Dimitriadis, K.; Moulias, A.; Katsanos, K.; et al. Thirty-days versus standard duration of dual antiplatelet treatment after percutaneous coronary interventions: A systematic review and meta-analysis. *Eur. Heart J.* **2022**, *43*, ehac544-2717. [CrossRef]
40. Tsigkas, G.; Apostolos, A.; Trigka, A.; Chlorogiannis, D.; Katsanos, K.; Toutouzas, K.; Alexopoulos, D.; Brilakis, E.S.; Davlouros, P. Very Short Versus Longer Dual Antiplatelet Treatment After Coronary Interventions: A Systematic Review and Meta-analysis. *Am. J. Cardiovasc. Drugs* **2022**, *23*, 35–46. [CrossRef]
41. Balanescu, D.V.; Aziz, M.K.; Donisan, T.; Palaskas, N.; Lopez-Mattei, J.; Hassan, S.; Kim, P.; Song, J.; Ntim, W.; Cilingiroglu, M.; et al. Cancer treatment resumption in patients with new generation drug eluting stents. *Coron. Artery Dis.* **2021**, *32*, 295–301. [CrossRef] [PubMed]
42. Ahmed, T.; Pacha, H.M.; Addoumieh, A.; Koutroumpakis, E.; Song, J.; Charitakis, K.; Boudoulas, K.D.; Cilingiroglu, M.; Marmagkiolis, K.; Grines, C.; et al. Percutaneous coronary intervention in patients with cancer using bare metal stents compared to drug-eluting stents. *Front. Cardiovasc. Med.* **2022**, *9*, 901431. [CrossRef] [PubMed]
43. Ozaki, Y.; Garcia-Garcia, H.M.; Mintz, G.S.; Waksman, R. Supporting evidence from optical coherence tomography for shortening dual antiplatelet therapy after drug-eluting stents implantation. *Expert Rev. Cardiovasc. Ther.* **2020**, *18*, 261–267. [CrossRef] [PubMed]
44. Apostolos, A.; Gerakaris, A.; Tsoni, E.; Pappelis, K.; Vasilagkos, G.; Bousoula, E.; Moulias, A.; Konstantinou, K.; Dimitriadis, K.; Karamasis, G.V.; et al. Imaging of Left Main Coronary Artery; Untangling the Gordian Knot. *Rev. Cardiovasc. Med.* **2023**, *24*, 26. [CrossRef]
45. Iliescu, C.A.; Cilingiroglu, M.; Giza, D.E.; Rosales, O.; Lebeau, J.; Guerrero-Mantilla, I.; Lopez-Mattei, J.; Song, J.; Silva, G.; Loyalka, P.; et al. "Bringing on the light" in a complex clinical scenario: Optical coherence tomography-guided discontinuation of antiplatelet therapy in cancer patients with coronary artery disease (PROTECT-OCT registry). *Am. Heart J.* **2017**, *194*, 83–91. [CrossRef]
46. Doolub, G.; Mamas, M.A. Percutaneous Coronary Angioplasty in Patients with Cancer: Clinical Challenges and Management Strategies. *J. Pers. Med.* **2022**, *12*, 1372. [CrossRef]
47. Bisceglia, I.; Canale, M.L.; Lestuzzi, C.; Parrini, I.; Russo, G.; Colivicchi, F.; Gabrielli, D.; Gulizia, M.M.; Iliescu, C.A. Acute coronary syndromes in cancer patients. *J. Cardiovasc. Med.* **2020**, *21*, 944–952. [CrossRef]

48. Gevaert, S.A.; Halvorsen, S.; Sinnaeve, P.R.; Sambola, A.; Gulati, G.; Lancellotti, P.; Van Der Meer, P.; Lyon, A.R.; Farmakis, D.; Lee, G.; et al. Evaluation and management of cancer patients presenting with acute cardiovascular disease: A Consensus Document of the Acute CardioVascular Care (ACVC) association and the ESC council of Cardio-Oncology-Part 1: Acute coronary syndromes and acute pericardial diseases. *Eur. Heart J. Acute Cardiovasc. Care* **2021**, *10*, 947–959.
49. Lucà, F.; Parrini, I.; Abrignani, M.G.; Rao, C.M.; Piccioni, L.; Di Fusco, S.A.; Ceravolo, R.; Bisceglia, I.; Riccio, C.; Gelsomino, S.; et al. Management of Acute Coronary Syndrome in Cancer Patients: It's High Time We Dealt with It. *J. Clin. Med.* **2022**, *11*, 1792. [CrossRef]
50. Tsigkas, G.; Apostolos, A.; Chlorogiannis, D.-D.; Bousoula, E.; Vasilagkos, G.; Tsalamandris, S.; Tsiafoutis, I.; Katsanos, K.; Toutouzas, K.; Aminian, A.; et al. Thirty-Days versus Longer Duration of Dual Antiplatelet Treatment after Percutaneous Coronary Interventions with Newer Drug-Eluting Stents: A Systematic Review and Meta-Analysis. *Life* **2023**, *13*, 666. [CrossRef]
51. Apostolos, A.; Chlorogiannis, D.; Vasilagkos, G.; Katsanos, K.; Toutouzas, K.; Aminian, A.; Alexopoulos, D.; Davlouros, P.; Tsigkas, G. Safety and efficacy of shortened dual antiplatelet therapy after complex percutaneous coronary intervention: A systematic review and meta-analysis. *Hell. J. Cardiol.* **2023**. [CrossRef] [PubMed]
52. Collet, J.P.; Thiele, H.; Barbato, E.; Bauersachs, J.; Dendale, P.; Edvardsen, T.; Gale, C.P.; Jobs, A.; Lambrinou, E.; Mehilli, J.; et al. 2020 ESC Guidelines for the management of acute coronary syndromes in patients presenting without persistent ST-segment elevation. *Eur. Heart J.* **2021**, *42*, 1289–1367. [CrossRef] [PubMed]
53. Neumann, F.J.; Sechtem, U.; Banning, A.P.; Bonaros, N.; Bueno, H.; Bugiardini, R.; Chieffo, A.; Crea, F.; Czerny, M.; Delgado, V.; et al. 2019 ESC Guidelines for the diagnosis and management of chronic coronary syndromes. *Eur. Heart J.* **2020**, *41*, 407–477.
54. Baran, D.A.; Grines, C.L.; Bailey, S.; Burkhoff, D.; Hall, S.A.; Henry, T.D.; Hollenberg, S.M.; Kapur, N.K.; O'Neill, W.; Ornato, J.P.; et al. SCAI clinical expert consensus statement on the classification of cardiogenic shock: This document was endorsed by the American College of Cardiology (ACC), the American Heart Association (AHA), the Society of Critical Care Medicine (SCCM), and the Society of Thoracic Surgeons (STS) in April 2019. *Catheter. Cardiovasc. Interv.* **2019**, *94*, 29–37.
55. McCarthy, C.P.; Steg, G.P.; Bhatt, D.L. The management of antiplatelet therapy in acute coronary syndrome patients with thrombocytopenia: A clinical conundrum. *Eur. Heart J.* **2017**, *38*, 3488–3492. [CrossRef] [PubMed]
56. Schiffer, C.A.; Bohlke, K.; Anderson, K.C. Platelet Transfusion for Patients With Cancer: American Society of Clinical Oncology Clinical Practice Guideline Update Summary. *J. Oncol. Pract.* **2018**, *14*, 129–133. [CrossRef] [PubMed]
57. Stucchi, M.; Cantoni, S.; Piccinelli, E.; Savonitto, S.; Morici, N. Anemia and acute coronary syndrome: Current perspectives. *Vasc. Health Risk Manag.* **2018**, *14*, 109–118. [CrossRef]
58. Ueki, Y.; Vögeli, B.; Karagiannis, A.; Zanchin, T.; Zanchin, C.; Rhyner, D.; Otsuka, T.; Praz, F.; Siontis, G.C.M.; Moro, C.; et al. Ischemia and Bleeding in Cancer Patients Undergoing Percutaneous Coronary Intervention. *Cardio Oncol.* **2019**, *1*, 145–155. [CrossRef]
59. Gerotziafas, G.T.; Taher, A.; Abdel-Razeq, H.; AboElnazar, E.; Spyropoulos, A.C.; El Shemmari, S.; Larsen, A.K.; Elalamy, I. A Predictive Score for Thrombosis Associated with Breast, Colorectal, Lung, or Ovarian Cancer: The Prospective COMPASS-Cancer-Associated Thrombosis Study. *Oncologist* **2017**, *22*, 1222–1231. [CrossRef]
60. Sibbing, D.; Aradi, D.; Alexopoulos, D.; ten Berg, J.; Bhatt, D.L.; Bonello, L.; Collet, J.P.; Cuisset, T.; Franchi, F.; Gross, L.; et al. Updated Expert Consensus Statement on Platelet Function and Genetic Testing for Guiding P2Y12 Receptor Inhibitor Treatment in Percutaneous Coronary Intervention. *JACC Cardiovasc. Interv.* **2019**, *12*, 1521–1537. [CrossRef]
61. Cooke, N.M.; Egan, K.; Mcfadden, S.; Grogan, L.; Breathnach, O.S.; O'Leary, J.; Hennessy, B.T.; Kenny, D. Increased platelet reactivity in patients with late-stage metastatic cancer. *Cancer Med.* **2013**, *2*, 564–570. [CrossRef] [PubMed]
62. Cowman, J.; Richter, L.; Walsh, R.; Keegan, N.; Tinago, W.; Ricco, A.J.; Hennessy, B.T.; Kenny, D.; Dunne, E. Dynamic platelet function is markedly different in patients with cancer compared to healthy donors. *Platelets* **2019**, *30*, 737–742. [CrossRef] [PubMed]
63. Just Vinholt, P.; Højrup Knudsen, G.; Sperling, S.; Frederiksen, H.; Nielsen, C. Platelet function tests predict bleeding in patients with acute myeloid leukemia and thrombocytopenia. *Am. J. Hematol.* **2019**, *94*, 891–901. [CrossRef] [PubMed]
64. Guha, A.; Dey, A.K.; Kalra, A.; Gumina, R.; Lustberg, M.; Lavie, C.J.; Sabik, J.F.; Addison, D. Coronary Artery Bypass Grafting in Cancer Patients: Prevalence and Outcomes in the United States. *Mayo Clin. Proc.* **2020**, *95*, 1865–1876. [CrossRef] [PubMed]
65. Valgimigli, M.; Bueno, H.; Byrne, R.A.; Collet, J.P.; Costa, F.; Jeppsson, A.; Kastrati, A.; Kolh, P.; Mauri, L.; Montalescot, G.; et al. 2017 ESC focused update on dual antiplatelet therapy in coronary artery disease developed in collaboration with EACTS: The Task Force for dual antiplatelet therapy in coronary artery disease of the European Society of Cardiology (ESC) and of the European Association for Cardio-Thoracic Surgery (EACTS). *Eur. Heart J.* **2018**, *39*, 213–254.
66. Kulik, A.; Ruel, M.; Jneid, H.; Ferguson, T.B.; Hiratzka, L.F.; Ikonomidis, J.S.; Lopez-Jimenez, F.; McNallan, S.M.; Patel, M.; Roger, V.L.; et al. Secondary prevention after coronary artery bypass graft surgery: A scientific statement from the American Heart Association. *Circulation* **2015**, *131*, 927–964. [CrossRef]
67. Sousa-Uva, M.; Head, S.J.; Milojevic, M.; Collet, J.P.; Landoni, G.; Castella, M.; Dunning, J.; Gudbjartsson, T.; Linker, N.J.; Sandoval, E.; et al. 2017 EACTS Guidelines on perioperative medication in adult cardiac surgery. *Eur. J. Cardiothorac. Surg.* **2018**, *53*, 5–33. [CrossRef]

68. Lawton, J.S.; Tamis-Holland, J.E.; Bangalore, S.; Bates, E.R.; Beckie, T.M.; Bischoff, J.M.; Bittl, J.A.; Cohen, M.G.; DiMaio, J.M.; Don, C.W.; et al. 2021 ACC/AHA/SCAI Guideline for Coronary Artery Revascularization: Executive Summary: A Report of the American College of Cardiology/American Heart Association Joint Committee on Clinical Practice Guidelines. *J. Am. Coll. Cardiol.* **2022**, *79*, 197–215. [CrossRef]
69. Vener, C.; Banzi, R.; Ambrogi, F.; Ferrero, A.; Saglio, G.; Pravettoni, G.; Sant, M. First-line imatinib vs second- and third-generation TKIs for chronic-phase CML: A systematic review and meta-analysis. *Blood Adv.* **2020**, *4*, 2723–2735. [CrossRef]
70. Capodanno, D.; Di Maio, M.; Greco, A.; Bhatt, D.L.; Gibson, C.M.; Goette, A.; Lopes, R.D.; Mehran, R.; Vranckx, P.; Angiolillo, D.J. Safety and Efficacy of Double Antithrombotic Therapy With Non-Vitamin K Antagonist Oral Anticoagulants in Patients With Atrial Fibrillation Undergoing Percutaneous Coronary Intervention: A Systematic Review and Meta-Analysis. *J. Am. Heart Assoc.* **2020**, *9*, e017212. [CrossRef]
71. Lopes, R.D.; Hong, H.; Harskamp, R.E.; Bhatt, D.L.; Mehran, R.; Cannon, C.P.; Granger, C.B.; Verheugt, F.W.A.; Li, J.; Ten Berg, J.M.; et al. Optimal Antithrombotic Regimens for Patients With Atrial Fibrillation Undergoing Percutaneous Coronary Intervention: An Updated Network Meta-analysis. *JAMA Cardiol.* **2020**, *5*, 582–589. [CrossRef]
72. Lopes, R.D.; Hong, H.; Harskamp, R.E.; Bhatt, D.L.; Mehran, R.; Cannon, C.P.; Granger, C.B.; Verheugt, F.W.A.; Li, J.; Ten Berg, J.M.; et al. Safety and Efficacy of Antithrombotic Strategies in Patients With Atrial Fibrillation Undergoing Percutaneous Coronary Intervention: A Network Meta-analysis of Randomized Controlled Trials. *JAMA Cardiol.* **2019**, *4*, 747–755. [CrossRef] [PubMed]
73. Golwala, H.B.; Cannon, C.P.; Steg, P.G.; Doros, G.; Qamar, A.; Ellis, S.G.; Oldgren, J.; Ten Berg, J.M.; Kimura, T.; Hohnloser, S.H.; et al. Safety and efficacy of dual vs. triple antithrombotic therapy in patients with atrial fibrillation following percutaneous coronary intervention: A systematic review and meta-analysis of randomized clinical trials. *Eur. Heart J.* **2018**, *39*, 1726–1735. [CrossRef] [PubMed]
74. Capodanno, D.; Huber, K.; Mehran, R.; Lip, G.Y.H.; Faxon, D.P.; Granger, C.B.; Vranckx, P.; Lopes, R.D.; Montalescot, G.; Cannon, C.P.; et al. Management of Antithrombotic Therapy in Atrial Fibrillation Patients Undergoing PCI: JACC State-of-the-Art Review. *J. Am. Coll. Cardiol.* **2019**, *74*, 83–99. [CrossRef]
75. Buccheri, S.; Angiolillo, D.J.; Capodanno, D. Evolving paradigms in antithrombotic therapy for anticoagulated patients undergoing coronary stenting. *Ther. Adv. Cardiovasc. Dis.* **2019**, *13*, 1753944719891688. [CrossRef]
76. Gibson, C.M.; Mehran, R.; Bode, C.; Halperin, J.; Verheugt, F.W.; Wildgoose, P.; Birmingham, M.; Ianus, J.; Burton, P.; van Eickels, M.; et al. Prevention of Bleeding in Patients with Atrial Fibrillation Undergoing PCI. *N. Engl. J. Med.* **2016**, *375*, 2423–2434. [CrossRef]
77. Cannon, C.P.; Bhatt, D.L.; Oldgren, J.; Lip, G.Y.H.; Ellis, S.G.; Kimura, T.; Maeng, M.; Merkely, B.; Zeymer, U.; Gropper, S.; et al. Dual Antithrombotic Therapy with Dabigatran after PCI in Atrial Fibrillation. *N. Engl. J. Med.* **2017**, *377*, 1513–1524. [CrossRef] [PubMed]
78. Lopes, R.D.; Heizer, G.; Aronson, R.; Vora, A.N.; Massaro, T.; Mehran, R.; Goodman, S.G.; Windecker, S.; Darius, H.; Li, J.; et al. Antithrombotic Therapy after Acute Coronary Syndrome or PCI in Atrial Fibrillation. *N. Engl. J. Med.* **2019**, *380*, 1509–1524. [CrossRef] [PubMed]
79. Vranckx, P.; Valgimigli, M.; Eckardt, L.; Tijssen, J.; Lewalter, T.; Gargiulo, G.; Batushkin, V.; Campo, G.; Lysak, Z.; Vakaliuk, I.; et al. Edoxaban-based versus vitamin K antagonist-based antithrombotic regimen after successful coronary stenting in patients with atrial fibrillation (ENTRUST-AF PCI): A randomised, open-label, phase 3b trial. *Lancet* **2019**, *394*, 1335–1343. [CrossRef]
80. Capodanno, D.; Angiolillo, D.J. Dual antithrombotic therapy for atrial fibrillation and PCI. *Lancet* **2019**, *394*, 1300–1302. [CrossRef]
81. Popma, J.J.; Deeb, G.M.; Yakubov, S.J.; Mumtaz, M.; Gada, H.; O'Hair, D.; Bajwa, T.; Heiser, J.C.; Merhi, W.; Kleiman, N.S.; et al. Transcatheter Aortic-Valve Replacement with a Self-Expanding Valve in Low-Risk Patients. *N. Engl. J. Med.* **2019**, *380*, 1706–1715. [CrossRef] [PubMed]
82. Mack, M.J.; Leon, M.B.; Thourani, V.H.; Makkar, R.; Kodali, S.K.; Russo, M.; Kapadia, S.R.; Malaisrie, S.C.; Cohen, D.J.; Pibarot, P.; et al. Transcatheter Aortic-Valve Replacement with a Balloon-Expandable Valve in Low-Risk Patients. *N. Engl. J. Med.* **2019**, *380*, 1695–1705. [CrossRef] [PubMed]
83. Schechter, M.; Balanescu, D.V.; Donisan, T.; Dayah, T.J.; Kar, B.; Gregoric, I.; Giza, D.E.; Song, J.; Lopez-Mattei, J.; Kim, P.; et al. An update on the management and outcomes of cancer patients with severe aortic stenosis. *Catheter. Cardiovasc. Interv.* **2019**, *94*, 438–445. [CrossRef] [PubMed]
84. Marmagkiolis, K.; Monlezun, D.J.; Cilingiroglu, M.; Grines, C.; Herrmann, J.; Toutouzas, K.P.; Ates, I.; Iliescu, C. TAVR in Cancer Patients: Comprehensive Review, Meta-Analysis, and Meta-Regression. *Front. Cardiovasc. Med.* **2021**, *8*, 641268. [CrossRef]
85. Karaduman, B.D.; Ayhan, H.; Keleş, T.; Bozkurt, E. Clinical outcomes after transcatheter aortic valve implantation in active cancer patients and cancer survivors. *Turk. J. Thorac. Cardiovasc. Surg.* **2021**, *29*, 45–51. [CrossRef] [PubMed]
86. Drakopoulou, M.; Soulaidopoulos, S.; Oikonomou, G.; Stathogiannis, K.; Latsios, G.; Synetos, A.; Tousoulis, D.; Toutouzas, K. Novel Perspective for Antithrombotic Therapy in TAVI. *Curr. Pharm. Des.* **2020**, *26*, 2789–2803. [CrossRef] [PubMed]
87. Capodanno, D.; Collet, J.P.; Dangas, G.; Montalescot, G.; ten Berg, J.M.; Windecker, S.; Angiolillo, D.J. Antithrombotic Therapy After Transcatheter Aortic Valve Replacement. *JACC. Cardiovasc. Interv.* **2021**, *14*, 1688–1703. [CrossRef]
88. Kuno, T.; Takagi, H.; Sugiyama, T.; Ando, T.; Miyashita, S.; Valentin, N.; Shimada, Y.J.; Kodaira, M.; Numasawa, Y.; Kanei, Y.; et al. Antithrombotic strategies after transcatheter aortic valve implantation: Insights from a network meta-analysis. *Catheter. Cardiovasc. Interv.* **2020**, *96*, E177–E186. [CrossRef]

89. Kobari, Y.; Inohara, T.; Tsuruta, H.; Yashima, F.; Shimizu, H.; Fukuda, K.; Naganuma, T.; Mizutani, K.; Yamawaki, M.; Tada, N.; et al. No Antithrombotic Therapy After Transcatheter Aortic Valve Replacement: Insight From the OCEAN-TAVI Registry. *JACC. Cardiovasc. Interv.* **2023**, *16*, 79–91. [CrossRef]
90. Drakopoulou, M.; Soulaidopoulos, S.; Stathogiannis, K.; Oikonomou, G.; Papanikolaou, A.; Toutouzas, K.; Tousoulis, D. Antiplatelet and Antithrombotic Therapy After Patent Foramen Oval and Atrial Septal Defect Closure. *Curr. Pharm. Des.* **2020**, *26*, 2769–2779. [CrossRef]
91. Apostolos, A.; Drakopoulou, M.; Gregoriou, S.; Synetos, A.; Trantalis, G.; Tsivgoulis, G.; Deftereos, S.; Tsioufis, K.; Toutouzas, K. Nickel Hypersensitivity to Atrial Septal Occluders: Smoke Without Fire? *Clin. Rev. Allergy Immunol.* **2022**, *62*, 476–483. [CrossRef] [PubMed]
92. Apostolos, A.; Drakopoulou, M.; Toutouzas, K. New migraines after atrial septal defect occlusion. Is the nickel hypersensitivity the start of everything? *Med. Hypotheses* **2021**, *146*, 110442. [CrossRef] [PubMed]
93. Pristipino, C.; Sievert, H.; D'Ascenzo, F.; Louis Mas, J.; Meier, B.; Scacciatella, P.; Hildick-Smith, D.; Gaita, F.; Toni, D.; Kyrle, P.; et al. European position paper on the management of patients with patent foramen ovale. General approach and left circulation thromboembolism. *Eur. Heart J.* **2019**, *40*, 3182–3195. [CrossRef] [PubMed]
94. Mas, J.-L.; Derumeaux, G.; Guillon, B.; Massardier, E.; Hosseini, H.; Mechtouff, L.; Arquizan, C.; Béjot, Y.; Vuillier, F.; Detante, O.; et al. Patent Foramen Ovale Closure or Anticoagulation vs. Antiplatelets after Stroke. *N. Engl. J. Med.* **2017**, *377*, 1011–1021. [CrossRef]
95. Lee, P.H.; Song, J.K.; Kim, J.S.; Heo, R.; Lee, S.; Kim, D.H.; Song, J.M.; Kang, D.H.; Kwon, S.U.; Kang, D.W.; et al. Cryptogenic Stroke and High-Risk Patent Foramen Ovale: The DEFENSE-PFO Trial. *J. Am. Coll. Cardiol.* **2018**, *71*, 2335–2342. [CrossRef] [PubMed]
96. Meier, B. Closure of patent foramen ovale: Technique, pitfalls, complications, and follow up. *Heart* **2005**, *91*, 444–448. [CrossRef]
97. Hobohm, L.; von Bardeleben, R.S.; Ostad, M.A.; Wenzel, P.; Münzel, T.; Gori, T.; Keller, K. 5-Year Experience of In-Hospital Outcomes After Percutaneous Left Atrial Appendage Closure in Germany. *JACC. Cardiovasc. Interv.* **2019**, *12*, 1044–1052. [CrossRef]
98. Chen, S.; Weise, F.K.; Chun, K.R.J.; Schmidt, B. Antithrombotic strategies after interventional left atrial appendage closure: An update. *Expert Rev. Cardiovasc. Ther.* **2018**, *16*, 675–678. [CrossRef]
99. Ledwoch, J.; Sievert, K.; Boersma, L.V.A.; Bergmann, M.W.; Ince, H.; Kische, S.; Pokushalov, E.; Schmitz, T.; Schmidt, B.; Gori, T.; et al. Initial and long-term antithrombotic therapy after left atrial appendage closure with the WATCHMAN. *Europace* **2020**, *22*, 1036–1043. [CrossRef]
100. Patti, G.; Sticchi, A.; Verolino, G.; Pasceri, V.; Vizzi, V.; Brscic, E.; Casu, G.; Golino, P.; Russo, V.; Rapacciuolo, A.; et al. Safety and Efficacy of Single Versus Dual Antiplatelet Therapy After Left Atrial Appendage Occlusion. *Am. J. Cardiol.* **2020**, *134*, 83–90. [CrossRef]
101. Aboyans, V.; Ricco, J.B.; Bartelink, M.L.E.L.; Björck, M.; Brodmann, M.; Cohnert, T.; Collet, J.P.; Czerny, M.; De Carlo, M.; Debus, S.; et al. 2017 ESC Guidelines on the Diagnosis and Treatment of Peripheral Arterial Diseases, in collaboration with the European Society for Vascular Surgery (ESVS): Document covering atherosclerotic disease of extracranial carotid and vertebral, mesenteric, renal, upper and lower extremity arteriesEndorsed by: The European Stroke Organization (ESO)The Task Force for the Diagnosis and Treatment of Peripheral Arterial Diseases of the European Society of Cardiology (ESC) and of the European Society for Vasc. *Eur. Heart J.* **2018**, *39*, 763–816.
102. Katan, M.; Luft, A. Global Burden of Stroke. *Semin. Neurol.* **2018**, *38*, 208–211. [CrossRef] [PubMed]
103. Fonseca, A.C.; Merwick, Á.; Dennis, M.; Ferrari, J.; Ferro, J.M.; Kelly, P.; Lal, A.; Ois, A.; Olivot, J.M.; Purroy, F. European Stroke Organisation (ESO) guidelines on management of transient ischaemic attack. *Eur. Stroke J.* **2021**, *6*, CLXIII–CLXXXVI. [CrossRef] [PubMed]
104. Bhatia, K.; Jain, V.; Aggarwal, D.; Vaduganathan, M.; Arora, S.; Hussain, Z.; Uberoi, G.; Tafur, A.; Zhang, C.; Ricciardi, M.; et al. Dual Antiplatelet Therapy Versus Aspirin in Patients With Stroke or Transient Ischemic Attack: Meta-Analysis of Randomized Controlled Trials. *Stroke* **2021**, *52*, E217–E223. [CrossRef] [PubMed]
105. Zaorsky, N.G.; Zhang, Y.; Tchelebi, L.T.; Mackley, H.B.; Chinchilli, V.M.; Zacharia, B.E. Stroke among cancer patients. *Nat. Commun.* **2019**, *10*, 5172. [CrossRef]
106. Bang, O.Y.; Chung, J.W.; Lee, M.J.; Seo, W.K.; Kim, G.M.; Ahn, M.J. Cancer-Related Stroke: An Emerging Subtype of Ischemic Stroke with Unique Pathomechanisms. *J. Stroke* **2020**, *22*, 1–10. [CrossRef]

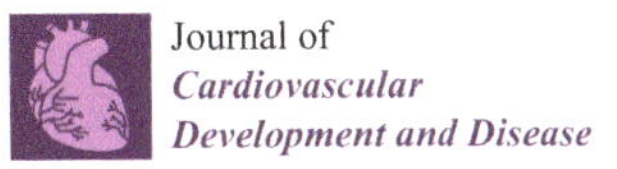
Journal of
Cardiovascular Development and Disease

MDPI

Review

Dual Antiplatelet Therapy with Parenteral $P2Y_{12}$ Inhibitors: Rationale, Evidence, and Future Directions

Giulia Alagna [1], Paolo Mazzone [2], Marco Contarini [2] and Giuseppe Andò [1,*]

1 Department of Clinical and Experimental Medicine, University of Messina, 98124 Messina, Italy; giulia.alagna@icloud.com
2 Cardiology Unit, "Umberto I" Hospital, 96100 Siracusa, Italy; mazzonepaolo89@gmail.com (P.M.); m.contarini@asp.sr.it (M.C.)
* Correspondence: giuseppe.ando@unime.it

Abstract: Dual antiplatelet therapy (DAPT), consisting of the combination of aspirin and an inhibitor of the platelet $P2Y_{12}$ receptor for ADP, remains among the most investigated treatments in cardiovascular medicine. While a substantial amount of research initially stemmed from the observations of late and very late stent thrombosis events in the first-generation drug-eluting stent (DES) era, DAPT has been recently transitioning from a purely stent-related to a more systemic secondary prevention strategy. Oral and parenteral platelet $P2Y_{12}$ inhibitors are currently available for clinical use. The latter have been shown to be extremely suitable in drug-naïve patients with acute coronary syndrome (ACS), mainly because oral $P2Y_{12}$ inhibitors are associated with delayed efficacy in patients with STEMI and because pre-treatment with $P2Y_{12}$ inhibitors is discouraged in NSTE-ACS, and in patients with recent DES implantation and in need of urgent cardiac and non-cardiac surgery. More definitive evidence is needed, however, about optimal switching strategies between parenteral and oral $P2Y_{12}$ inhibitors and about newer potent subcutaneous agents that are being developed for the pre-hospital setting.

Keywords: dual antiplatelet therapy; $P2Y_{12}$ inhibitors; acute coronary syndrome; clopidogrel; prasugrel; ticagrelor; cangrelor; selatogrel; zalunfiban

Citation: Alagna, G.; Mazzone, P.; Contarini, M.; Andò, G. Dual Antiplatelet Therapy with Parenteral $P2Y_{12}$ Inhibitors: Rationale, Evidence, and Future Directions. *J. Cardiovasc. Dev. Dis.* **2023**, *10*, 163. https://doi.org/10.3390/jcdd10040163

Academic Editor: Krishnaraj Sinhji Rathod

Received: 12 March 2023
Revised: 4 April 2023
Accepted: 7 April 2023
Published: 9 April 2023

1. Introduction

Dual antiplatelet therapy (DAPT) consists of the combination of aspirin and an inhibitor of the platelet $P2Y_{12}$ receptor for adenosine diphosphate (ADP). At the end of the 1990s, two randomized trials definitively established DAPT with aspirin and ticlopidine as the gold standard therapy after percutaneous coronary intervention (PCI) with stent implantation, in comparison to aspirin or to aspirin and anticoagulant therapy [1,2]. Ticlopidine was soon replaced by clopidogrel at the beginning of the 2000s. DAPT has proven to be among the most investigated treatments in cardiovascular medicine. Such necessity of research initially arose from the observations of late and very late stent thrombosis (ST) events occurring after first-generation drug-eluting stent (DES) implantation, highlighting lack of efficacy of clopidogrel as one of the possible drivers of thrombotic events [3] and paving the way to development of potent oral agents such as prasugrel [4] and ticagrelor [5]. More recent evidence in high-risk patients has suggested that DAPT reduces the long-term risk of cardiovascular death, spontaneous myocardial infarction (MI), stroke and major adverse cardiac events (MACE) [6,7]. After decades of research, DAPT has been moving from a stent-related to a systemic treatment among other secondary prevention strategies such as lipid-lowering therapy and control of diabetes and hypertension. Most evidence remains largely based on post-PCI patients [8], while patients that are either medically managed (e.g., those with MINOCA [9], spontaneous coronary artery dissection [10], or takotsubo syndrome [11]) or undergoing coronary artery bypass grafting (CABG) [12] remain underrepresented in clinical trials. On this background, we will discuss the role, indications, and utilization of cangrelor, the only parenteral $P2Y_{12}$ inhibitor available so far, its recommendations as a bridging antiplatelet agent for cardiac and non-cardiac surgery and the future directions of DAPT with new parenteral agents.

2. $P2Y_{12}$ Inhibitor Antiplatelet Agents

2.1. Oral $P2Y_{12}$ Inhibitors

While ticlopidine was the first $P2Y_{12}$ inhibitor to be associated with low-dose aspirin for DAPT, its unfavorable safety profile made it obsolete after the introduction of clopidogrel. Clopidogrel is a second-generation thienopyridine and an irreversible $P2Y_{12}$ receptor antagonist that is administered as an inactive pro-drug and requires enzymatic liver conversion into its active metabolite by a series of cytochrome P450 (CYP) enzymes. After activation, clopidogrel irreversibly binds to $P2Y_{12}$, an ADP receptor, on the surface of platelets, resulting in an inactivation of the glycoprotein (GP) IIb/IIIa receptor and destabilization of the platelet aggregate [6]. The recommended regimen is a loading dose of 600 mg followed by a maintenance dose of 75 mg once daily. No dose adjustment is required in CKD patients. The onset of action is particularly delayed and variable, ranging from 2 to 6 h and the offset of effect ranges from 3 to 10 days. The evidence provided by the landmark CURE trial established DAPT with clopidogrel as the standard of care after acute coronary syndrome (ACS) and after coronary stent implantation [13]. However, clopidogrel has too much inter-individual variability in platelet inhibition and has significant non responsiveness and resistance in some patients. The enzymatic liver conversion is one of the main causes of variability of clopidogrel action. CYP2C19 is one of the most important polymorphic CYP enzymes across different populations and this is associated with worse outcomes, for instance, in those with the CYP2C19*2 variant [14]. Likewise, all comedications that are inhibitors of CYP2C19 suppress clopidogrel bioactivation (e.g., some proton pump inhibitors, statins and calcium channel blockers) [15]. Moreover, poor intestinal absorption can delay the onset of action of clopidogrel, which can be worsened by concomitant administration of opioids for angina relief. Inadequate $P2Y_{12}$ inhibition, especially in the setting of ACS, contributes to more frequent periprocedural complications such as need for recurrent revascularization, MI, and ST. This highlighted the need for a more potent and consistent platelet inhibition that was introduced with novel generation $P2Y_{12}$ inhibitors.

Prasugrel is thienopyridine as well and an irreversible $P2Y_{12}$ receptor antagonist that is administered as an inactive pro-drug and requires an enzymatic liver activation. Differently than clopidogrel, it gains a faster, greater, and more consistent degree of platelet inhibition [16]. The recommended regimen is a loading dose of 60 mg followed by a maintenance dose of 10 mg once daily, reduced to 5 mg in patients $\geq$75 years old or <60 kg. No dose adjustment is required in CKD patients. The onset of action is rapid, ranging from 0.5 to 4 h and the offset of effect ranges from 5 to 10 days. The TRITON-TIMI 38 trial compared prasugrel versus clopidogrel in $P2Y_{12}$ inhibitor-naïve ACS patients referred to PCI [4]. Prasugrel determined a reduction in primary ischemic endpoint compared to clopidogrel, counterbalanced by a significant increase in the rate of major bleeding. Prasugrel was also compared to ticagrelor, the other potent $P2Y_{12}$ inhibitor, in the recent ISAR-REACT 5 randomized trial. Prasugrel was superior in reducing the rate of death, MI, and stroke without any increase in bleeding complications [17]. Thus, prasugrel is the recommended $P2Y_{12}$ inhibitor in ACS patients without high bleeding risk proceeding to PCI [18].

Ticagrelor is a direct oral reversible $P2Y_{12}$ receptor inhibitor, which belongs to a novel chemical class, the cyclopentyl triazolopyrimidine. Following intestinal absorption, ticagrelor does not need to be metabolized for platelet inhibition. The recommended dose is a loading dose of 180 mg followed by a maintenance dose of 90 mg twice a day. No dose adjustment is required in CKD patients. The onset of action is rapid as well, ranging from 0.5 to 2 h and the offset of effect ranges from 3 to 4 days. The PLATO trial proved the superiority of ticagrelor compared to clopidogrel in ACS patients regarding the rate of death from vascular causes, MI, or stroke, without significant difference in major bleeding rates [5]. Nevertheless, ticagrelor also led to more patients stopping medication because of side effects, mainly dyspnea. As it is not associated with pulmonary or cardiac dysfunction, alterations in the mechanisms and the neurological pathways of the sensation of dyspnea may be involved in its pathogenesis [19].

2.2. Drawbacks of Oral $P2Y_{12}$ Inhibitors

Despite potent $P2Y_{12}$ inhibitors (prasugrel and ticagrelor) provide lower rates of ischemic events compared to clopidogrel, significant concerns remain about their onset of action. Moreover, their administration does not counterbalance the high residual platelet reactivity (HRPR) up to 4–6 h after the standard loading dose [20–22]. For this reason, strategies have been tested to increase the bioavailability of oral $P2Y_{12}$ inhibitors, such as crushing or chewing tablets. However, pharmacokinetic and pharmacodynamic data remain limited [23–25]. So far, clopidogrel remains the $P2Y_{12}$ inhibitor recommended in stable coronary artery disease (CAD) patients, unless specific high-risk procedural characteristics are present, such as complex left main or multivessel stenting, suboptimal stent deployment, or other conditions associated with high risk of stent thrombosis; in such cases, initial treatment with either prasugrel or ticagrelor may be considered according to European guidelines [26] if the tradeoff between risk of ischemia and bleeding is favorable [27]. All these therapies are limited by their need to be absorbed in the gastrointestinal (GI) tract before becoming available and this leads to an inevitable delay between drug intake and time of reaching effective platelet inhibition. Gastric emptying, intestinal motility, blood perfusion of the mucosa and its permeability are all factors influencing the absorption rate of medications [28]. Moreover, it has been reported that the velocity of platelet inhibition after oral intake was influenced by the clinical presentation: faster for stable CAD undergoing PCI, slower for NSTE-ACS patients, and the slowest for STEMI patients [20,29]. This phenomenon can be explained by a decreased cardiac output in ACS patients, which leads to a sympathetic system activation, and a vasoconstriction of the peripheral arteries that shunts the blood to vital organs, impairing gastric emptying, intestinal motility, and permeability of the hypo-perfused mucosa [30]. Elevated central pressure due to reduced cardiac output also leads to the release of atrial natriuretic peptide, which inhibits intestinal permeability and motility [31]. In acute settings, nausea and vomiting are common, reducing drug absorption as well. Finally, concomitant treatment with morphine, an opioid analgesic usually used to alleviate chest pain, delays gastric emptying, reduces intestinal peristalsis, and itself induces nausea and vomiting. Another barrier concerns the inability for oral administration of medications in intubated or unconscious patients. A new formulation of ticagrelor in orodispersible tablets that promptly releases its components upon contact with the oral cavity has recently become available and has been tested in a prospective trial of high-risk ACS patients. Although a superior grade of platelet inhibition was not obtained as compared with standard ticagrelor tablets, the trial confirmed the feasibility and safety of administration of ticagrelor without the need of swallowing water, that may prove to be convenient in critical ACS patients [32].

That said, following intake of oral $P2Y_{12}$ inhibitors there is a variable timeframe of hours of inadequate antiplatelet protection. While the risk for ST is low with new generation stents, the delayed antiplatelet effects may still increase the risk of peri-procedural MI and impaired coronary/myocardial reperfusion, translating into worse clinical outcomes. Pretreatment whenever possible could reduce this delay, but most recent ESC guidelines do not recommend (class III) pre-treatment with oral $P2Y_{12}$ inhibitors in NSTE-ACS patients, because several trials showed no ischemic benefits and more bleeding complications [18]. In addition, treatment of stable CAD patients does not include a $P2Y_{12}$ inhibitor before coronary angiography. These observations underscore the need to define strategies that can bridge the gap in platelet inhibitory effects following intake of oral $P2Y_{12}$ inhibitors.

2.3. Parenteral $P2Y_{12}$ Inhibitors

Parenteral administration of a $P2Y_{12}$ inhibitor allows for immediate antiplatelet effects, skipping the delay and variability in intestinal absorption velocity and providing an enhanced platelet inhibition during the time window of inadequate response to oral agents. This is notable especially in high-risk patients undergoing PCI, who require an immediate platelet inhibition.

Cangrelor is an adenosine triphosphate-analog that is a highly specific and a direct reversible antagonist for the $P2Y_{12}$ receptor on the surface of platelets. This leads to blockage of ADP-induced GP IIb/IIIa receptors and inhibition of platelet aggregation. After administration, cangrelor does not need bioactivation and is immediately ready for platelet inhibition. It is available as a lyophilized powder and it is administered initially as a 30 mcg/kg intravenous bolus prior to PCI and then continued with a 4 mcg/kg/min infusion for at least 2 h or for the duration of PCI, whichever is longer. It reaches an immediate (~2 min) onset of action and has a very short offset with a rapid (30–60 min) restoration of platelet function after its discontinuation. There is neither dosage adjustment required for renal or hepatic impairment, nor for age. It has a short plasma half-life of 3–5 min as it is rapidly inactivated via dephosphorylation by nucleotidases in the blood and the major metabolite is considered inactive. Cangrelor allows high levels of platelet inhibition (>95%) and provides further decrease in platelet aggregation in patients treated than with the more potent oral $P2Y_{12}$ inhibitors [33]. This reduces the risk of periprocedural and early postprocedural complications such as MI, repeat coronary revascularization and ST. Cangrelor is the only parenteral $P2Y_{12}$ receptor inhibitor that has received approval. In 2015, both the US FDA and the EMA approved it in $P2Y_{12}$ naïve patients undergoing PCI, both with ACS and with CAD. A large RCT showed faster and enhanced platelet inhibition in the peri-PCI period, translating into reduced ischemic events leading to clinical approval of the drug [34]. We will discuss later the CHAMPION program and more recent randomized clinical trials that have been designed to compare cangrelor vs. the more potent $P2Y_{12}$ inhibitors (prasugrel and ticagrelor).

Some parenteral antithrombotic drugs that interact with multiple pathways are currently being developed for the treatment of ACS, with the aim of further reducing ischemic events without significantly increasing bleeding complications [35]. Selatogrel is a reversible binding $P2Y_{12}$ inhibitor formulated for subcutaneous (SC) administration. Its molecular structure derives from incorporation of the pyrimidine group of ticagrelor into a family of compounds previously studied as $P2Y_{12}$ receptor antagonists [36,37]. Preclinical studies have suggested that selatogrel is potent and selective, but also that it may have a broader therapeutic index than clopidogrel or ticagrelor with regards to increased bleeding risk while maintaining antithrombotic effect [38]. Selatogrel has a rapid onset and one study of the radiolabeled drug suggested that there were no significant plasma metabolites, and that elimination was largely fecal, predicting no significant drug–drug interactions [39]. Phase II trials in both ACS and stable, chronic CAD are now being reported with promising results. Selatogrel reliably and potently inhibits platelet reactivity within 30 min after subcutaneous administration and for approximately 8 h in patients with chronic coronary syndrome, the effect fading within 24 h [40]. In patients with AMI, a single subcutaneous injection of selatogrel rapidly induced a profound and dose-dependent inhibition of platelet activity, independently from age, sex or clinical presentation, without major bleeding events and with short-term dyspnea as the only relevant adverse event [41]. The clinical context in which selatogrel may find its place remains to be determined; however, as it provides potent, rapid and reversible $P2Y_{12}$ inhibition without the need for intravenous access or infusion, it could represent a promising pre-treatment option for early prehospital administration by healthcare professionals or even from self-administration by patients during a suspected re-infarction [42]. A large-scale clinical outcomes trial (SOS-AMI, Selatogrel Outcome Study in Suspected Acute Myocardial Infarction) in patients with a recent history of AMI, employing an autoinjector for early and convenient subcutaneous self-administration of selatogrel by the patient him/herself, is now ongoing (ClinicalTrials.gov Identifier: NCT04957719).

RUC-4 (zalunfiban) is a second-generation GP IIb/IIIa inhibitor (GPI) which has shown a good safety profile and a high and limited-duration antiplatelet efficacy in both stable [43] and STEMI [44] patients. Zalunfiban is now being investigated in a large-scale Phase 3 RCT testing pre-hospital subcutaneous injection in STEMI patients (CELEBRATE, A Phase 3 Study of Zalunfiban in Subjects with ST-elevation MI, ClinicalTrials.gov Identifier: NCT04825743).

3. Efficacy and Safety of Cangrelor: Main Evidence Available

3.1. The CHAMPION Program

The pharmacologic profile of cangrelor makes it not only an attractive agent for protection of ischemic events in patients undergoing PCI, but also a safe one in case of procedural complications, such as bleeding or need for emergent surgery, given its fast offset of effects, obviating the need for an antidote for reversal [45–47]. The efficacy and safety of cangrelor in the setting of PCI were evaluated in three large randomized controlled, double-blind, phase III trials (Table 1):

Table 1. Overview of the CHAMPION Program trials.

	CHAMPION PLATFORM	CHAMPION PCI	CHAMPION PHOENIX
Years	2007–2009	2007–2009	2010–2012
Patients (n)	5362	8877	11,145
Diagnosis	NSTE-ACS (94.8%); stable angina (5.2%)	STEMI (11.2%); NSTE-ACS (73.8%); stable angina (1.5%)	STEMI (18%); NSTE-ACS (25.7%); stable angina (62.3%)
Antiplatelet therapy	Clopidogrel naïve	Clopidogrel	Clopidogrel naïve
Treatment	Cangrelor: 30 µg/kg bolus, 4 µg/kg/min infusion	Cangrelor: 30 µg/kg bolus, 4 µg/kg/min infusion	Cangrelor: 30 µg/kg bolus, 4 µg/kg/min infusion
Transition to clopidogrel	Clopidogrel 600 mg at the end of cangrelor infusion	Clopidogrel 600 mg at the end of cangrelor infusion	Clopidogrel 600 mg at the end of cangrelor infusion
Control arm	Placebo	Clopidogrel 600 mg	Clopidogrel 600 mg or 300 mg
Definition of myocardial infarction	Clinical	Clinical	Universal definition
Primary composite endpoint	Death, MI, IDR at 48 h	Death, MI, IDR at 48 h	Death, MI, IDR at 48 h
Results	OR 0.87 (95% CI 0.71–1.07; $p = 0.17$)	OR 1.05 (95% CI 0.88–1.24; $p = 0.59$)	OR 0.78 (95% CI 0.66–0.93; $p = 0.005$)

The CHAMPION-PLATFORM trial enrolled 5362 patients with stable angina, unstable angina or NSTE-ACS undergoing PCI [48]. Patients were randomized to either cangrelor or placebo, bolus and infusion initiated during PCI, followed by 600 mg of clopidogrel at the end of the cangrelor infusion or at the end of the PCI for the placebo group. The primary endpoint of a composite of death, MI or ischemia-driven revascularization at 48 h was not significatively different between cangrelor or placebo (7.0 vs. 8.0%; $p = 0.17$) but cangrelor, had significantly lower rate of ST (0.2 vs. 0.6%; $p = 0.02$) and death from any cause (0.2 vs. 0.7%; $p = 0.02$) at 48 h. Cangrelor had no differences compared to placebo for major or minor bleeding according to the TIMI criteria and for severe or moderate bleeding according to the GUSTO study [49]. There was only a difference in major bleeding according to the ACUITY criteria, due to an excess of groin hematomas in the cangrelor group. However, the rates of blood transfusion were not significantly different. The CHAMPION-PCI trial (n = 8877) had a similar design to the prior trial but clopidogrel was given at the start of the placebo infusion, before PCI. The trial population was basically the same but also included ST-segment elevation myocardial infarction (STEMI) patients undergoing primary PCI (pPCI). The primary and secondary endpoints were the same as for CHAMPION-PLATFORM. However, in CHAMPION-PCI there were no statistically significant differences between the cangrelor and clopidogrel groups for any endpoint. The incidence of bleeding was significatively higher in the cangrelor group only by ACUITY minor (17.6 vs. 15.2%; $p = 0.003$) or GUSTO mild (19.6 vs. 16.9%; $p = 0.001$) criteria [50]. These discouraging results could be explained by the MI definition used in these trials which was considered obsolete and did not appropriately discriminate periprocedural MI especially from the first MI in ACS patients, being based mainly on CK and CKMB assays [51]. An analysis of these two trials using the universal MI definition demonstrated that the primary endpoint of a composite of death, MI and ischemia-driven revascularization was significantly reduced with cangrelor compared with the control (3.1 vs. 3.8%; $p = 0.037$). This difference was seen early, within the prior 6 h, according to the cangrelor time of action. Even acute ST was lower with cangrelor compared to placebo (0.2 vs. 0.4%; $p = 0.018$). In addition, cangrelor caused more rate of major and minor bleeding by ACUITY criteria and more hematomas, though they did not need more blood transfusions, according to trial results [52]. The benefit of cangrelor in ischemic endpoints seen in this analysis led to conduct of a similar trial, incorporating the universal definition of MI, the CHAMPION PHOENIX [34]. Enrolled patients (n=11,145), who were $P2Y_{12}$ inhibitor naïve, underwent PCI for stable angina, NSTEMI or STEMI. They received cangrelor and a loading dose of clopidogrel (600 mg) at the end of infusion, or placebo and a loading dose of clopidogrel (300 or 600 mg) before or after PCI. Cangrelor led to a significantly lower rate of the primary endpoint (composite of death, MI, ischemia-driven revascularization or ST at 48 h) (4.7 vs. 5.9%; $p = 0.005$), particularly driven by a reduction in the periprocedural MIs; it led also to a significantly lower rate of the secondary endpoint of intraprocedural ST at 48 h [53]. Bleeding outcomes defined as GUSTO major and moderate criteria were not significantly different between the cangrelor group and the clopidogrel group. Bleeding measured using the more sensitive ACUITY criteria was consistently increased with cangrelor relative to clopidogrel in both stable and ACS patients. However, the need for blood transfusions was similar between the groups [53]. A post-hoc analysis combined the primary efficacy and safety endpoints to provide a composite of net adverse clinical events. Cangrelor compared with clopidogrel consistently reduced net adverse clinical events, in both ST and ACS subsets, both early at 48 h and at 30 days. These results were confirmed, at 48 h and 30 days, by a pooled analysis of all three CHAMPION trials [54].

3.2. *Use of Cangrelor in Combination with Potent Oral $P2Y_{12}$ Inhibitors*

As already mentioned, the first trials of cangrelor mainly involved patients with stable or unstable CAD and a limited proportion of patients with STEMI. Thus, there was an urgent need for clinical and pharmacodynamic information on the wide use of cangrelor in combination with ticagrelor, the fastest oral formulation of $P2Y_{12}$ inhibitors, for patients who have STEMI treated with pPCI. The CANTIC Study (Platelet Inhibition with Cangrelor and Crushed TICagrelor in STEMI Patients Undergoing Primary Percutaneous Coronary Intervention) was the first prospective randomized study designed in patients undergoing pPCI to explore the occurrence of drug–drug interaction (DDI) when cangrelor or placebo are concomitantly administered with ticagrelor [55]. Fifty STEMI patients scheduled for pPCI were randomized into two groups: one group received blinded 2 h cangrelor bolus followed by infusion, the other group received placebo. Additionally, both groups received

180 mg of crushed ticagrelor. Platelet reactivity was measured with VerifyNow $P2Y_{12}$ point-of-care testing as $P2Y_{12}$ reaction units (PRU) and vasodilator-stimulated phosphoprotein (VASP). Following PCI, all patients were prescribed aspirin indefinitely and ticagrelor 90 mg twice daily for at least 12 months. PRU levels were significantly lower in patients randomized to cangrelor than in those randomized to placebo as early as 5 min after the bolus ($p < 0.001$). PRU levels at 30 min (primary endpoint) were significantly lower with cangrelor versus placebo (63 vs. 214; $p < 0.001$) and remained significantly lower in the cangrelor group until completion of the 2 h infusion. In the placebo group, PRU levels decreased over time, with significant differences from baseline observed only 1 h after drug administration ($p < 0.001$), which became more marked after 2 h ($p < 0.001$). At the end of the infusion, there was an increase in PRU levels in the cangrelor group with significant differences at 1 h ($p = 0.001$) and 2 h ($p = 0.027$) after the infusion. In the placebo group, PRU levels continued to decrease at 1 h ($p = 0.059$) and 2 h ($p = 0.007$) after the infusion and remained similar at 1 and 2 h after having stopped the infusion. Rates of high platelet reactivity (HPR), as defined by PRU > 208, were significantly higher with placebo than with cangrelor at any study time point during the infusions (Table 2). A DDI during concomitant administration of cangrelor and ticagrelor was therefore ruled out, since no differences in PRU levels were found between the two groups after drug infusion was stopped. Indeed, patients in the cangrelor group did not have HPR, differently than placebo group where HPR status was reduced but still present in already half of the individuals at the end of the PCI and in one-third of the patients at the end of the placebo infusion. HPR levels were low overall and similar between groups after discontinuation of drug infusion. This consideration is consistent with the absence of DDI between cangrelor and ticagrelor. Despite several limitations, including the limited number of patients, the results were consistent with another nonrandomized pharmacodynamic study of the combination of cangrelor and ticagrelor for pPCI [56] and with a smaller open-label randomized trial [57]. The study demonstrated that in patients undergoing pPCI, the combination of cangrelor and ticagrelor results in a more rapid and potent platelet inhibitory effect compared to ticagrelor alone, with important implications for clinical practice such as a more versatile use of ticagrelor with respect to timing of its administration in patients treated with cangrelor.

Table 2. Characteristics of patients randomized in the CANTIC trial [55].

	Cangrelor Group	Placebo Group	*p*-Value
Patients, n	22	22	
Diagnosis	STEMI	STEMI	
Treatment	Cangrelor: 30 µg/kg 2 h bolus, 4 µg/kg/min infusion	Placebo	
Time from bolus to end of PCI, min (SD)	39 (18–51)	33 (26–60)	
Transition to ticagrelor	Crushed ticagrelor 180 mg	Crushed ticagrelor 180 mg	
HPR at baseline, n	15 (68%)	15 (68%)	NS
HPR during cangrelor			
5 min, n (%)	0 (0%)	15 (71%)	<0.001
30 min, n (%)	0 (0%)	12 (57%)	<0.001
End of PCI, n (%)	0 (0%)	13 (62%)	<0.001
1 h, n (%)	0 (0%)	8 (38%)	0.003
2 h, n (%)	0 (0%)	6 (33%)	0.007
HPR post cangrelor			
1 h, n (%)	2 (10%)	2 (12%)	NS
2 h, n (%)	1 (5%)	1 (6%)	NS

The findings of the CANTIC study were confirmed by the recently published results of the prospective, randomized, double-blind, placebo-controlled, crossover, pharmacokinetic (PK) and pharmacodynamic (PD) SWAP-5 (Pharmacodynamic and Pharmacokinetic Profiles of Switching Between Cangrelor and Ticagrelor Following Ticagrelor Pre-treatment: The Switching Antiplatelet-5 Study) trial, which aimed to rule out DDI among cangrelor-treated patients who were pre-treated with ticagrelor [58]. Indeed, many patients in real-world clinical practice, in whom there may be the desire to use cangrelor to achieve enhanced $P2Y_{12}$ inhibitory effects during PCI, are pre-treated with ticagrelor [59]. This may include patients in whom the full antiplatelet effects of ticagrelor may be delayed by several hours due to impaired absorption such as in patients presenting with ACS, especially STEMI, or treated with opioids [60,61]. In ticagrelor-pretreated patients there was a significant reduction in PRU at 30 min and 1 h after the start of the cangrelor infusion compared to the placebo group. At 2 h after stopping the cangrelor or placebo infusion, PRUs were low and similar in both groups (16.9 vs. 12.6), satisfying the primary endpoint of non-inferiority. No differences were found in PK/PD profiles such as plasma levels of ticagrelor and its metabolite between the two groups after drug infusion discontinuation, thus the absence of a DDI was also confirmed [58]. SWAP-5 Study was conducted in patients with stable CAD and not in patients with ACS undergoing PCI. Hence, the magnitude of the PK/PD findings observed may not be reflective of those in the acute setting. Several other studies are ongoing and will provide further insights into the use of cangrelor in patients undergoing pPCI. More data on transition to potent oral $P2Y_{12}$ receptor inhibitors is desirable, for instance for patients who require a fast-acting intravenous agent such as cangrelor in emergency situations, such as cardiac arrest or cardiogenic shock, or for those who have been preloaded with oral antiplatelet agents and have angiographic findings requiring an additional antiplatelet agent.

The first results of the CAMEO Registry, aimed at retrospectively addressing optimal platelet inhibition during early management of patients with MI prior to coronary angiography or coronary artery bypass grafting, demonstrated inter-hospital variability in how cangrelor was administered and switched to an oral $P2Y_{12}$ inhibitor [62]. These findings highlight opportunities for optimization of cangrelor dosing, infusion duration, and the transition of care from the catheterization lab to the coronary intensive care unit. Data from recently published Cangrelor OHCA (Out-of-Hospital Cardiac Arrest) Study showed that in comatose survivors of OHCA undergoing PCI and target temperature management, cangrelor safely induced immediate and profound platelet inhibition without significant DDI with ticagrelor; nevertheless the study is a single-center and non-placebo-controlled trial [63]. Furthermore, the ongoing multicenter, randomized, double blind trial DAPT-SHOCK-AMI (Dual Antiplatelet Therapy for Shock Patients with Acute Myocardial Infarction; ClinicalTrials.gov Identifier: NCT03551964) will provide results on the comparison between the combination of cangrelor and crushed ticagrelor versus ticagrelor alone in patients with AMI complicated by initial cardiogenic shock and treated with pPCI. The ARCANGELO (Italian Prospective Study on Cangrelor) is a recently published multicenter, observational, prospective cohort study that included patients with ACS undergoing PCI who had not received an oral $P2Y_{12}$ inhibitor before the PCI procedure and in whom oral therapy with $P2Y_{12}$ inhibitors was not feasible or desirable; this study aimed to assess the safety of cangrelor in daily practice [64]. The primary endpoint is the incidence of any hemorrhage, according to Bleeding Academic Research Consortium (BARC) criteria, in the 30 days following the PCI, calculated as the ratio between the number of patients experiencing at least one event during the 30-day observation period and the total number of evaluable patients. The different types of bleedings according to the GUSTO criteria and MACE at various timeframes (from 48 h to 30 days) were investigated, too. The preliminary results showed that all bleedings were classified as BARC Type 1–2, BARC Grade 3a bleeding occurred in one (0.3%) patient, while more severe bleedings were not reported. A total of 17 bleedings were observed in the 320 patients who completed the study. MACE was observed in four patients (two AMI, one sudden cardiac death, one non-cardiovascular death). None bleeding was classified as related to cangrelor. The final analysis of data will assess a more precise evaluation of the study endpoints; however, the use of cangrelor in patients with ACS undergoing PCI does not appear to be associated with severe bleedings. The ongoing SWAP-6 (Pharmacodynamic and Pharmacokinetic Profiles on Switching from Cangrelor to Prasugrel in Patients with Acute Coronary Syndrome Undergoing Percutaneous Coronary Intervention: The Switching Antiplatelet-6 Study; ClinicalTrials.gov Identifier: NCT04668144) trial will further clarify pharmacodynamic effects to rule out a DDI when cangrelor and prasugrel are concomitantly administered in patients undergoing coronary stenting. Currently, a single study has suggested that prasugrel can be administered at the beginning of the cangrelor infusion, with no evidence of drug interactions [65]. Whether this evidence applies to patients with STEMI is unknown and this treatment strategy remains off-label [66].

FABOLUS-FASTER (Facilitation through Aggrastat or Cangrelor Bolus and Infusion over Prasugrel: A Multicenter Randomized Open-Label Trial in Patients with ST-Elevation Myocardial Infarction Referred for Primary Percutaneous Intervention) is a trial that compared, for the first time, the pharmacodynamic effects of cangrelor with the GPI inhibitor tirofiban and the pharmacodynamic and pharmacokinetic effects of prasugrel 60 mg in chewed or whole tablets in patients with STEMI undergoing pPCI [67]. Patients were randomly assigned (1:1:1) to cangrelor (n = 40), tirofiban (n = 40) (both given as a bolus and 2 h infusion followed by a loading dose of 60 mg prasugrel at the time of infusion interruption) or prasugrel 60 mg loading dose (n = 42). Patients in the prasugrel group underwent further 1:1 sub-randomization to oral administration of the loading dose as chewed (n = 21) or whole (n = 21) tablets. Briefly, the aim of the study was to test three primary hypotheses: non-inferiority of cangrelor versus tirofiban, superiority of both tirofiban and cangrelor versus chewed prasugrel, and superiority of chewed prasugrel versus whole prasugrel. Cangrelor did not reach non-inferiority as compared to tirofiban in terms of ADP-induced platelet aggregation (Table 3) due to a lower platelet aggregation in patients treated with tirofiban than cangrelor or chewed prasugrel up to 2 h. Interestingly, residual platelet reactivity was lower with cangrelor compared to chewed prasugrel within the first hour, but higher thereafter.

Table 3. Rates of high residual platelet reactivity (>59%) at Light Transmittance Aggregometry (LTA) after ADP 20 µmol/L stimulation in FABOLUS FASTER trial [67].

	Rates				***p*-Values**			
	Tirofiban	**Cangrelor**	**Chewed Prasugrel**	**Integral Prasugrel**	**Tirofiban vs. Cangrelor**	**Tirofiban vs. Chewed Prasugrel**	**Cangrelor vs. Chewed Prasugrel**	**Chewed Prasugrel vs. Integral Prasugrel**
>59% LTA with ADP 20 µmol/L								
15 min	0.0%	57.5%	100.0%	95.2%	<0.001	<0.001	<0.001	NS
30 min	0.0%	55.0%	90.5%	95.2%	<0.001	<0.001	0.012	NS
1 h	0.0%	55.0%	66.7%	81.0%	<0.001	<0.001	NS	NS
2 h	0.0%	50.0%	38.1%	52.4%	<0.001	<0.001	NS	NS
3 h	7.5%	81.6%	28.6%	19.0%	<0.001	0.030	<0.001	NS
4 to 6 h	7.5%	68.4%	33.3%	19.0%	<0.001	0.014	0.009	NS

Tirofiban was associated with lower TRAP-induced platelet aggregation than cangrelor or chewed prasugrel ($p < 0.001$ at any time point for both comparisons) whereas there was no difference between cangrelor and chewed prasugrel or between the two prasugrel groups. The FABOLUS-FASTER study strengthened the notion of the superiority of parenteral over oral antiplatelet drugs in the acute phase of STEMI treatment in terms of platelet inhibition; however, the observed superiority of tirofiban versus cangrelor remains a mechanistic observation, and whether it could be translated into better clinical outcomes without impairing risk of bleeding remains to be elucidated. Large-scale studies re-evaluating the comparative risks and benefits of a short infusion of parenteral platelet inhibitors such as cangrelor or GPI versus the newer oral $P2Y_{12}$ receptor blockers alone in contemporary pPCI practice remain desirable. Based on the observations from the FABOLUS-FASTER that cangrelor followed by prasugrel is associated with a rebound in platelet activation over 2 to 4 h and on the data of CANTIC Study [55] and Alexopoulos [20] showing some HRPR during and after the cangrelor infusion, it could be hypothesized that when cangrelor is used, ticagrelor may be the preferred oral $P2Y_{12}$ inhibitor.

4. Current Recommendations for the Transition from Cangrelor to Oral $P2Y_{12}$ Inhibitors

At the end of cangrelor infusion, which should be prolonged at least for two hours, patients who underwent PCI with stent implantation should receive a loading dose of an oral $P2Y_{12}$ inhibitor, beyond aspirin. The timing for the $P2Y_{12}$ inhibitor loading dose is related to the pharmacology of the specific drug. Clopidogrel active metabolite is rapidly degraded if it does not bind $P2Y_{12}$ receptor. So, if the receptor is already occupied by cangrelor, a more potent $P2Y_{12}$ inhibitor, clopidogrel active metabolite is degraded, getting no platelet inhibition following the cangrelor infusion cessation. Therefore, loading dose of 600 mg clopidogrel must be administered only after cangrelor infusion cessation. This is also widely supported by CHAMPION platelet sub-study, where there was no apparent significant pharmacodynamic interaction when clopidogrel was administered at the end of the cangrelor infusion [68].

Prasugrel is a thienopyridine requiring activation with similar pharmacodynamics to clopidogrel. Therefore, the loading dose administration of prasugrel should be administered at the end of cangrelor infusion as well. A study examining the transition from cangrelor to thienopyridines showed a transient recovery of platelet reactivity during the switch and found the optimal administration time of prasugrel, to limit the recovery of platelet function, at 30 min prior to cangrelor cessation [69]. This is in line with the more potent binding power to $P2Y_{12}$ receptor of prasugrel compared to clopidogrel. In accordance with this study, the EMA recommends the administration of a prasugrel loading dose (60 mg) either 30 min prior to the cangrelor cessation or immediately after; the FDA recommends it only immediately after cangrelor cessation.

Ticagrelor is a reversible $P2Y_{12}$ inhibitor, and it binds a different site of the receptor compared to cangrelor. Previous studies have demonstrated that there are no DDI between ticagrelor and cangrelor, suggesting that ticagrelor can be given at any time during cangrelor infusions or at the end of it [70]. Both the FDA and EMA have recommended the administration of a ticagrelor loading dose (180 mg) either during the cangrelor infusion or immediately after the infusion cessation.

For clopidogrel and prasugrel, the recommended transitions from cangrelor may result in a brief inadequate $P2Y_{12}$ inhibition, due to the delayed onset of action of clopidogrel and prasugrel. This is consistent with the results of a recent observational pharmacodynamic registry confirming that the switch from cangrelor to clopidogrel could expose patients to a variable period of inadequate platelet inhibition, while ticagrelor given as early as possible after starting cangrelor infusion may avoid any rebound effect in platelet reactivity [71]. Therefore, it is reasonable to prefer ticagrelor as the maintenance $P2Y_{12}$ inhibitor in oral DAPT, as it can be started prior to the cessation of cangrelor.

5. Antiplatelet Bridging for CABG and Non-Cardiac Surgery

Patients treated with a $P2Y_{12}$ inhibitor, who require a major cardiac or non-cardiac surgery, have worse outcomes due to an increased risk for peri- and post-operative bleedings, reoperation and need for blood transfusions. The European guidelines recommended to delay, if it is possible, a non-emergent surgery after PCI with DES implantation until completion of the full course of DAPT, or at least after one month of DAPT [72]. In cases when surgery cannot be delayed for a longer period, a minimum of 1 month of DAPT should be considered, because the higher risk of adverse cardiac events is within the first 30 days after PCI. In any case of patients who need earlier surgery, it is recommended to withhold $P2Y_{12}$ inhibitor at least 7 days for prasugrel, 5 days for clopidogrel and 3 days for ticagrelor before surgery (Figure 1). However, cessation of DAPT in the setting of recent ACS or PCI with stent implantation is associated with a time-dependent increased risk for worse outcomes. It is particularly true for ACS patients with high ischemic risk features, who need at least 6 months of DAPT.

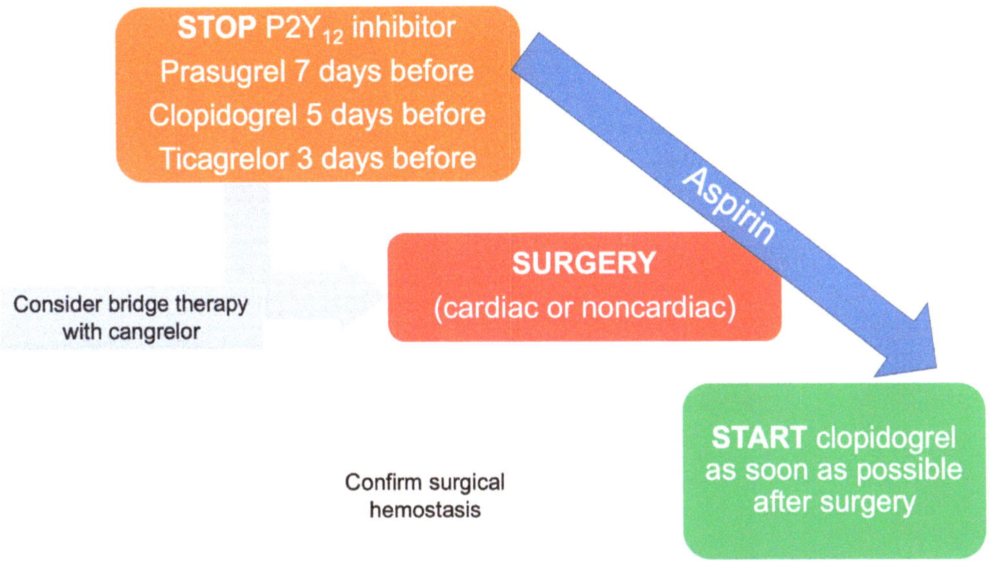

Figure 1. Time frames of $P2Y_{12}$ inhibitor discontinuation and restarting in patients undergoing cardiac or non cardiac surgery. Elaborated from Valgimigli et al. [72].

For patients with a very high risk of ST who cannot delay surgery, bridging therapy with intravenous, reversible platelet inhibitor may be considered (Figure 1). Due to its profile of being a rapid onset/offset, potent and reversible $P2Y_{12}$ inhibitor, cangrelor was tested as a $P2Y_{12}$ inhibitor 'bridge' after discontinuation of thienopyridines, in patients undergoing surgery (BRIDGE trial). In this trial, 210 participants who were planned to undergo non-emergent CABG and had received an oral $P2Y_{12}$ inhibitor were randomized to either cangrelor as bridge therapy or placebo. Aspirin therapy was maintained. Cangrelor, compared with placebo, resulted in higher proportions of suppressed platelet activity, without a significant increase in CABG-related bleeding (11.8 vs. 10.4%; $p = 0.76$), despite some participants receiving cangrelor infusion for up to 7 days [73].

Based on the BRIDGE trial protocol, a recent consensus document standardized management of antithrombotic therapy in patients treated with coronary stents in various types of surgery [74]. It is recommended to stop prasugrel 7 days, clopidogrel and ticagrelor 5 days before surgery. Cangrelor as bridge therapy should be started within 72 h from $P2Y_{12}$ discontinuation, at the dose of 0.75 µg/kg/min without bolus and continued until 1–6 h before skin incision. Clopidogrel should be started, with a new loading dose of 300 or 600 mg, as soon as possible after surgery (within 1–6 h). If oral administration is not possible due to intubation, cangrelor should be restarted. Prasugrel and ticagrelor are discouraged. The MONET BRIDGE study was designed to assess the use of cangrelor as a platelet-inhibiting bridge for patients who discontinue DAPT before cardiac and non-cardiac surgery within 12 months from coronary stent implantation [75]. It demonstrated that perioperative bridging therapy with cangrelor is a feasible approach for patients with DES at high thrombotic risk and undergoing surgery requiring interruption of DAPT: no ischemic outcomes occurred after surgery and up to 30-days follow-up. Moreover, the mean hemoglobin drop was <2 g/dL; nine patients received blood transfusions consistent with the type of surgery, but no life-threatening or fatal bleeding occurred. More studies are warranted to support the efficacy and safety of a standardized bridging strategy by identifying the patient population that would receive the maximum clinical benefit from bridge therapy. In addition to MONET BRIDGE, the MARS (Management of Antiplatelet Regimen During Surgical Procedures; ClinicalTrials.gov Identifier: NCT03981835) trial is currently studying the area of perioperative antiplatelet therapy management through a multi-center, observational US national registry designed to collect preoperative, intraoperative and postoperative clinical strategies, therapeutic interventions, and 30-day outcomes data of ~1500 patients post-PCI scheduled to undergo cardiac or noncardiac surgery.

6. Future Directions

The current available oral $P2Y_{12}$ has a relatively slow onset of action, so drug-naïve patients, and especially those with ACS, undergoing PCI lack the protection conferred by antiplatelet therapy for a too long period and may be exposed to a greater thrombotic risk. Cangrelor proved its effectiveness in drug-naïve patients undergoing PCI, both in the stable and the acute setting, by reducing early and 30-day ischemic outcomes, with particular emphasis for ischemia driven revascularization and early ST. Cangrelor appears to be a very safe drug with a low rate of bleeding and specifically of major (BARC 3–5) events. The results of ongoing randomized trials with new short-acting and potent parenteral antiplatelet agents will be likely to open a new debate about the optimal choice and timing to administer parenteral DAPT in patients with STEMI, since we could have available, at the same time, a drug to be self-administered at home (selatogrel), a drug to be administered in the ambulance (zalunfiban) and a drug to be administered in the hospital (cangrelor). Further RCTs are needed about the combination of parenteral and potent oral $P2Y_{12}$ inhibitors in patients with ACS and about the optimal switching strategies. Available studies so far support the most adopted practice to administer ticagrelor at the same time of cangrelor bolus or as soon as possible after initiation of cangrelor infusion.

Funding: This research received no external funding.

Institutional Review Board Statement: This study did not require ethical approval.

Informed Consent Statement: Not applicable.

Data Availability Statement: Not applicable.

Conflicts of Interest: The authors declare no conflict of interest.

References

1. Leon, M.B.; Baim, D.S.; Popma, J.J.; Gordon, P.C.; Cutlip, D.E.; Ho, K.K.; Giambartolomei, A.; Diver, D.J.; Lasorda, D.M.; Williams, D.O.; et al. A clinical trial comparing three antithrombotic-drug regimens after coronary-artery stenting. Stent Anticoagulation Restenosis Study Investigators. *N. Engl. J. Med.* **1998**, *339*, 1665–1671. [CrossRef] [PubMed]
2. Schömig, A.; Neumann, F.J.; Kastrati, A.; Schühlen, H.; Blasini, R.; Hadamitzky, M.; Walter, H.; Zitzmann-Roth, E.M.; Richardt, G.; Alt, E.; et al. A randomized comparison of antiplatelet and anticoagulant therapy after the placement of coronary-artery stents. *N. Engl. J. Med.* **1996**, *334*, 1084–1089. [CrossRef]
3. McFadden, E.P.; Stabile, E.; Regar, E.; Cheneau, E.; Ong, A.T.; Kinnaird, T.; Suddath, W.O.; Weissman, N.J.; Torguson, R.; Kent, K.M.; et al. Late thrombosis in drug-eluting coronary stents after discontinuation of antiplatelet therapy. *Lancet* **2004**, *364*, 1519–1521. [CrossRef]
4. Wiviott, S.D.; Braunwald, E.; McCabe, C.H.; Montalescot, G.; Ruzyllo, W.; Gottlieb, S.; Neumann, F.J.; Ardissino, D.; De Servi, S.; Murphy, S.A.; et al. Prasugrel versus clopidogrel in patients with acute coronary syndromes. *N. Engl. J. Med.* **2007**, *357*, 2001–2015. [CrossRef]
5. Wallentin, L.; Becker, R.C.; Budaj, A.; Cannon, C.P.; Emanuelsson, H.; Held, C.; Horrow, J.; Husted, S.; James, S.; Katus, H.; et al. Ticagrelor versus clopidogrel in patients with acute coronary syndromes. *N. Engl. J. Med.* **2009**, *361*, 1045–1057. [CrossRef] [PubMed]
6. Bonaca, M.P.; Bhatt, D.L.; Cohen, M.; Steg, P.G.; Storey, R.F.; Jensen, E.C.; Magnani, G.; Bansilal, S.; Fish, M.P.; Im, K.; et al. Long-term use of ticagrelor in patients with prior myocardial infarction. *N. Engl. J. Med.* **2015**, *372*, 1791–1800. [CrossRef]

7. Yeh, R.W.; Kereiakes, D.J.; Steg, P.G.; Windecker, S.; Rinaldi, M.J.; Gershlick, A.H.; Cutlip, D.E.; Cohen, D.J.; Tanguay, J.F.; Jacobs, A.; et al. Benefits and Risks of Extended Duration Dual Antiplatelet Therapy After PCI in Patients With and Without Acute Myocardial Infarction. *J. Am. Coll. Cardiol.* **2015**, *65*, 2211–2221. [CrossRef]
8. Ando, G.; De Santis, G.A.; Greco, A.; Pistelli, L.; Francaviglia, B.; Capodanno, D.; De Caterina, R.; Capranzano, P. P2Y(12) Inhibitor or Aspirin Following Dual Antiplatelet Therapy After Percutaneous Coronary Intervention: A Network Meta-Analysis. *JACC Cardiovasc. Interv.* **2022**, *15*, 2239–2249. [CrossRef]
9. Reynolds, H.R.; Maehara, A.; Kwong, R.Y.; Sedlak, T.; Saw, J.; Smilowitz, N.R.; Mahmud, E.; Wei, J.; Marzo, K.; Matsumura, M.; et al. Coronary Optical Coherence Tomography and Cardiac Magnetic Resonance Imaging to Determine Underlying Causes of Myocardial Infarction With Nonobstructive Coronary Arteries in Women. *Circulation* **2021**, *143*, 624–640. [CrossRef] [PubMed]
10. Waterbury, T.M.; Tarantini, G.; Vogel, B.; Mehran, R.; Gersh, B.J.; Gulati, R. Non-atherosclerotic causes of acute coronary syndromes. *Nat. Rev. Cardiol.* **2020**, *17*, 229–241. [CrossRef]
11. Ando, G.; Trio, O.; de Gregorio, C. Transient left ventricular dysfunction in patients with neurovascular events. *Acute Card Care* **2010**, *12*, 70–74. [CrossRef]
12. Tarantini, G.; Mojoli, M.; Varbella, F.; Caporale, R.; Rigattieri, S.; Ando, G.; Cirillo, P.; Pierini, S.; Santarelli, A.; Sganzerla, P.; et al. Timing of Oral P2Y(12) Inhibitor Administration in Patients With Non-ST-Segment Elevation Acute Coronary Syndrome. *J. Am. Coll. Cardiol.* **2020**, *76*, 2450–2459. [CrossRef] [PubMed]
13. Yusuf, S.; Zhao, F.; Mehta, S.R.; Chrolavicius, S.; Tognoni, G.; Fox, K.K. Effects of clopidogrel in addition to aspirin in patients with acute coronary syndromes without ST-segment elevation. *N. Engl. J. Med.* **2001**, *345*, 494–502. [CrossRef] [PubMed]
14. Gurbel, P.A.; Tantry, U.S. Drug insight: Clopidogrel nonresponsiveness. *Nat. Clin. Pract. Cardiovasc. Med.* **2006**, *3*, 387–395. [CrossRef] [PubMed]
15. Jiang, X.L.; Samant, S.; Lesko, L.J.; Schmidt, S. Clinical pharmacokinetics and pharmacodynamics of clopidogrel. *Clin Pharmacokinet* **2015**, *54*, 147–166. [CrossRef]
16. Brandt, J.T.; Payne, C.D.; Wiviott, S.D.; Weerakkody, G.; Farid, N.A.; Small, D.S.; Jakubowski, J.A.; Naganuma, H.; Winters, K.J. A comparison of prasugrel and clopidogrel loading doses on platelet function: Magnitude of platelet inhibition is related to active metabolite formation. *Am. Heart J.* **2007**, *153*, 66.e9–66.e16. [CrossRef]
17. Schüpke, S.; Neumann, F.J.; Menichelli, M.; Mayer, K.; Bernlochner, I.; Wöhrle, J.; Richardt, G.; Liebetrau, C.; Witzenbichler, B.; Antoniucci, D.; et al. Ticagrelor or Prasugrel in Patients with Acute Coronary Syndromes. *N. Engl. J. Med.* **2019**, *381*, 1524–1534. [CrossRef]
18. Collet, J.P.; Thiele, H.; Barbato, E.; Barthélémy, O.; Bauersachs, J.; Bhatt, D.L.; Dendale, P.; Dorobantu, M.; Edvardsen, T.; Folliguet, T.; et al. 2020 ESC Guidelines for the management of acute coronary syndromes in patients presenting without persistent ST-segment elevation. *Eur. Heart J.* **2021**, *42*, 1289–1367. [CrossRef]
19. Cattaneo, M.; Faioni, E.M. Why does ticagrelor induce dyspnea? *Thromb. Haemost.* **2012**, *108*, 1031–1036. [CrossRef]
20. Alexopoulos, D.; Xanthopoulou, I.; Gkizas, V.; Kassimis, G.; Theodoropoulos, K.C.; Makris, G.; Koutsogiannis, N.; Damelou, A.; Tsigkas, G.; Davlouros, P.; et al. Randomized assessment of ticagrelor versus prasugrel antiplatelet effects in patients with ST-segment-elevation myocardial infarction. *Circ. Cardiovasc. Interv.* **2012**, *5*, 797–804. [CrossRef]
21. Valgimigli, M.; Tebaldi, M.; Campo, G.; Gambetti, S.; Bristot, L.; Monti, M.; Parrinello, G.; Ferrari, R. Prasugrel versus tirofiban bolus with or without short post-bolus infusion with or without concomitant prasugrel administration in patients with myocardial infarction undergoing coronary stenting: The FABOLUS PRO (Facilitation through Aggrastat By drOpping or shortening Infusion Line in patients with ST-segment elevation myocardial infarction compared to or on top of PRasugrel given at loading dOse) trial. *JACC Cardiovasc. Interv.* **2012**, *5*, 268–277. [CrossRef]
22. Parodi, G.; Valenti, R.; Bellandi, B.; Migliorini, A.; Marcucci, R.; Comito, V.; Carrabba, N.; Santini, A.; Gensini, G.F.; Abbate, R.; et al. Comparison of prasugrel and ticagrelor loading doses in ST-segment elevation myocardial infarction patients: RAPID (Rapid Activity of Platelet Inhibitor Drugs) primary PCI study. *J. Am. Coll. Cardiol.* **2013**, *61*, 1601–1606. [CrossRef]
23. Parodi, G.; Xanthopoulou, I.; Bellandi, B.; Gkizas, V.; Valenti, R.; Karanikas, S.; Migliorini, A.; Angelidis, C.; Abbate, R.; Patsilinakos, S.; et al. Ticagrelor crushed tablets administration in STEMI patients: The MOJITO study. *J. Am. Coll. Cardiol.* **2015**, *65*, 511–512. [CrossRef]
24. Rollini, F.; Franchi, F.; Hu, J.; Kureti, M.; Aggarwal, N.; Durairaj, A.; Park, Y.; Seawell, M.; Cox-Alomar, P.; Zenni, M.M.; et al. Crushed Prasugrel Tablets in Patients With STEMI Undergoing Primary Percutaneous Coronary Intervention: The CRUSH Study. *J. Am. Coll. Cardiol.* **2016**, *67*, 1994–2004. [CrossRef] [PubMed]
25. Venetsanos, D.; Sederholm Lawesson, S.; Swahn, E.; Alfredsson, J. Chewed ticagrelor tablets provide faster platelet inhibition compared to integral tablets: The inhibition of platelet aggregation after administration of three different ticagrelor formulations (IPAAD-Tica) study, a randomised controlled trial. *Thromb. Res.* **2017**, *149*, 88–94. [CrossRef] [PubMed]
26. Knuuti, J.; Wijns, W.; Saraste, A.; Capodanno, D.; Barbato, E.; Funck-Brentano, C.; Prescott, E.; Storey, R.F.; Deaton, C.; Cuisset, T.; et al. 2019 ESC Guidelines for the diagnosis and management of chronic coronary syndromes. *Eur. Heart J.* **2020**, *41*, 407–477. [CrossRef] [PubMed]
27. Orme, R.C.; Parker, W.A.E.; Thomas, M.R.; Judge, H.M.; Baster, K.; Sumaya, W.; Morgan, K.P.; McMellon, H.C.; Richardson, J.D.; Grech, E.D.; et al. Study of Two Dose Regimens of Ticagrelor Compared with Clopidogrel in Patients Undergoing Percutaneous Coronary Intervention for Stable Coronary Artery Disease (STEEL-PCI). *Circulation* **2018**, *138*, 1290–1300. [CrossRef]

28. Serenelli, M.; Pavasini, R.; Vitali, F.; Tonet, E.; Bilotta, F.; Parodi, G.; Campo, G. Efficacy and safety of alternative oral administrations of P2Y12-receptor inhibitors: Systematic review and meta-analysis. *J. Thromb. Haemost.* **2019**, *17*, 944–950. [CrossRef]
29. Capranzano, P.; Angiolillo, D.J. Tackling the gap in platelet inhibition with oral antiplatelet agents in high-risk patients undergoing percutaneous coronary intervention. *Expert. Rev. Cardiovasc. Ther.* **2021**, *19*, 519–535. [CrossRef]
30. Reilly, P.M.; Wilkins, K.B.; Fuh, K.C.; Haglund, U.; Bulkley, G.B. The mesenteric hemodynamic response to circulatory shock: An overview. *Shock* **2001**, *15*, 329–343. [CrossRef]
31. Wakabayashi, S.; Kitahara, H.; Nishi, T.; Sugimoto, K.; Nakayama, T.; Fujimoto, Y.; Ariyoshi, N.; Kobayashi, Y. Platelet inhibition after loading dose of prasugrel in patients with ST-elevation and non-ST-elevation acute coronary syndrome. *Cardiovasc. Interv. Ther.* **2018**, *33*, 239–246. [CrossRef]
32. Parodi, G.; Talanas, G.; Mura, E.; Canonico, M.E.; Siciliano, R.; Guarino, S.; Marini, A.; Dossi, F.; Franca, P.; Raccis, M.; et al. Orodispersible Ticagrelor in Acute Coronary Syndromes: The TASTER Study. *J. Am. Coll. Cardiol.* **2021**, *78*, 292–294. [CrossRef] [PubMed]
33. Schilling, U.; Dingemanse, J.; Ufer, M. Pharmacokinetics and Pharmacodynamics of Approved and Investigational P2Y12 Receptor Antagonists. *Clin. Pharmacokinet.* **2020**, *59*, 545–566. [CrossRef] [PubMed]
34. Bhatt, D.L.; Stone, G.W.; Mahaffey, K.W.; Gibson, C.M.; Steg, P.G.; Hamm, C.W.; Price, M.J.; Leonardi, S.; Gallup, D.; Bramucci, E.; et al. Effect of platelet inhibition with cangrelor during PCI on ischemic events. *N. Engl. J. Med.* **2013**, *368*, 1303–1313. [CrossRef]
35. Zwart, B.; Parker, W.A.E.; Storey, R.F. New Antithrombotic Drugs in Acute Coronary Syndrome. *J. Clin. Med.* **2020**, *9*, 2059. [CrossRef]
36. Caroff, E.; Meyer, E.; Treiber, A.; Hilpert, K.; Riederer, M.A. Optimization of 2-phenyl-pyrimidine-4-carboxamides towards potent, orally bioavailable and selective P2Y(12) antagonists for inhibition of platelet aggregation. *Bioorg. Med. Chem. Lett.* **2014**, *24*, 4323–4331. [CrossRef]
37. Parker, W.A.E.; Storey, R.F. Novel approaches to P2Y(12) inhibition and aspirin dosing. *Platelets* **2021**, *32*, 7–14. [CrossRef] [PubMed]
38. Rey, M.; Kramberg, M.; Hess, P.; Morrison, K.; Ernst, R.; Haag, F.; Weber, E.; Clozel, M.; Baumann, M.; Caroff, E.; et al. The reversible P2Y(12) antagonist ACT-246475 causes significantly less blood loss than ticagrelor at equivalent antithrombotic efficacy in rat. *Pharmacol. Res. Perspect.* **2017**, *5*, e00338. [CrossRef]
39. Ufer, M.; Huynh, C.; van Lier, J.J.; Caroff, E.; Fischer, H.; Dingemanse, J. Absorption, distribution, metabolism and excretion of the P2Y12 receptor antagonist selatogrel after subcutaneous administration in healthy subjects. *Xenobiotica* **2020**, *50*, 427–434. [CrossRef]
40. Storey, R.F.; Gurbel, P.A.; Ten Berg, J.; Bernaud, C.; Dangas, G.D.; Frenoux, J.M.; Gorog, D.A.; Hmissi, A.; Kunadian, V.; James, S.K.; et al. Pharmacodynamics, pharmacokinetics, and safety of single-dose subcutaneous administration of selatogrel, a novel P2Y12 receptor antagonist, in patients with chronic coronary syndromes. *Eur. Heart J.* **2020**, *41*, 3132–3140. [CrossRef]
41. Sinnaeve, P.; Fahrni, G.; Schelfaut, D.; Spirito, A.; Mueller, C.; Frenoux, J.M.; Hmissi, A.; Bernaud, C.; Ufer, M.; Moccetti, T.; et al. Subcutaneous Selatogrel Inhibits Platelet Aggregation in Patients With Acute Myocardial Infarction. *J. Am. Coll. Cardiol.* **2020**, *75*, 2588–2597. [CrossRef]
42. Valgimigli, M.; Landi, A. Subcutaneous RUC-4 for acute myocardial infarction: A new treatment on the horizon for pre-hospital care? *EuroIntervention* **2021**, *17*, e362–e363. [CrossRef]
43. Kereiakes, D.J.; Henry, T.D.; DeMaria, A.N.; Bentur, O.; Carlson, M.; Seng Yue, C.; Martin, L.H.; Midkiff, J.; Mueller, M.; Meek, T.; et al. First Human Use of RUC-4: A Nonactivating Second-Generation Small-Molecule Platelet Glycoprotein IIb/IIIa (Integrin alphaIIbbeta3) Inhibitor Designed for Subcutaneous Point-of-Care Treatment of ST-Segment-Elevation Myocardial Infarction. *J. Am. Heart Assoc.* **2020**, *9*, e016552. [CrossRef]
44. Bor, W.L.; Zheng, K.L.; Tavenier, A.H.; Gibson, C.M.; Granger, C.B.; Bentur, O.; Lobatto, R.; Postma, S.; Coller, B.S.; van 't Hof, A.W.J.; et al. Pharmacokinetics, pharmacodynamics, and tolerability of subcutaneous administration of a novel glycoprotein IIb/IIIa inhibitor, RUC-4, in patients with ST-segment elevation myocardial infarction. *EuroIntervention* **2021**, *17*, e401–e410. [CrossRef] [PubMed]
45. Angiolillo, D.J.; Bhatt, D.L.; Steg, P.G.; Stone, G.W.; White, H.D.; Gibson, C.M.; Hamm, C.W.; Price, M.J.; Prats, J.; Liu, T.; et al. Impact of cangrelor overdosing on bleeding complications in patients undergoing percutaneous coronary intervention: Insights from the CHAMPION trials. *J. Thromb. Thrombolysis* **2015**, *40*, 317–322. [CrossRef]
46. Franchi, F.; Rollini, F.; Park, Y.; Angiolillo, D.J. A Safety Evaluation of Cangrelor in Patients Undergoing PCI. *Expert Opin. Drug Saf.* **2016**, *15*, 275–285. [CrossRef]
47. Leonardi, S.; Bhatt, D.L. Practical considerations for cangrelor use in patients with acute coronary syndromes. *Eur. Heart J. Acute Cardiovasc. Care* **2019**, *8*, 39–44. [CrossRef]
48. Bhatt, D.L.; Lincoff, A.M.; Gibson, C.M.; Stone, G.W.; McNulty, S.; Montalescot, G.; Kleiman, N.S.; Goodman, S.G.; White, H.D.; Mahaffey, K.W.; et al. Intravenous platelet blockade with cangrelor during PCI. *N. Engl. J. Med.* **2009**, *361*, 2330–2341. [CrossRef] [PubMed]
49. Mehran, R.; Rao, S.V.; Bhatt, D.L.; Gibson, C.M.; Caixeta, A.; Eikelboom, J.; Kaul, S.; Wiviott, S.D.; Menon, V.; Nikolsky, E.; et al. Standardized bleeding definitions for cardiovascular clinical trials: A consensus report from the Bleeding Academic Research Consortium. *Circulation* **2011**, *123*, 2736–2747. [CrossRef]

50. Harrington, R.A.; Stone, G.W.; McNulty, S.; White, H.D.; Lincoff, A.M.; Gibson, C.M.; Pollack, C.V., Jr.; Montalescot, G.; Mahaffey, K.W.; Kleiman, N.S.; et al. Platelet inhibition with cangrelor in patients undergoing PCI. *N. Engl. J. Med.* **2009**, *361*, 2318–2329. [CrossRef] [PubMed]
51. Feng, K.Y.; Mahaffey, K.W. Cangrelor in clinical use. *Future Cardiol.* **2020**, *16*, 89–102. [CrossRef]
52. White, H.D.; Chew, D.P.; Dauerman, H.L.; Mahaffey, K.W.; Gibson, C.M.; Stone, G.W.; Gruberg, L.; Harrington, R.A.; Bhatt, D.L. Reduced immediate ischemic events with cangrelor in PCI: A pooled analysis of the CHAMPION trials using the universal definition of myocardial infarction. *Am. Heart J.* **2012**, *163*, 182–190.e184. [CrossRef]
53. Abtan, J.; Steg, P.G.; Stone, G.W.; Mahaffey, K.W.; Gibson, C.M.; Hamm, C.W.; Price, M.J.; Abnousi, F.; Prats, J.; Deliargyris, E.N.; et al. Efficacy and Safety of Cangrelor in Preventing Periprocedural Complications in Patients With Stable Angina and Acute Coronary Syndromes Undergoing Percutaneous Coronary Intervention: The CHAMPION PHOENIX Trial. *JACC Cardiovasc. Interv.* **2016**, *9*, 1905–1913. [CrossRef] [PubMed]
54. Steg, P.G.; Bhatt, D.L.; Hamm, C.W.; Stone, G.W.; Gibson, C.M.; Mahaffey, K.W.; Leonardi, S.; Liu, T.; Skerjanec, S.; Day, J.R.; et al. Effect of cangrelor on periprocedural outcomes in percutaneous coronary interventions: A pooled analysis of patient-level data. *Lancet* **2013**, *382*, 1981–1992. [CrossRef]
55. Franchi, F.; Rollini, F.; Rivas, A.; Wali, M.; Briceno, M.; Agarwal, M.; Shaikh, Z.; Nawaz, A.; Silva, G.; Been, L.; et al. Platelet Inhibition With Cangrelor and Crushed Ticagrelor in Patients With ST-Segment-Elevation Myocardial Infarction Undergoing Primary Percutaneous Coronary Intervention. *Circulation* **2019**, *139*, 1661–1670. [CrossRef] [PubMed]
56. Mohammad, M.A.; Andell, P.; Koul, S.; James, S.; Scherstén, F.; Götberg, M.; Erlinge, D. Cangrelor in combination with ticagrelor provides consistent and potent P2Y12-inhibition during and after primary percutaneous coronary intervention in real-world patients with ST-segment-elevation myocardial infarction. *Platelets* **2017**, *28*, 414–416. [CrossRef] [PubMed]
57. Alexopoulos, D.; Pappas, C.; Sfantou, D.; Xanthopoulou, I.; Didagelos, M.; Kikas, P.; Ziakas, A.; Tziakas, D.; Karvounis, H.; Iliodromitis, E. Cangrelor in Ticagrelor-Loaded STEMI Patients Undergoing Primary Percutaneous Coronary Intervention. *J. Am. Coll. Cardiol.* **2018**, *72*, 1750–1751. [CrossRef]
58. Franchi, F.; Ortega-Paz, L.; Rollini, F.; Galli, M.; Been, L.; Ghanem, G.; Shalhoub, A.; Ossi, T.; Rivas, A.; Zhou, X.; et al. Cangrelor in Patients With Coronary Artery Disease Pretreated With Ticagrelor: The Switching Antiplatelet (SWAP)-5 Study. *JACC Cardiovasc. Interv.* **2023**, *16*, 36–46. [CrossRef]
59. Grimfjärd, P.; Lagerqvist, B.; Erlinge, D.; Varenhorst, C.; James, S. Clinical use of cangrelor: Nationwide experience from the Swedish Coronary Angiography and Angioplasty Registry (SCAAR). *Eur. Heart J. Cardiovasc. Pharmacother.* **2019**, *5*, 151–157. [CrossRef]
60. Franchi, F.; Rollini, F.; Angiolillo, D.J. Antithrombotic therapy for patients with STEMI undergoing primary PCI. *Nat. Rev. Cardiol.* **2017**, *14*, 361–379. [CrossRef]
61. Franchi, F.; Rollini, F.; Park, Y.; Hu, J.; Kureti, M.; Rivas Rios, J.; Faz, G.; Yaranov, D.; Been, L.; Pineda, A.M.; et al. Effects of Methylnaltrexone on Ticagrelor-Induced Antiplatelet Effects in Coronary Artery Disease Patients Treated With Morphine. *JACC Cardiovasc. Interv.* **2019**, *12*, 1538–1549. [CrossRef] [PubMed]
62. Rymer, J.A.; Bhatt, D.L.; Angiolillo, D.J.; Diaz, M.; Garratt, K.N.; Waksman, R.; Edwards, L.; Tasissa, G.; Salahuddin, K.; El-Sabae, H.; et al. Cangrelor Use Patterns and Transition to Oral P2Y(12) Inhibitors Among Patients With Myocardial Infarction: Initial Results From the CAMEO Registry. *J. Am. Heart Assoc.* **2022**, *11*, e024513. [CrossRef] [PubMed]
63. Kordis, P.; Bozic Mijovski, M.; Berden, J.; Steblovnik, K.; Blinc, A.; Noc, M. Cangrelor for comatose survivors of out-of-hospital cardiac arrest undergoing percutaneous coronary intervention: The CANGRELOR-OHCA study. *EuroIntervention* **2023**, *18*, 1269–1271. [CrossRef] [PubMed]
64. De Luca, L.; Calabrò, P.; Chirillo, F.; Rolfo, C.; Menozzi, A.; Capranzano, P.; Menichelli, M.; Nicolini, E.; Mauro, C.; Trani, C.; et al. Use of cangrelor in patients with acute coronary syndromes undergoing percutaneous coronary intervention: Study design and interim analysis of the ARCANGELO study. *Clin. Cardiol.* **2022**, *45*, 913–920. [CrossRef]
65. Hochholzer, W.; Kleiner, P.; Younas, I.; Valina, C.M.; Loffelhardt, N.; Amann, M.; Bomicke, T.; Ferenc, M.; Hauschke, D.; Trenk, D.; et al. Randomized Comparison of Oral P2Y(12)-Receptor Inhibitor Loading Strategies for Transitioning From Cangrelor: The ExcelsiorLOAD2 Trial. *JACC Cardiovasc. Interv.* **2017**, *10*, 121–129. [CrossRef]
66. Rollini, F.; Franchi, F.; Angiolillo, D.J. Drug-Drug Interactions When Switching Between Intravenous and Oral P2Y(12) Receptor Inhibitors: How Real Is It? *JACC Cardiovasc. Interv.* **2017**, *10*, 130–132. [CrossRef]
67. Gargiulo, G.; Esposito, G.; Avvedimento, M.; Nagler, M.; Minuz, P.; Campo, G.; Gragnano, F.; Manavifar, N.; Piccolo, R.; Tebaldi, M.; et al. Cangrelor, Tirofiban, and Chewed or Standard Prasugrel Regimens in Patients With ST-Segment-Elevation Myocardial Infarction: Primary Results of the FABOLUS-FASTER Trial. *Circulation* **2020**, *142*, 441–454. [CrossRef]
68. Angiolillo, D.J.; Schneider, D.J.; Bhatt, D.L.; French, W.J.; Price, M.J.; Saucedo, J.F.; Shaburishvili, T.; Huber, K.; Prats, J.; Liu, T.; et al. Pharmacodynamic effects of cangrelor and clopidogrel: The platelet function substudy from the cangrelor versus standard therapy to achieve optimal management of platelet inhibition (CHAMPION) trials. *J. Thromb. Thrombolysis* **2012**, *34*, 44–55. [CrossRef]
69. Schneider, D.J.; Seecheran, N.; Raza, S.S.; Keating, F.K.; Gogo, P. Pharmacodynamic effects during the transition between cangrelor and prasugrel. *Coron Artery Dis.* **2015**, *26*, 42–48. [CrossRef]
70. Schneider, D.J.; Agarwal, Z.; Seecheran, N.; Keating, F.K.; Gogo, P. Pharmacodynamic effects during the transition between cangrelor and ticagrelor. *JACC Cardiovasc. Interv.* **2014**, *7*, 435–442. [CrossRef] [PubMed]

71. Gargiulo, G.; Marenna, A.; Sperandeo, L.; Manzi, L.; Avvedimento, M.; Simonetti, F.; Canonico, M.E.; Paolillo, R.; Spinelli, A.; Borgia, F.; et al. Pharmacodynamic effects of cangrelor in elective complex PCI: Insights from the POMPEII Registry. *EuroIntervention* **2023**, *18*, 1266–1268. [CrossRef]
72. Valgimigli, M.; Bueno, H.; Byrne, R.A.; Collet, J.P.; Costa, F.; Jeppsson, A.; Jüni, P.; Kastrati, A.; Kolh, P.; Mauri, L.; et al. 2017 ESC focused update on dual antiplatelet therapy in coronary artery disease developed in collaboration with EACTS: The Task Force for dual antiplatelet therapy in coronary artery disease of the European Society of Cardiology (ESC) and of the European Association for Cardio-Thoracic Surgery (EACTS). *Eur. Heart J.* **2018**, *39*, 213–260. [CrossRef]
73. Angiolillo, D.J.; Firstenberg, M.S.; Price, M.J.; Tummala, P.E.; Hutyra, M.; Welsby, I.J.; Voeltz, M.D.; Chandna, H.; Ramaiah, C.; Brtko, M.; et al. Bridging antiplatelet therapy with cangrelor in patients undergoing cardiac surgery: A randomized controlled trial. *JAMA* **2012**, *307*, 265–274. [CrossRef]
74. Rossini, R.; Tarantini, G.; Musumeci, G.; Masiero, G.; Barbato, E.; Calabrò, P.; Capodanno, D.; Leonardi, S.; Lettino, M.; Limbruno, U.; et al. A Multidisciplinary Approach on the Perioperative Antithrombotic Management of Patients With Coronary Stents Undergoing Surgery: Surgery After Stenting 2. *JACC Cardiovasc. Interv.* **2018**, *11*, 417–434. [CrossRef]
75. Rossini, R.; Masiero, G.; Fruttero, C.; Passamonti, E.; Calvaruso, E.; Cecconi, M.; Carlucci, C.; Mojoli, M.; Guido, P.; Talanas, G.; et al. Antiplatelet Therapy with Cangrelor in Patients Undergoing Surgery after Coronary Stent Implantation: A Real-World Bridging Protocol Experience. *TH Open* **2020**, *4*, e437–e445. [CrossRef]

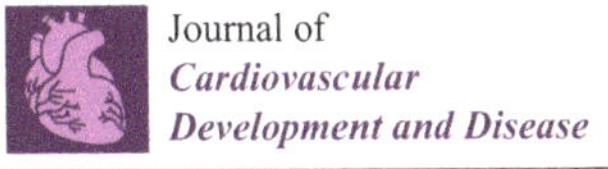
Journal of
Cardiovascular Development and Disease

MDPI

Review

Novel Antithrombotic Agents in Ischemic Cardiovascular Disease: Progress in the Search for the Optimal Treatment

Ignacio Barriuso [1,2,3], Fernando Worner [1] and Gemma Vilahur [2,4,*]

1 Hospital Universitario Arnau de Vilanova, Institut de Recerca Biomèdica de Lleida, 25198 Lleida, Spain
2 Institut de Recerca, Hospital Santa Creu i Sant Pau, IIB Sant Pau, 08025 Barcelona, Spain
3 Department of Medicine, Autonomous University of Barcelona, 08193 Barcelona, Spain
4 Centro de Investigaciones Biomédicas En Red de enfermedades CardioVasculares (CiberCV), 28029 Madrid, Spain
* Correspondence: gvilahur@santpau.cat; Tel.: +34-935537100

Abstract: Ischemic cardiovascular diseases have a high incidence and high mortality worldwide. Therapeutic advances in the last decades have reduced cardiovascular mortality, with antithrombotic therapy being the cornerstone of medical treatment. Yet, currently used antithrombotic agents carry an inherent risk of bleeding associated with adverse cardiovascular outcomes and mortality. Advances in understanding the pathophysiology of thrombus formation have led to the discovery of new targets and the development of new anticoagulants and antiplatelet agents aimed at preventing thrombus stabilization and growth while preserving hemostasis. In the following review, we will comment on the key limitation of the currently used antithrombotic regimes in ischemic heart disease and ischemic stroke and provide an in-depth and state-of-the-art overview of the emerging anticoagulant and antiplatelet agents in the pipeline with the potential to improve clinical outcomes.

Keywords: cardiovascular diseases; novel antithrombotic agents; antiplatelet drugs; anticoagulants; hemostasis

Citation: Barriuso, I.; Worner, F.; Vilahur, G. Novel Antithrombotic Agents in Ischemic Cardiovascular Disease: Progress in the Search for the Optimal Treatment. *J. Cardiovasc. Dev. Dis.* **2022**, *9*, 397. https://doi.org/10.3390/jcdd9110397

Academic Editor: Qingping Feng

Received: 3 October 2022
Accepted: 11 November 2022
Published: 16 November 2022

1. Introduction

Cardiovascular diseases (CVDs) remain the leading cause of death worldwide. In 2016, 17.9 million people died from all causes of CVDs [1,2]. There were approximately 8.9 million deaths due to ischemic heart disease (IHD) worldwide, remaining the first cause of death; a less prevalent disease was ischemic stroke, with an incidence of 7.6 million globally [3–5].

Therapeutic advances in the last decades have reduced CVD mortality, with antithrombotic therapy being the cornerstone of medical treatment. Several antithrombotic drugs are currently used to either block platelet activation (Figure 1), prevent the activation of the coagulation cascade, or induce fibrinolysis once the clot is formed (Figure 2) [6–9]. Yet, although these antithrombotic agents have robustly demonstrated their effectiveness in preventing atherothrombotic events, they also carry an inherent risk of bleeding. Bleeding is associated with adverse cardiovascular outcomes and mortality; hence, there is a need to discover new targets and develop novel antithrombotic strategies to effectively inhibit thrombosis while preserving hemostasis.

In the following review, we will comment on the key limitations of the currently used antithrombotic regimes in ischemic heart disease and ischemic stroke and provide an in-depth and state-of-the-art overview of the emerging anticoagulant and antiplatelet agents in the pipeline with the potential to improve clinical outcomes.

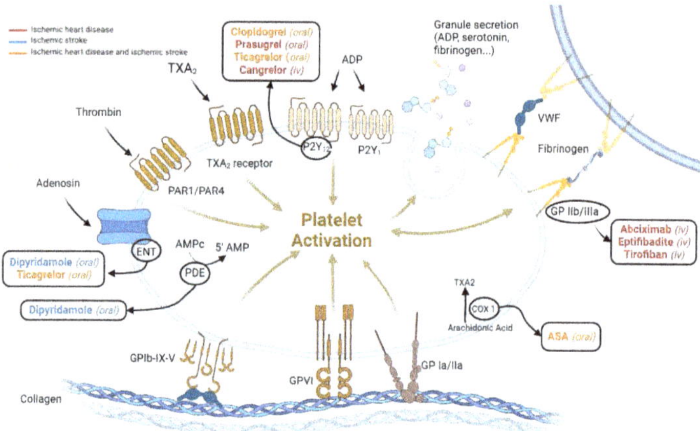

Figure 1. Antiplatelet drugs currently used to treat ischemic heart disease and ischemic stroke. ENT: equilibrative nucleoside transporter; ASA: acetylsalicylic acid; TXA2: thromboxane A2; VWF: Von Willebrand Factor; PDR: phosphodiesterase; COX: cyclooxygenase. Figure created with BioRender.com.

2. The Coagulation Cascade: Targeting the Intrinsic Coagulation Pathway

Anticoagulants are the treatment of choice to prevent cardioembolic stroke in patients with atrial fibrillation [10,11]. During the last decade, the development of direct oral anticoagulants (DOACs; Figure 2) has brought many advantages as compared to vitamin K antagonists, including a predictable pharmacokinetic profile, rapid onset and offset of action, and fixed dosing with no need for laboratory monitoring or dietary discretion [12]. Conversely, different reversal agents have also been developed to block the effect of anticoagulants in case of need (Table 1).

Table 1. Anticoagulant reversal agents in clinical use and preclinical/clinical development.

In Clinical Use [13–15]	
Agent	*Target*
Vitamin K	Warfarin, acenocumarol
Idarucimab	Dabigatran
Andexanet alfa	Apixaban, rivaroxavan, edoxaban
Protamine sulfate	Unfractionated heparin LMWH (partially)
Prothrombin complex concentrate, fresh frozen plasma	Non-specific prohemostatic agents
Preclinical/Clinical Development	
Agent	*Target*
Aripazine (ciraparantag/PER977) (NCT04593784) [16,17]	LMWH, fondaparinux, FXa inhibitors, dabigatran
γ-thrombine S195A [18]	Dabigatran
GDFXa-α2M complex [19]	Rivaroxaban, apixaban, dabigatran and heparins

Yet, important challenges still need to be addressed. As such, bleeding remains the most reported side effect of DOACs, and in certain sub-groups of patients, including patients with mechanical heart valves or triple-positive antiphospholipid disease syndrome, DOACs seem to be less effective than vitamin K antagonists and are not recommended [20–22].

Anticoagulants are also implemented in ischemic heart disease since patients who suffer an acute coronary event present an excess of thrombin generation that persists beyond the acute presentation [23]. So far, several trials have demonstrated the ability of anticoagulants to protect against cardiovascular events. As such, the addition of warfarin [24], rivaroxaban [25,26], or ximelagatran [27] to a standard antiplatelet regime has shown to significantly reduce ischemic events, though at the expense of increased bleeding risk. Ximelagatran was, however, withdrawn from the market due to hepatotoxicity and the only anticoagulant recommended by the guidelines for long-term secondary prevention is rivaroxaban, which may be administered at low doses on top of aspirin at 1-year post-MI [25].

Altogether, these trials have evidenced the need to discover new targets that effectively block thrombin generation without displaying hemorrhagic side effects. In recent years, special attention has focused on the main components of the intrinsic coagulation pathway, particularly factor (F)XII, FXI, and FIX [28].

2.1. Targeting Factor XII

FXII has been associated with thrombosis, hereditary angioedema, and (neuro) inflammation. On the other hand, FXII deficiency (i.e., Hageman factor deficiency) is a rare genetic blood disorder that is entirely asymptomatic, showing prolonged active partial thromboplastin times (aPTT) as the only alteration on coagulation tests [29]. FXII circulates in plasma in a zymogen form, and its activation is brought about by the interaction with negatively charged molecules that induce a conformational change in zymogen FXII leading to activated protease FXIIa followed by activation of the enzyme precursors FXI and FIX (Figure 2) [30]. Hence, FXII inhibitors are expected to be particularly efficient in patients whose blood is exposed to non-physiological surfaces of medical devices such as vascular catheters, hemodialysis circuit tubes and membranes, and mechanical valves or stents, that expose negative charge molecules [31]. Alternatively, contact system proteins FXII, high-molecular-weight kininogen (HK), and plasma kallikrein (PK) may assemble on cell surface proteoglycans of various cardiovascular cells. Contact with surface-exposed moieties and plasma-borne soluble contact activators induces FXII activation, which initiates the intrinsic coagulation pathway and activates PK leading to the release of the proinflammatory mediator bradykinin (BK) by PK-mediated cleavage of HK. FXII inhibitors are also being evaluated as a potential treatment for hereditary angioedema, a BK-mediated life-threatening inherited swelling disorder where Serpin C1 esterase inhibitor (a major plasma inhibitor of FXII and and PK) is dysfunctional or deficient [16].

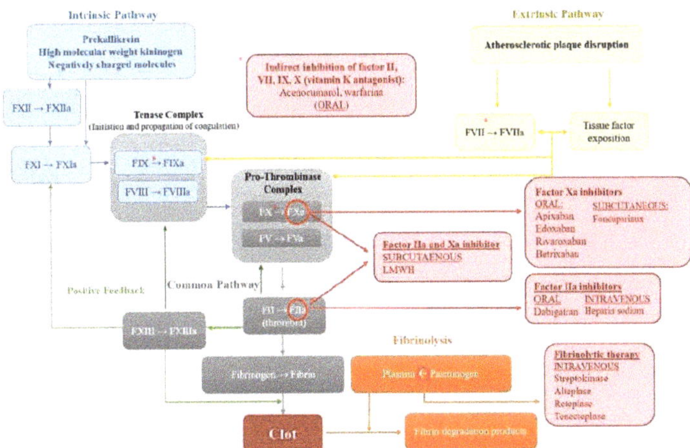

Figure 2. Diagram of the coagulation and fibrinolytic pathways and the different anticoagulants and fibrinolytic agents used in the clinical setting. LMWH: low molecular weight heparin. * Indirect inhibition of factors II, VII, IX and X.

Multiple prototypes have been discovered within the last years, including monoclonal antibodies, small interfering RNAs, antisense oligonucleotides, and serine protease inhibitors which are currently being tested at a preclinical level (details are provided in Table 2) [32]. However, each one of these strategies confers different pharmacological properties, which may limit their indications. Antibodies and approaches to silence gene expression require parenteral delivery by subcutaneous or intravenous injection, whereas small molecules can be delivered orally or parenterally (Table 2). On the other hand, small interfering RNAs and antisense oligonucleotides have a slow onset of action requiring about four weeks to achieve therapeutic levels. Although they are not optimal for use in acute settings, their effect extends over time which may enable once-monthly administration. On the other hand, however, they may also require the development of antidotes. In contrast, antibodies and serine protease inhibitors have a rapid onset of action and an expected half-live of <24 h, thereby limiting the need to develop reversal strategies [33].

Table 2. Factor XIIa inhibitors currently under development. This table includes the emerging FXIIa inhibitors and details the studies conducted so far to assess their efficacy and safety.

Factor XIIa Inhibitors	Type	Phase	Studies Conducted So Far
Garadacimab ***(subcutaneous)***	Antibody	III	Tested in patients with C1-esterase inhibitor-deficient hereditary angioedema showing a significant reduction of angioedema attacks. A dose-dependent increase in aPTT with no change in prothrombin time was also observed without increasing of bleeding events [34,35]. Currently ongoing phase III trials (NCT04656418, NCT04739059).
3F7 ***(intravenous)***	Antibody	Preclinical	Thromboprotection in ECMO without impairing the hemostatic capacity or increasing bleeding [36,37].
9A2 and 15H8 ***(intravenous)***	Antibody	I	Both antibodies have been shown to protect against ferric chloride-induced arterial thrombosis. 15H8 prolonged the aPTT time in non-human primates and humans and reduced fibrin formation in collagen-coated vascular grafts inserted into arteriovenous shunts in non-human primates [38].
5C12 ***(intravenous)***	Antibody	Preclinical	Thromboprotection in ECMO in non-human primates [39].
Ir-CPI ***(intravenous)***	Kunitz-type serine protease inhibitor	Preclinical	It has demonstrated antithrombotic activity in: (1) venous and arterial in vitro thrombosis models; (2) arteriovenous shunt rabbit models; and (3) extracorporeal circuit [40,41]. It can interact with factors XIIa, XIa, and Kallikrein [42].
FXII-ASO ***(subcutaneous)***	Antisense oligonucleotide	Preclinical	Prolonged the time to catheter thrombotic occlusion (implanted in jugular vein) compared to control in a rabbit model of thrombosis [43].
ALN-F12 ***(subcutaneous)***	Interfering RNA	Preclinical	Dose-dependently reduced platelet and fibrin deposition in mice models of venous and arterial thrombosis models [44].
rHA-Infestin-4 ***(intravenous)***	Kazal-type serine protease inhibitor	Preclinical	Protects against arterial and venous thrombosis in mouse and rabbit models. Reduces infarct size and brain edema formation leading to better neurological scores and survival in a mouse model of stroke [45–47].

aPTT: activated partial thromboplastin time; ECMO: extracorporeal membrane oxygenation.

Garadacimab, a monoclonal antibody, has been the sole FXIIa inhibitor to reach phase III clinical trials (NCT04656418) in patients with hereditary angioedema, showing promising preliminary data after a 6-month follow-up. Another phase III trial (NCT04739059) is ongoing to evaluate its benefits in a longer term (32 months). Based on its proven safety profile and the outcome of both trials, garadacimab may be considered a promising strategy for other indications, including CVDs.

2.2. Targeting Factor XI

Factor XI congenital deficiency has been proven to protect against arterial and venous thrombosis reducing the incidence of deep-vein thrombosis, ischemic stroke, myocardial infarction, and vascular graft occlusion [48–51]. Most importantly, FXI-deficient patients do not generally exhibit spontaneous bleeding, and the bleeding associated with injury or surgery tends to be mild [52,53]. These observations have supported the development of multiple FXI inhibitors, most of which have reached Phase II testing. Table 3 details the studies conducted so far as per FXI inhibitors.

Table 3. Factor XIa inhibitors currently under development. This table includes the emerging FXIa inhibitors and details the studies conducted so far to assess their efficacy and safety. VTE: venous thromboembolism; AF: atrial fibrilation; aPTT: activated partial thromboplastin time; MI: myocardial infarction.

Factor XIa Inhibitors	Type	Phase	Studies Conducted So Far
Osocimab *(subcutaneous)*	Antibody	II	Effective in thromboprophylaxis in patients undergoing knee arthroplasty [54].
Abelacimab *(intravenous)*	Antibody	III	Effective in preventing venous thromboembolism and is associated with a low bleeding risk [55]. There are ongoing phase III trials in cancer patients to compare the effect of abelacimab relative to apixaban (NCT05171049) or dalteparin (NCT05171075) in VTE recurrence and bleeding.
AB023 (Xisomab) *(intravenous)*	Antibody	II	Effective and secure in patients with end-stage renal disease [56]. Ongoing phase II trial to test xisomab for the prevention of catheter-associated thrombosis in patients with cancer receiving chemotherapy (NCT04465760).
14E11 *(subcutaneous)*	Antibody	Preclinical	In mice, 14E11 has been shown to prevent arterial occlusion induced by ferric chloride to a similar degree as that accomplished by total FXI deficiency. In baboons, it has been shown to reduce platelet-rich thrombus growth in collagen-coated grafts inserted into arteriovenous shunts [57].
FXI-175, FXI-203 *(intravenous)*	Antibody	Preclinical	Ferric chloride-induced thrombosis was reduced in mice treated with FXI-175 and FX-203 compared to placebo-treated mice. Neither antibody caused severe blood loss assessed through the tail bleeding assay [58].
Frunexian EP-7041a *(intravenous)*	Small molecule $C_{19}H_{27}ClN_4O_4$	II	EP-7041 was safe and well tolerated in healthy volunteers with rapid onset and offset of action and predictable dose-related increases of aPTT [59]. In addition, there is an ongoing trial in thromboprophylaxis in COVID-19 patients (NCT05040776).
Milvexian (BMS-986177) *(oral)*	Small molecule $C_{28}H_{23}Cl_2F_2N_9O_2$	II	Prevention of venous thromboembolism with low risk of bleeding (phase II) [60]. In rabbits, it has demonstrated effective antithrombotic potential with limited impact on hemostasis, even when combined with aspirin [61]. A recent phase II trial (AXIOMATIC-SSP) has shown it is safe in secondary stroke prevention [62].
Asundexian *(oral)*	Small molecule $C_{26}H_{21}ClF_4N_6O_4$	II b	In patients with AF, it has shown low rates of bleeding as compared with apixaban [63]. It has also shown no increase in bleeding events in MI [64] and stroke [65] patients. New phase III clinical trials have been announced to test its efficacy in patients with AF (OCEAN-AF) and in secondary prevention of stroke (OCEAN-STROKE).
BMS-962212 *(intravenous)*	Small molecule $C_{32}H_{28}ClFN_8O_5$	I	Tested in healthy subjects showing good tolerance, no signs of bleeding and significant changes in aPTT and FXI clotting activity [66].
ONO-7684 *(oral)*	Small molecule $C_{23}H1_6ClF_2N_9O$	I	It strongly inhibited factor XI coagulation activity and increased activated partial thromboplastin time [67].
BMS-654457 *(intravenous)*	Small molecule $C_{36}H_{37}N_5O_{4)}$	Preclinical	It has been shown in vitro to increase aPTT without altering prothrombin time or ADP-, arachidonic acid-, or collagen-induced platelet aggregation. In rabbit models, it has shown equivalent antithrombotic effect to that achieved by standard doses of reference anticoagulants (warfarin and dabigatran) and antiplatelet agents (clopidogrel and prasugrel) in addition to reducing bleeding time [68].
ONO-5450598 *(oral)*	Small molecule	Preclinical	It provided a significant reduction of thrombus formation as compared to rivaroxaban in a non-human primate arteriovenous shunt model of thrombosis [69].
BMS-262084 *(intravenous)*	Small molecule $C_{18}H_{31}N_7O_5$	Preclinical	Evaluated in rabbits, where it displayed antithrombotic potential in an arteriovenous-shunt model of thrombosis, and in an electrolytic-mediated carotid arterial thrombosis [70].
FXI-ASO (ISIS416858) *(subcutaneous)*	Antisense oligonucleotide	II	Effective in thromboprophylaxis in patients undergoing knee arthroplasty [71].

2.3. Targeting Factor IX

Factor IX is another potential drug target currently under intensive research because of its efficacy and safety profile [72]. Factor IX is activated by both the intrinsic and extrinsic pathways (Figure 2). In the intrinsic pathway, FXIa induces FIX activation, whereas, in the extrinsic coagulation pathway, FIX is activated by the tissue factor (TF)–VIIa complex. FIXa forms a complex with FVIIIa that binds to platelets serving as a very potent activator of FX [73].

As for FXIIa and FXIa, multiple FIX inhibitors have been developed, most of them in the preclinical development phase, and only a few have reached clinical trials (Table 4) [74]. One that raised great interest is pegnivacogin, a RNA-aptamer based FIXa inhibitor featuring a reversal agent, anivamersen [73]. However, both phase II trials where pegnivacogin has been tested have not resulted in the expected positive outcome. The RADAR trial in NSTEMI patients undergoing cardiac catheterization [75] did not show differences between pegnivacogin and heparin, and the REGULATE-PCI trial performed in patients undergoing PCI [76] had to be prematurely terminated due to the presence of severe allergic reactions.

Table 4. Factor IXa inhibitors currently under development. This table includes the emerging FIXa inhibitors and details the studies conducted so far to assess their efficacy and safety.

Factor IXa Inhibitors	Type	Phase	Studies Conducted So Far
Pegnivacogin ***(intravenous)***	RNA aptamer	II	- Phase II trial in NSTEMI patients undergoing cardiac catheterization did not show significant differences compared with heparin [75]. - A randomized clinical trial in patients undergoing percutaneous coronary intervention had to be terminated early due to severe allergic reactions [76]. - It has decreased platelet activation and aggregation in vitro [77].
SB249417 ***(intravenous)***	Antibody	I	It has demonstrated prolongation of coagulation measures in humans [78].
TTP889 ***(oral)***	Small molecule	II	It has not been shown to be effective for the extended prevention of venous thromboembolism [79].

3. Targeting the Platelet: What Is in the Pipeline for Novel Antiplatelet Agents?

Antiplatelet agents are used in treating both ischemic heart disease and ischemic stroke (Figure 1). The currently available antiplatelet agents either: (1) target intraplatelet enzymes (COX-1 inhibition by ASA and PDE inhibition by dipyridamole and cilostazol), preventing the formation of thromboxane A2 (TXA_2) or the degradation of AMPc, respectively; or (2) block platelet membrane receptors ($P2Y_{12}$ receptor antagonists, GPIIb/IIIa inhibitors, and PAR antagonist) preventing their downstream signaling activation (Figure 1) [80].

In the setting of ischemic heart disease, antiplatelet agents are used both in acute and chronic coronary syndromes and after stent implantation to prevent stent-related thrombosis [81–84]. A double antiplatelet regime with a combination of ASA and a $P2Y_{12}$ inhibitor is recommended during the first year after an acute myocardial event [85–87]. Among the $P2Y_{12}$ inhibitors, clopidogrel, a second-generation thienopyridine, is a pro-drug that requires a two-enzyme-mediated transformation to become active and irreversibly block the $P2Y_{12}$ platelet ADP receptors. Yet, clopidogrel exhibits high individual variability because of differences in the activity of cytochrome P450 2C19. Prasugrel, a third-generation thienopyridine, is also a pro-drug but requires fewer hepatic steps to be converted into an active metabolite [88] and hence is less affected by variation in CYP enzymes and exerts a higher degree of platelet inhibition as compared to clopidogrel. Finally, ticagrelor, the first of a new class of $P2Y_{12}$ inhibitors named cyclopentyl-triazole-pyrimidines, is a reversible $P2Y_{12}$ receptor inhibitor that does not need hepatic metabolism and accordingly has a more predictable metabolic pathway resulting in a better inter-individual consistency among patients and clinical efficacy. Both prasugrel and ticagrelor have demonstrated greater efficacy than clopidogrel [86,87] and accordingly are recommended over clopidogrel in clinical guidelines in patients with no high bleeding risk [83].

Another known antiplatelet target is the GPIIb/IIIa, the most abundant platelet receptor mainly involved in platelet aggregation [89]. Two GPIIb/IIIa receptor blockers have been approved for intravenous clinical use in STEMI patients, including tirofiban (tyrosine-derived non-peptide derivative) and eptifibatide (heptapeptide). Both antagonize the GPIIb/IIIa receptor preventing fibrinogen and Von Willebrand factor (VWF) from binding to the receptor [90].

In the setting of strokes, antiplatelet therapy is used in secondary prevention in patients with non-cardioembolic transient ischemic attack or stroke. Single antiplatelet therapy with ASA or clopidogrel, or the combination of low dose ASA and dipyridamole or cilostazol, is usually recommended for secondary prevention. In some patients, a combination of ASA and clopidogrel is recommended for up to 90 days to reduce early recurrences [91]. Recent data have demonstrated that ticagrelor on top of ASA reduces the total burden of disability owing to ischemic stroke recurrence compared to ASA alone [92]. Based on these recent findings, the combination of ASA and ticagrelor for up to one month might be considered in patients at risk of ischemic stroke [91]. Cilostazol may also be used for secondary stroke prevention, particularly in Asian patients [93], since randomized clinical trials are still needed to determine its usefulness in non-Asian populations.

There are no reversal agents for the antiplatelet drugs presently used in the clinical setting. However, this might change in short/medium term for ticagrelor. Bentracimab (PB2452) is a recombinant human monoclonal antibody antigen-binding fragment with a dual mechanism of action; it binds both to free ticagrelor and to its major active metabolite (AR-C124910XX) [94]. Bentracimab is currently being tested in a phase III clinical trial (NCT04286438) in patients with uncontrolled major or life-threatening bleeding or requiring urgent survey or invasive procedure.

Despite the currently available antiplatelet armamentarium, recurrent thrombotic events still occur, and enhanced bleeding risk remains a challenge that needs to be addressed. These limitations have stimulated research interest in identifying and developing new antiplatelet targets (Figure 3).

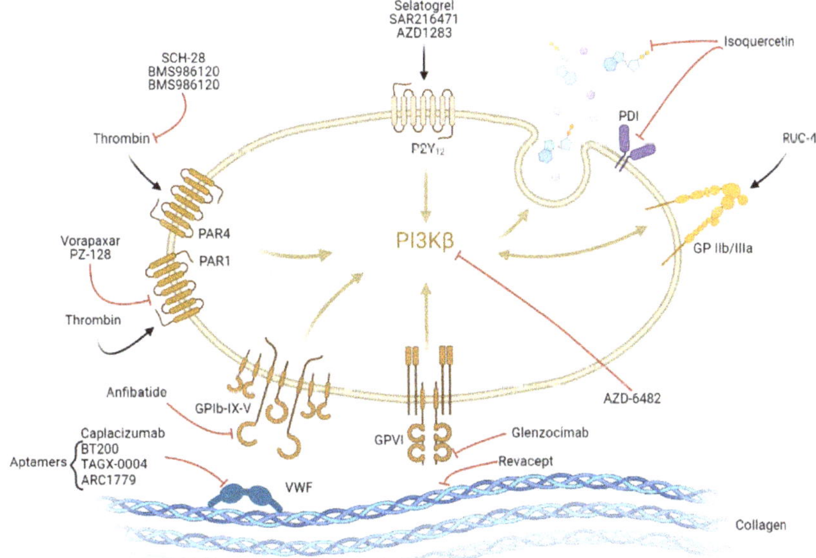

Figure 3. Emerging antiplatelet targets and drugs. PDI: phosphodiesterase; VWF: Von Willebrand factor; GP: glycoprotein; PAR: protease activator receptor.

3.1. Targeting Platelet Adhesion

3.1.1. Inhibition of Von Willebrand Factor-Glycoprotein 1bα-Mediated Platelet Activation

VWF is synthesized by endothelial cells and megakaryocytes. VWF activity depends on the size of the multimer being ultra-large VWF multimers highly reactive with platelets. The monomeric VWF displays a multi-domain structure which includes an A3 domain (interacts with exposed vascular collagen) and an A1 domain (binds to platelet GPIbα). A1 interaction with GPIbα favours platelet rolling and adhesion, especially under high share rate conditions. GPVI and integrin α2β1 further support tight platelet adhesion.

The resultant platelet activation induces the conformational change of GPIIb/IIIa, which favours platelet–platelet interaction (i.e., platelet aggregation) by binding to fibrinogen (primary ligand) or to the C1 domain of VWF. Although VWF also plays a pivotal role in platelet aggregation by serving as an intercellular bridge between platelets, efforts have mainly focused on the discovery of pharmacological agents able to interfere with VWF-mediated platelet adhesion either by blocking the VWF-collagen or the VWF-GPIbα interaction. Promising preclinical and proof-of-concept clinical trials have supported their antithrombotic potential, as described below [95].

Anfibatide is a direct GPIb antagonist purified from snake (*Deinagkistrodon acutus*) venom that prevents GPIb interaction with VWF. Intravenous administration of anfibatide in NSTEMI patients (phase Ib/IIa study) proved feasible and safe and markedly inhibited platelet aggregation without increasing the risk of bleeding [96]. A phase II trial is currently assessing its safety and efficacy in STEMI patients before primary PCI being the primary endpoint TIMI myocardial perfusion grades (NCT02495012). In the field of stroke, administering anfibatide after cerebral ischemia/reperfusion injury in rats has been shown to significantly improve ischemic lesions alleviating inflammation and apoptosis in a dose-dependent manner [97] and preserving blood-brain barrier integrity [98]. These observations further support the contribution of platelets to inflammation and immune responses in ischemic damage beyond their function in hemostasis [99].

Caplacizumab (formerly ALX-0081) is a humanized monoclonal nanobody that targets the A1 domain of VWF, preventing its interaction with GPIb. After promising results in phase I studies (healthy subjects and stable angina patients undergoing PIC) [100], a phase II study in high-risk patients with ACS undergoing PCI (NCT01020383) is currently underway and aims to compare the safety and efficacy of caplacizumab vs. abciximab on top of standard antithrombotic therapy (ASA, clopidogrel, and heparin).

Aptamers have also been developed to block the A1 domain (Table 5). As such, BT200 has been shown to effectively block VWF activity in both ACS [101] and stroke [102] patients by binding to the VWF-A1 domain and is currently being tested in healthy volunteers (phase I, NCT04103034).

3.1.2. Glycoprotein VI: Inhibition of Collagen-Mediated Platelet Activation

GPVI is a platelet- and megakaryocyte-specific 60–65 kDa immunoglobulin-like transmembrane receptor. It is expressed at the platelet surface and is associated with the FcR γ (Fc receptor γ)-chain, which is responsible for the signaling via its immunoreceptor-tyrosine-based-activation motif. GPVI is considered the main collagen receptor in platelets, although it also binds to other substrates, including fibrin, fibrinogen, fibronectin, galectin-3, or laminin [103]. Activation of the GPVI–FcRγ complex initiates intracellular signaling through a tyrosine kinase-based signaling pathway [104] that eventually triggers calcium mobilization and the resultant platelet activation [105]. Several experimental studies have supported that GPVI seems to have little or no impact on hemostasis. As such, patients lacking functional GPVI have shown mild bleeding diathesis [106] unless they have moderate to severe thrombocytopenia [107]. Likewise, a mutation in the GPVI gene identified in the Chilean population that prevents GPVI surface expression has not been associated with a significant increase in bleeding and has been hypothesized to confer a protective benefit against CVD [108]. Overall, the fact that GPVI is uniquely expressed in platelets and megakaryocytes and has reported minor involvement in hemostasis [109,110] has made GPVI inhibition a promising approach to prevent thrombosis while limiting bleeding risk.

Revacept, a competitive antagonist to GPVI-collagen signaling, is one of the most studied drugs. Revacept is a dimeric, soluble fusion protein composed of the extracellular domain of the GPVI receptor and the human Fc-fragment. It competes with endogenous platelet GPVI for binding to exposed collagen fibers preventing platelet activation [105]. Since revacept targets the exposed vascular collagen, it does not interfere with circulating platelets beyond the atherosclerotic lesion, exerting a little effect on systemic hemostasis or bleeding as suggested in experimental models and a phase I clinical trial [111]. Revacept has been tested in phase II clinical trials [112] in patients with stable coronary heart disease undergoing PCI. Yet, no significant differences were observed in the primary endpoint (death or myocardial injury) or bleeding between the treated and placebo arm. Future studies are being planned to address its efficacy in patients at higher risk of ischemic events (e.g., in the context of ACS), where collagen-induced platelet activation may play a more important role.

In the setting of ischemic stroke, revacept is currently being tested in a phase II clinical trial (NCT01645306) in patients with symptomatic carotid artery stenosis (history of ischemic stroke, transitory ischemic attack or amaurosis fugax within the last 30 days) to check its efficacy in secondary prevention of thromboembolic ischemic events.

Several monoclonal antibodies against GPVI have also been developed, as detailed in Table 5, the most important being glenzocimab (ACT017). This monoclonal antibody binds to human GPVI and has inhibited platelet adhesion, aggregation, and thrombus formation onto collagen surface under arterial flow conditions [113]. Glenzocimab has a short plasma half-life requiring to be infused intravenously for 6 or 12 h to maintain the necessary duration of effect [114]. Glenzocimab has been demonstrated to sufficiently block collagen-induced platelet aggregation in a phase I study [115] with an excellent safety profile (no evidence of thrombocytopenia or excess bleeding). Glenzocimab is being tested in a phase II/III trial to evaluate the safety and efficacy of a single dose of glenzocimab used in combination with standard of care (thrombolysis and thrombectomy) for acute ischemic stroke (ACTIVASE NCT05070260).

Table 5. Novel antiplatelets in the preclinical phase.

Antiplatelet	Type	Mechanism of Action	Studies Conducted So Far
TAGX-0004 (*studies* in vitro)	Aptamer	VWF inhibition	It has excellent affinity with VWF-A1 domain and a superior antithrombotic potential than ARC1779 [116].
ARC1779 (*intravenous*)	Aptamer	VWF inhibition	In a phase II trial, it reduced cerebral thromboembolism in patients undergoing carotid endarterectomy [117]. However, the study was terminated due to a lack of funding and associated increased bleeding risk. Further development of ARC1779 was halted.
AJW200 (*intravenous*)	Monoclonal antibody	VWF inhibition	Tested as adjunctive therapy with tPA in a mouse model of embolic stroke where it showed a synergistic effect and improved behavioural function [118]. In monkeys, it has been shown to inhibit high-shear-stress-induced platelet adhesion, aggregation, and thrombin generation [119].
82D6A3 (*intravenous*)	Monoclonal antibody (A3 domain)	VWF inhibition	It has been tested in baboons, showing potent antithrombotic activities without significantly prolonging the bleeding time [120].
Caplacizumab (*intravenous/subcutaneous*)	Nanobody	VWF inhibition	Approved for the treatment of immune mediated thrombotic thrombocytopenic purpura [121].
h6B4-Fab (*intravenous*)	Monoclonal antibody	GPIb inhibition	Reduce thrombus formation in baboons with minimal effect on bleeding time [122].
SZ2 (*intravenous*)	Monoclonal antibody	GPIb inhibition	In vitro, functional studies revealed that it prevents platelet adhesion to VWF under high-shear stress and inhibits ristocetin-induced platelet aggregation in a dose-dependent manner [123].
JAQ1 (*Intravenous*)	Monoclonal antibody	GPVI inhibition	It protects against lethal thromboembolism in mice with minimal impact on hemostasis [124,125].
SCH-28 (*studies* in vitro)	Small molecule	PAR4 inhibition	It inhibits PAR-4-mediated platelet activation and aggregation by blocking the thrombin exosite II binding domain [126].
HPW-RX40 (*intravenous*)	Small molecule	PDI inhibition	Reduces thrombus formation in whole human blood under flow conditions and protects mice from ferric chloride-induced thrombus formation [127].
ML359 (*studies* in vitro)	Small molecule	PDI Inhibition	It exerts no cytotoxicity in three human cell lines and inhibits platelet aggregation [128].
ML355 (*oral*)	Small molecule	12-Lipoxygenase inhibition	It reduces thrombus growth and vessel occlusion in a mouse model of arterial thrombosis with minimal impact on hemostasis [129].

Table 5. *Cont.*

Antiplatelet	Type	Mechanism of Action	Studies Conducted So Far
MIPS-9922 *(intravenous)*	Small molecule	PI3Kβ inhibition	It prevents arterial thrombus formation in an in vivo electrolytic mouse model of thrombosis with minimal impact on hemostasis [130].
scFv *(intravenous)*	Antibody	GPIIb/IIIa inhibition	It has demonstrated comparable antithrombotic efficacy to currently used GPIIb/IIIa inhibitors (tirofiban and eptifibatide) in a mice model of ferric chloride-induced thrombosis with minimal impact on hemostasis [131].
mP$_6$ *(intravenous)*	Péptide	GPIIb/IIIa inhibition	It has proven superior to aspirin and is similar to ticagrelor in a mice model of ferric chloride-induced thrombosis with minimal effects on hemostasis [132].
SAR216471 *(oral)*	Small molecule	P2Y$_{12}$ Inhibition	It has shown potent antithrombotic activity in a rat arterio-venous shunt model with no effect on hemostasia [133].
AZD1283 *(oral)*	Small molecule	P2Y$_{12}$ Inhibition	It has shown potent antithrombotic efficacy in a rat model of ferric chloride-induced thrombosis and lowers bleeding risk compared to clopidogrel [134].
BMS-884775 *(oral)*	Small molecule	P2Y$_1$ Inhibition	It has demonstrated, in a rabbit model of thrombosis, similar efficacy to prasugrel with less bleeding risk [135].
MRS2500 *(intravenous)*	Small molecule	P2Y$_1$ Inhibition	It prevents carotid artery thrombosis in monkey models of electrolytic-mediated arterial thrombosis with a concomitant mild prolongation in bleeding time [136].
GLS-409 *(intravenous)*	Small molecule	P2Y$_1$ and P2Y$_{12}$ Inhibition	A It attenuates thrombosis in a canine model of unstable angina and reduces platelet aggregation to a comparable extent to cangrelor or the combination of cangrelor with a selective P2Y1 inhibitor [137].
Troα6 and Troα10 *(intravenous)*	Peptides	GPVI inhibition	It inhibits collagen-induced platelet aggregation and thrombus formation in a ferric chloride-induced thrombosis model without prolonging bleeding time [138].
BI1002494 *(oral)*	Peptide	GPVI inhibition	It reduces infarct sizes and improves neurological outcomes in a mouse model of cerebral ischemia without affecting hemostasis [139].

Tyrosine kinase inhibitors are also being developed to prevent downstream signaling initiated by activation of the GPVI-FcRγ complex. As such, platelet activation through GPVI relies on a potent protein tyrosine kinase cascade culminating in the activation of the tyrosine kinase Syk (spleen-associated tyrosine kinase). Tyrosine kinase inhibitors have been shown to exert antiplatelet effects in cancer patients (e.g., pazopanib in patients with renal cell carcinoma) [140] and short-term studies with ibrutinib analogs Btki (Bruton's tyrosine kinase inhibitors) 43607 and Btki 43761 have shown a dramatic reduction in collagen-induced platelet aggregation in non-human primates without measurable effects on plasma clotting times or bleeding risk [141]. In addition, the Syk inhibitors PRT-060318 and BI1002494 have been shown to reduce thrombus stability in vitro [142] and thrombosis in a mouse model of cerebral ischemia [139], respectively. Finally, ibrutinib has also been shown to block CLEC-2-mediated platelet activation [143]. CLEC-2 is a platelet-activating type II transmembrane receptor which has a function similar to that of GPVI in activating Syk [144].

3.2. Targeting Platelet Activation

3.2.1. PAR1 and PAR4: Inhibition of Thrombin-Mediated Platelet Activation

Thrombin activates human platelets through the protease-activated receptor (PAR)-1 and PAR-4. PARs are G protein-coupled receptors whose activation by thrombin depends on proteolytic cleavage of the N-terminal domain of the receptor, generating a new amino terminus that acts as a tethered ligand to activate the receptor. PAR-4 has shown to interact with PAR-1 and P2Y$_{12}$, inducing sustained platelet activation, whereas PAR-1 does not interact with ADP receptors leading to an acute platelet response. Hence, blockage of the P2Y$_{12}$ receptor may suppress PAR-4-mediated platelet aggregation, while PAR-1-mediated effects remain unaltered [145].

PAR-1 has shown a high affinity for thrombin, whereas higher thrombin levels are required to activate PAR-4. Hence, PAR-1 has become the focus of intense research as a therapeutic antiplatelet target. Vorapaxar is a competitive PAR-1 antagonist that irreversibly binds to the ligand-binding pocket on the extracellular surface of PAR-1. Based on two large phases III clinical trials (TRA 2°P–TIMI 50 [146] and TRACER [147] vorapaxar may be used on top of standard antiplatelet therapy in the secondary prevention of ischemic events in patients with a history of MI or symptomatic peripheral artery disease. Yet, vorapaxar is contraindicated in patients with a history of stroke or transient ischemic attack because it has been associated with increased intracranial bleeding. However, subgroup analyses of both trials have found that vorapaxar might be potentially beneficial in patients with previous MI, diabetes, coronary artery bypass grafting, and ischemic stroke [148,149]. So far, vorapaxar has been shown to reduce thrombus formation in post-MI patients treated with potent $P2Y_{12}$ inhibitors [150].

PZ-128 is a pepducin inhibitor of PAR1 for patients with CAD/ACS undergoing coronary interventions. Pepducins are lipidated peptides which target the cytoplasmic surface of their cognate receptor, not affecting the ligand-binding. PZ-128 has experimentally been demonstrated to reduce acute arterial thrombosis and atherosclerotic plaque burden [151]. Furthermore, PZ-128 was recently tested in a phase II trial in NSTEMI or stable angina patients undergoing PCI and appeared to be safe, well-tolerated, and potentially reduce periprocedural myonecrosis when administered on top of standard antiplatelet therapy [152].

As per PAR-4 inhibitors, BMS-986141 has been demonstrated to reduce platelet-rich thrombus formation under a high shear rate [153] and is currently being tested in a phase IIa trial (NCT05093790). A similar drug, BMS-986120, has been recently tested with success in humans (phase I) [154] after encouraging data from preclinical studies where it has shown robust antithrombotic activity and a low bleeding profile [155,156].

3.2.2. Inhibition of Phosphoinositide 3-Kinase Beta (PI3Kβ)

PI3Kβ is a lipid kinase that acts as an important mediator in the signal transduction downstream of the activation of $P2Y_{12}$, GPIIb/IIIa, GPVI, PAR, and GPIb and plays a pivotal role in platelet aggregation and thrombus stability.

Based on the specific PI3Kβ inhibitor, TGX-221, which has only been tested in preclinical studies, a new molecule with better pharmacological properties has been developed, AZD-6482. This drug has shown in a phase I trial to moderately inhibit ADP- and collagen-induced platelet aggregation, particularly under high shear stress conditions with only mild prolonged bleeding time [157]. In another phase I study, the combination of AZD-6482 with ASA provided greater platelet inhibition compared to DAPT with ASA and clopidogrel without translating into prolonged bleeding times [158]. A new phase II trial (STARS) is planned to test the safety and tolerability of this drug in reperfusion for stroke (NCT05363397).

3.2.3. Selatogrel: The New Antagonist of the $P2Y_{12}$ Receptor

Selatogrel (ACT-246475) is a new potent, reversible, and selective inhibitor of the $P2Y_{12}$ platelet receptor. Its efficacy and safety have already been confirmed in phase I and II clinical trials. In contrast to the currently used $P2Y_{12}$ inhibitors (i.e., oral or intravenous administration), selatogrel is administered subcutaneously, overcoming potential pharmacokinetic limitations of other $P2Y_{12}$ inhibitors, including the delay of absorption and lack of enteral access for administration with oral formulations; the need for intravenous access with cangrelor; or the need for metabolization (e.g., clopidogrel and prasugrel) to be ideal in the critical 3-h window during an ACS [159]. Additionally, selatogrel seems to have a lower bleeding risk profile than clopidogrel or ticagrelor. A study performed in mice showed that the stability of hemostatic seals was undisturbed in the presence of selatogrel, unlike clopidogrel or ticagrelor. The authors suggested that the mechanism underlying the differences in blood loss profiles among these $P2Y_{12}$ receptor antagonists was related to off-target interference with endothelial and neutrophil cells and fibrin-mediated stabilization of hemostatic seals [160]. Subsequently, phase I and phase II clinical trials have confirmed that selatogrel provides sustained and reversible $P2Y_{12}$ platelet inhibition with an acceptable safety profile [159]. A phase III clinical trial is currently underway (NCT04957719).

3.2.4. New $P2Y_1$ Receptor Antagonists

Besides the $P2Y_{12}$ receptor, human platelets express another purinergic ADP receptor named $P2Y_1$. The binding of ADP to $P2Y_1$ initiates platelet aggregation response which may be reverted, while $P2Y_{12}$ activation leads to irreversible platelet aggregation. Therefore, complete platelet aggregation requires a complex interplay and coactivation of both $P2Y_1$ and $P2Y_{12}$ receptors [161]. Following this assumption, several $P2Y_1$ inhibitors have been developed, though so far, they have only been tested in animal models, as detailed in Table 5 [162].

3.3. Targeting Platelet Aggregation and Thrombus Propagation

3.3.1. New Inhibitor of GP IIb/IIIa

Zalunfiban (RUC-4) is a second-generation small-molecule platelet GPIIb/IIIa inhibitor that blocks the receptor in its inactive conformation. This blockade avoids the drug-induced thrombocytopenia associated with other GPIIb/IIIa inhibitors since it prevents the exposition of epitopes that are potential targets for thrombocytopenia-related antibodies. Subcutaneous administration of RUC-4 in healthy subjects and stable coronary artery disease patients on ASA (Phase I trial) has shown a rapid (<15 min), potent (>80% reduction of platelet aggregation), and reversible (platelet function is restored after 1–2 h) platelet inhibitory effect [163]. These observations were confirmed in a phase IIa trial in the setting of STEMI [164]. Currently, RUC-4 is being tested in phase IIb trial in STEMI patients undergoing primary PCI (NCT04825743). Other GPIIb/IIIa inhibitors have been developed and are currently being tested in the preclinical setting, as detailed in Table 5.

3.3.2. Inhibition of Protein Disulfide Isomerase (PDI)

PDI is an enzyme in the endoplasmic reticulum that catalyzes the modification of thiol-disulfide bonds during protein synthesis and is also expressed on the surface of multiple cells, including platelets. Four members of the PDI family of enzymes, including PDI, ERp57, ERp72, and ERp5, are secreted from activated platelets and endothelial cells at the site of vascular injury. The mechanisms by which extracellular PDI regulates platelet function remain to be determined. However, it is thought to interact with prothrombotic components, including GPIIb/IIIa, $\alpha 2\beta 1$, vWF, GPIbα, and TF supporting, and thus, platelet activation, aggregation, and coagulation [165–167].

Quercetin flavonoids (mainly isoquercetin) are potent PDI inhibitors present in fruits and vegetables. They have been tested primarily in the field of cancer and venous thromboembolism where it has been shown, in phase II trial, to improve hypercoagulability in advanced cancer [168]. However, its potential role in the context of CVD has yet to be established [169,170]. So far, other PDI inhibitors are in the pipeline since they have been shown to exert antithrombotic effects in vitro and in experimental animal models (Table 5).

4. Conclusions

Despite the major advances in antithrombotic therapy accomplished over the last decades, atherothrombotic events remain a leading cause of death worldwide. The secondary prevention of both ischemic heart disease and ischemic stroke requires effective antiplatelets and anticoagulants without bleeding side effects. Research conducted over the last years has led to a deeper understanding of the molecular mechanisms regulating atherothrombosis and hemostasis, providing new targets for intervention [5,6,171–173]. New antithrombotic strategies have been developed and assessed in preclinical animal models, and some have already reached clinical testing. As per the coagulation cascade, new anticoagulants have focused on the intrinsic coagulation pathway to prevent ischemic coronary and cerebral events. In this regard, although the long journey from animal studies to randomized clinical trials has just started, hopefully, some of these promising strategies will reach routine clinical use, providing the patient with optimal protection against arterial thrombosis inhibition while preserving hemostasis.

Author Contributions: I.B. and G.V. wrote the paper. F.W. provided an in-depth and constructive review of the content. All authors have read and agreed to the published version of the manuscript.

Funding: This work was supported by Grant PID2021-128891OB-I00 (to G.V.) and PLEC2021- 007664-NextGenerationEU (to G.V.) funded by MCIN/AEI/10.13039/501100011033 and Fondo Europeo de Desarrollo Regional (FEDER) A way of making Europe; and the SEC/FEC-INV-TRL 20/015 funded by the Spanish Society of Cardiology (to G.V.) We thank the Fundación Investigación Cardiovascular—Fundación Jesus Serra for their continuous support.

Institutional Review Board Statement: Not applicable.

Informed Consent Statement: Not applicable.

Data Availability Statement: Not applicable.

Acknowledgments: This work is part of the Autonomous University of Barcelona requirement for the Doctorate in Medicine (Ignacio Barriuso).

Conflicts of Interest: The authors declare no conflict of interest.

References

1. Cardiovascular Diseases (CVDs) [Internet]. Available online: https://www.who.int/news-room/fact-sheets/detail/cardiovascular-diseases-(cvds) (accessed on 24 September 2022).
2. Roth, G.A.; Forouzanfar, M.H.; Moran, A.E.; Barber, R.; Nguyen, G.; Feigin, V.L.; Naghavi, M.; Mensah, G.A.; Murray, C.J. Demographic and Epidemiologic Drivers of Global Cardiovascular Mortality. *N. Engl. J. Med.* **2015**, *372*, 1333–1341. [CrossRef] [PubMed]
3. Global Burden of Disease Collaborative Network. *Global Burden of Disease Study 2019 (GBD 2019) Results*; Institute for Health Metrics and Evaluation (IHME): Seattle, WA, USA, 2019. Available online: http://ghdx.healthdata.org/gbd-results-tool (accessed on 15 September 2022).
4. Vilahur, G.; Badimon, J.J.; Bugiardini, R.; Badimon, L. Perspectives: The burden of cardiovascular risk factors and coronary heart disease in Europe and worldwide. *Eur. Heart J. Suppl.* **2014**, *16*, A7–A11. [CrossRef]
5. Badimon, L.; Storey, R.F.; Vilahur, G. Update on lipids, inflammation and atherothrombosis. *Thromb. Haemost.* **2011**, *105*, 34–42.
6. Patrono, C.; Morais, J.; Baigent, C.; Collet, J.P.; Fitzgerald, D.; Halvorsen, S.; Rocca, B.; Siegbahn, A.; Storey, R.F.; Vilahur, G. Antiplatelet Agents for the Treatment and Prevention of Coronary Atherothrombosis. *J. Am. Coll. Cardiol.* **2017**, *70*, 1760–1776. [CrossRef] [PubMed]
7. Badimon, L.; Vilahur, G. Platelets, arterial thrombosis and cerebral ischemia. *Cerebrovasc. Dis.* **2007**, *24*, 30–39. [CrossRef] [PubMed]
8. Badimon, L.; Padró, T.; Vilahur, G. Atherosclerosis, platelets and thrombosis in acute ischaemic heart disease. *Eur. Heart J. Acute Cardiovasc. Care* **2012**, *1*, 60–74. [CrossRef]
9. Parker, W.A.E.; Gorog, D.A.; Geisler, T.; Vilahur, G.; Sibbing, D.; Rocca, B.; Storey, R.F. Prevention of stroke in patients with chronic coronary syndromes or peripheral arterial disease. *Eur. Heart J. Suppl.* **2020**, *22*, M26–M34. [CrossRef]
10. Hindricks, G.; Potpara, T.; Dagres, N.; Arbelo, E.; Bax, J.J.; Blomström-Lundqvist, C.; Boriani, G.; Castella, M.; Dan, G.-A.; Dilaveris, P.E.; et al. 2020 ESC Guidelines for the diagnosis and management of atrial fibrillation developed in collaboration with the European Association for Cardio-Thoracic Surgery (EACTS): The Task Force for the diagnosis and management of atrial fibrillation of the European Society of Cardiology (ESC) Developed with the special contribution of the European Heart Rhythm Association (EHRA) of the ESC. *Eur. Heart J.* **2021**, *42*, 373–498.
11. Díaz-Guzmán, J.; Freixa-Pamias, R.; García-Alegría, J.; Cabeza, A.-I.P.; Roldán-Rabadán, I.; Antolin-Fontes, B.; Rebollo, P.; Llorac, A.; Genís-Gironés, M.; Escobar-Cervantes, C. Epidemiology of atrial fibrillation-related ischemic stroke and its association with DOAC uptake in Spain: First national population-based study 2005 to 2018. *Rev. Esp. Cardiol.* **2022**, *75*, 496–505. [CrossRef]
12. Cha, M.-J.; Choi, E.-K.; Han, K.-D.; Lee, S.-R.; Lim, W.-H.; Oh, S.; Lip, G.Y.H. Effectiveness and Safety of Non-Vitamin K Antagonist Oral Anticoagulants in Asian Patients with Atrial Fibrillation. *Stroke* **2017**, *48*, 3040–3048. [CrossRef]
13. Dhakal, P.; Rayamajhi, S.; Verma, V.; Gundabolu, K.; Bhatt, V.R. Reversal of Anticoagulation and Management of Bleeding in Patients on Anticoagulants. *Clin. Appl. Thromb. Hemost.* **2017**, *23*, 410–415. [CrossRef]
14. Desai, N.R.; Cornutt, D. Reversal agents for direct oral anticoagulants: Considerations for hospital physicians and intensivists. *Hosp. Pract.* **2019**, *47*, 113–122. [CrossRef]

15. Jourdi, G.; Le Bonniec, B.; Gouin-Thibault, I. Strategies of neutralization of the direct oral anticoagulants effect: Review of the literature. *Ann. Biol. Clin.* **2019**, *77*, 67–78. [CrossRef]
16. Ansell, J.; Laulicht, B.E.; Bakhru, S.H.; Burnett, A.; Jiang, X.; Chen, L.; Baker, C.; Villano, S.; Steiner, S. Ciraparantag, an anticoagulant reversal drug: Mechanism of action, pharmacokinetics, and reversal of anticoagulants. *Blood* **2021**, *137*, 115–125. [CrossRef]
17. Ansell, J.E.; Bakhru, S.H.; Laulicht, B.E.; Steiner, S.S.; Grosso, M.; Brown, K.; Dishy, V.; Noveck, R.J.; Costin, J.C. Use of PER977 to reverse the anticoagulant effect of edoxaban. *N. Engl. J. Med.* **2014**, *371*, 2141–2142. [CrossRef]
18. Sheffield, W.P.; Lambourne, M.D.; Eltringham-Smith, L.J.; Bhakta, V.; Arnold, D.M.; Crowther, M.A. γT-S195A thrombin reduces the anticoagulant effects of dabigatran in vitro and in vivo. *J. Thromb. Haemost.* **2014**, *12*, 1110–1115. [CrossRef]
19. Jourdi, G.; Abdoul, J.; Siguret, V.; Decleves, X.; Frezza, E.; Pailleret, C.; Gouin-Thibault, I.; Gandrille, S.; Neveux, N.; Samama, C.M.; et al. Induced forms of α2-macroglobulin neutralize heparin and direct oral anticoagulant effects. *Int. J. Biol. Macromol.* **2021**, *184*, 209–217. [CrossRef]
20. Chan, N.; Sobieraj-Teague, M.; Eikelboom, J.W. Direct oral anticoagulants: Evidence and unresolved issues. *Lancet* **2020**, *396*, 1767–1776. [CrossRef]
21. Eikelboom, J.W.; Connolly, S.J.; Brueckmann, M.; Granger, C.B.; Kappetein, A.P.; Mack, M.J.; Blatchford, J.; Devenny, K.; Friedman, J.; Guiver, K.; et al. Dabigatran versus warfarin in patients with mechanical heart valves. *N. Engl. J. Med.* **2013**, *369*, 1206–1214. [CrossRef]
22. Pengo, V.; Denas, G.; Zoppellaro, G.; Jose, S.P.; Hoxha, A.; Ruffatti, A.; Andreoli, L.; Tincani, A.; Cenci, C.; Prisco, D.; et al. Rivaroxaban vs warfarin in high-risk patients with antiphospholipid syndrome. *Blood* **2018**, *132*, 1365–1371. [CrossRef]
23. Merlini, P.A.; Bauer, K.A.; Oltrona, L.; Ardissino, D.; Cattaneo, M.; Belli, C.; Mannucci, P.M.; Rosenberg, R.D. Persistent activation of coagulation mechanism in unstable angina and myocardial infarction. *Circulation* **1994**, *90*, 61–68. [CrossRef] [PubMed]
24. Rothberg, M.B.; Celestin, C.; Fiore, L.D.; Lawler, E.; Cook, J.R. Warfarin plus aspirin after myocardial infarction or the acute coronary syndrome: Meta-analysis with estimates of risk and benefit. *Ann. Intern. Med.* **2005**, *143*, 241–250. [CrossRef] [PubMed]
25. Eikelboom, J.W.; Connolly, S.J.; Bosch, J.; Dagenais, G.R.; Hart, R.G.; Shestakovska, O.; Diaz, R.; Alings, M.; Lonn, E.M.; Anand, S.S.; et al. Rivaroxaban with or without Aspirin in Stable Cardiovascular Disease. *N. Engl. J. Med.* **2017**, *377*, 1319–1330. [CrossRef] [PubMed]
26. Mega, J.L.; Braunwald, E.; Wiviott, S.D.; Bassand, J.P.; Bhatt, D.L.; Bode, C.; Burton, P.; Cohen, M.; Cook-Bruns, N.; Fox, K.A.; et al. Rivaroxaban in patients with a recent acute coronary syndrome. *N. Engl. J. Med.* **2012**, *366*, 9–19. [CrossRef] [PubMed]
27. Wallentin, L.; Wilcox, R.G.; Weaver, W.D.; Emanuelsson, H.; Goodvin, A.; Nyström, P.; Bylock, A.; ESTEEM Investigators. Oral ximelagatran for secondary prophylaxis after myocardial infarction: The ESTEEM randomised controlled trial. *Lancet* **2003**, *362*, 789–797. [CrossRef]
28. Wheeler, A.P.; Gailani, D. The Intrinsic Pathway of Coagulation as a Target for Antithrombotic Therapy. *Hematol. Oncol. Clin. N. Am.* **2016**, *30*, 1099–1114. [CrossRef]
29. Schmaier, A.H.; Stavrou, E.X. Factor XII—What's important but not commonly thought about. *Res. Pract. Thromb. Haemost.* **2019**, *3*, 599–606. [CrossRef]
30. Heestermans, M.; Naudin, C.; Mailer, R.K.; Konrath, S.; Klaetschke, K.; Jämsä, A.; Frye, M.; Deppermann, C.; Pula, G.; Kuta, P.; et al. Identification of the factor XII contact activation site enables sensitive coagulation diagnostics. *Nat. Commun.* **2021**, *12*, 5596. [CrossRef]
31. Srivastava, P.; Gailani, D. The rebirth of the contact pathway: A new therapeutic target. *Curr. Opin. Hematol.* **2020**, *27*, 311–319. [CrossRef]
32. Kalinin, D.V. Factor XII(a) inhibitors: A review of the patent literature. *Expert Opin. Ther. Pat.* **2021**, *31*, 1155–1176. [CrossRef]
33. Fredenburgh, J.C.; Weitz, J.I. New anticoagulants: Moving beyond the direct oral anticoagulants. *J. Thromb. Haemost.* **2021**, *19*, 20–29. [CrossRef]
34. Craig, T.; Magerl, M.; Levy, D.S.; Reshef, A.; Lumry, W.R.; Martinez-Saguer, I.; Jacobs, J.S.; Yang, W.H.; Ritchie, B.; Aygören-Pürsün, E.; et al. Prophylactic use of an anti-activated factor XII monoclonal antibody, garadacimab, for patients with C1-esterase inhibitor-deficient hereditary angioedema: A randomised, double-blind, placebo-controlled, phase 2 trial. *Lancet* **2022**, *399*, 945–955. [CrossRef]
35. McKenzie, A.; Roberts, A.; Malandkar, S.; Feuersenger, H.; Panousis, C.; Pawaskar, D. A phase I, first-in-human, randomized dose-escalation study of anti-activated factor XII monoclonal antibody garadacimab. *Clin. Transl. Sci.* **2022**, *15*, 626–637. [CrossRef]
36. Larsson, M.; Rayzman, V.; Nolte, M.W.; Nickel, K.F.; Björkqvist, J.; Jämsä, A.; Hardy, M.P.; Fries, M.; Schmidbauer, S.; Hedenqvist, P.; et al. A factor XIIa inhibitory antibody provides thromboprotection in extracorporeal circulation without increasing bleeding risk. *Sci. Transl. Med.* **2014**, *6*, 222ra17. [CrossRef]
37. Worm, M.; Köhler, E.C.; Panda, R.; Long, A.; Butler, L.M.; Stavrou, E.X.; Nickel, K.F.; Fuchs, T.A.; Renné, T. The factor XIIa blocking antibody 3F7: A safe anticoagulant with anti-inflammatory activities. *Ann. Transl. Med.* **2015**, *3*, 247. [CrossRef]
38. Matafonov, A.; Leung, P.Y.; Gailani, A.E.; Grach, S.L.; Puy, C.; Cheng, Q.; Sun, M.F.; McCarty, O.J.; Tucker, E.I.; Kataoka, H.; et al. Factor XII inhibition reduces thrombus formation in a primate thrombosis model. *Blood* **2014**, *123*, 1739–1746. [CrossRef]
39. Wallisch, M.; Lorentz, C.U.; Lakshmanan, H.H.S.; Johnson, J.; Carris, M.R.; Puy, C.; Gailani, D.; Hinds, M.T.; McCarty, O.J.T.; Gruber, A.; et al. Antibody inhibition of contact factor XII reduces platelet deposition in a model of extracorporeal membrane oxygenator perfusion in nonhuman primates. *Res. Pract. Thromb. Haemost.* **2020**, *4*, 205–216. [CrossRef]

40. Pireaux, V.; Tassignon, J.; Demoulin, S.; Derochette, S.; Borenstein, N.; Ente, A.; Fiette, L.; Douxfils, J.; Lancellotti, P.; Guyaux, M.; et al. Anticoagulation with an Inhibitor of Factors XIa and XIIa During Cardiopulmonary Bypass. *J. Am. Coll. Cardiol.* **2019**, *74*, 2178–2189. [CrossRef]
41. Demoulin, S.; Godfroid, E.; Hermans, C. Dual inhibition of factor XIIa and factor XIa as a therapeutic approach for safe thromboprotection. *J. Thromb. Haemost.* **2021**, *19*, 323–329. [CrossRef]
42. Decrem, Y.; Rath, G.; Blasioli, V.; Cauchie, P.; Robert, S.; Beaufays, J.; Frère, J.M.; Feron, O.; Dogné, J.M.; Dessy, C.; et al. Ir-CPI, a coagulation contact phase inhibitor from the tick Ixodes ricinus, inhibits thrombus formation without impairing hemostasis. *J. Exp. Med.* **2009**, *206*, 2381–2395. [CrossRef]
43. Yau, J.W.; Liao, P.; Fredenburgh, J.C.; Stafford, A.R.; Revenko, A.S.; Monia, B.P.; Weitz, J.I. Selective depletion of factor XI or factor XII with antisense oligonucleotides attenuates catheter thrombosis in rabbits. *Blood* **2014**, *123*, 2102–2107. [CrossRef] [PubMed]
44. Liu, J.; Cooley, B.C.; Akinc, A.; Butler, J.; Borodovsky, A. Knockdown of liver-derived factor XII by GalNAc-siRNA ALN-F12 prevents thrombosis in mice without impacting hemostatic function. *Thromb. Res.* **2020**, *196*, 200–205. [CrossRef] [PubMed]
45. May, F.; Krupka, J.; Fries, M.; Thielmann, I.; Pragst, I.; Weimer, T.; Panousis, C.; Nieswandt, B.; Stoll, G.; Dickneite, G.; et al. FXIIa inhibitor rHA-Infestin-4: Safe thromboprotection in experimental venous, arterial and foreign surface-induced thrombosis. *Br. J. Haematol.* **2016**, *173*, 769–778. [CrossRef] [PubMed]
46. Krupka, J.; May, F.; Weimer, T.; Pragst, I.; Kleinschnitz, C.; Stoll, G.; Panousis, C.; Dickneite, G.; Nolte, M.W. The Coagulation Factor XIIa Inhibitor rHA-Infestin-4 Improves Outcome after Cerebral Ischemia/Reperfusion Injury in Rats. *PLoS ONE* **2016**, *11*, e0146783. [CrossRef] [PubMed]
47. Hopp, S.; Albert-Weissenberger, C.; Mencl, S.; Bieber, M.; Schuhmann, M.K.; Stetter, C.; Nieswandt, B.; Schmidt, P.M.; Monoranu, C.M.; Alafuzoff, I.; et al. Targeting coagulation factor XII as a novel therapeutic option in brain trauma. *Ann. Neurol.* **2016**, *79*, 970–982. [CrossRef]
48. Salomon, O.; Steinberg, D.M.; Zucker, M.; Varon, D.; Zivelin, A.; Seligsohn, U. Patients with severe factor XI deficiency have a reduced incidence of deep-vein thrombosis. *Thromb. Haemost.* **2011**, *105*, 269–273. [CrossRef]
49. Tucker, E.I.; Marzec, U.M.; White, T.C.; Hurst, S.; Rugonyi, S.; McCarty, O.J.; Gailani, D.; Gruber, A.; Hanson, S.R. Prevention of vascular graft occlusion and thrombus-associated thrombin generation by inhibition of factor XI. *Blood* **2009**, *113*, 936–944. [CrossRef]
50. Salomon, O.; Steinberg, D.M.; Koren-Morag, N.; Tanne, D.; Seligsohn, U. Reduced incidence of ischemic stroke in patients with severe factor XI deficiency. *Blood* **2008**, *111*, 4113–4117. [CrossRef]
51. Preis, M.; Hirsch, J.; Kotler, A.; Zoabi, A.; Stein, N.; Rennert, G.; Saliba, W. Factor XI deficiency is associated with lower risk for cardiovascular and venous thromboembolism events. *Blood* **2017**, *129*, 1210–1215. [CrossRef]
52. James, P.; Salomon, O.; Mikovic, D.; Peyvandi, F. Rare bleeding disorders—Bleeding assessment tools, laboratory aspects and phenotype and therapy of FXI deficiency. *Haemophilia* **2014**, *20*, 71–75. [CrossRef]
53. Nourse, J.; Danckwardt, S. A novel rationale for targeting FXI: Insights from the hemostatic microRNA targetome for emerging anticoagulant strategies. *Pharmacol. Ther.* **2021**, *218*, 107676. [CrossRef]
54. Weitz, J.I.; Bauersachs, R.; Becker, B.; Berkowitz, S.D.; Freitas, M.C.S.; Lassen, M.R.; Metzig, C.; Raskob, G.E. Effect of Osocimab in Preventing Venous Thromboembolism Among Patients Undergoing Knee Arthroplasty: The FOXTROT Randomized Clinical Trial. *JAMA* **2020**, *323*, 130–139. [CrossRef]
55. Verhamme, P.; Yi, B.A.; Segers, A.; Salter, J.; Bloomfield, D.; Büller, H.R.; Raskob, G.E.; Weitz, J.I. Abelacimab for Prevention of Venous Thromboembolism. *N. Engl. J. Med.* **2021**, *385*, 609–617. [CrossRef]
56. Lorentz, C.U.; Tucker, E.I.; Verbout, N.G.; Shatzel, J.J.; Olson, S.R.; Markway, B.D.; Wallisch, M., Ralle, M.; Hinds, M.T.; McCarty, O.J.T.; et al. The contact activation inhibitor AB023 in heparin-free hemodialysis: Results of a randomized phase 2 clinical trial. *Blood* **2021**, *138*, 2173–2184. [CrossRef]
57. Cheng, Q.; Tucker, E.I.; Pine, M.S.; Sisler, I.; Matafonov, A.; Sun, M.F.; White-Adams, T.C.; Smith, S.A.; Hanson, S.R.; McCarty, O.J.; et al. A role for factor XIIa-mediated factor XI activation in thrombus formation in vivo. *Blood* **2010**, *116*, 3981–3989. [CrossRef]
58. Van Montfoort, M.L.; Knaup, V.L.; Marquart, J.A.; Bakhtiari, K.; Castellino, F.J.; Hack, C.E.; Meijers, J.C. Two novel inhibitory anti-human factor XI antibodies prevent cessation of blood flow in a murine venous thrombosis model. *Thromb. Haemost.* **2013**, *110*, 1065–1073. [CrossRef]
59. Hayward, N.J.; Goldberg, D.I.; Morrel, E.M.; Friden, P.M.; Bokesch, P.M. Abstract 13747: Phase 1a/1b Study of EP-7041: A novel, potent, selective, small molecule FXIa inhibitor. *Circulation* **2017**, *136*, A13747.
60. Weitz, J.I.; Strony, J.; Ageno, W.; Gailani, D.; Hylek, E.M.; Lassen, M.R.; Mahaffey, K.W.; Notani, R.S.; Roberts, R.; Segers, A.; et al. Milvexian for the Prevention of Venous Thromboembolism. *N. Engl. J. Med.* **2021**, *385*, 2161–2172. [CrossRef]
61. Wong, P.C.; Crain, E.J.; Bozarth, J.M.; Wu, Y.; Dilger, A.K.; Wexler, R.R.; Ewing, W.R.; Gordon, D.; Luettgen, J.M. Milvexian, an orally bioavailable, small-molecule, reversible, direct inhibitor of factor XIa: In vitro studies and in vivo evaluation in experimental thrombosis in rabbits. *J. Thromb. Haemost.* **2022**, *20*, 399–408. [CrossRef]
62. Bristol Myers Squibb—Late-Breaking Results from Phase 2 AXIOMATIC-SSP Study of Milvexian, an Investigational OralFactor XIa Inhibitor, Show Favorable Antithrombotic Profile in Combination with Dual Antiplatelet Therap. Available online: https://news.bms.com/news/details/2022/Late-Breaking-Results-From-Phase-2-AXIOMATIC-SSP-Study-of-Milvexian-an-Investigational-Oral-Factor-XIa-Inhibitor-Show-Favorable-Antithrombotic-Profile-in-Combination-With-Dual-Antiplatelet-Therapy/default.aspx (accessed on 29 September 2022).

63. Piccini, J.P.; Caso, V.; Connolly, S.J.; Fox, K.A.A.; Oldgren, J.; Jones, W.S.; Gorog, D.A.; Durdil, V.; Viethen, T.; Neumann, C.; et al. Safety of the oral factor XIa inhibitor asundexian compared with apixaban in patients with atrial fibrillation (PACIFIC-AF): A multicentre, randomised, double-blind, double-dummy, dose-finding phase 2 study. *Lancet* **2022**, *399*, 1383–1390. [CrossRef]
64. Rao, S.V.; Kirsch, B.; Bhatt, D.L.; Budaj, A.; Coppolecchia, R.; Eikelboom, J.; James, S.K.; Jones, W.S.; Merkely, B.; Keller, L.; et al. A Multicenter, Phase 2, Randomized, Placebo-Controlled, Double-Blind, Parallel-Group, Dose-Finding Trial of the Oral Factor XIa Inhibitor Asundexian to Prevent Adverse Cardiovascular Outcomes Following Acute Myocardial Infarction. *Circulation* **2022**, *146*, 1196–1206. [CrossRef] [PubMed]
65. Shoamanesh, A.; Mundl, H.; Smith, E.E.; Masjuan, J.; Milanov, I.; Hirano, T.; Agafina, A.; Campbell, B.; Caso, V.; Mas, J.-L.; et al. Factor XIa inhibition with asundexian after acute non-cardioembolic ischaemic stroke (PACIFIC-Stroke): An international, randomised, double-blind, placebo-controlled, phase 2b trial. *Lancet* **2022**, *400*, 997–1007. [CrossRef]
66. Perera, V.; Luettgen, J.M.; Wang, Z.; Frost, C.E.; Yones, C.; Russo, C.; Lee, J.; Zhao, Y.; LaCreta, F.P.; Ma, X.; et al. First-in-human study to assess the safety, pharmacokinetics and pharmacodynamics of BMS-962212, a direct, reversible, smallmolecule factor XIa inhibitor in non-Japaneseand Japanese healthy subjects. *Br. J. Clin. Pharmacol.* **2018**, *84*, 876–887. [CrossRef] [PubMed]
67. Beale, D.; Dennison, J.; Boyce, M.; Mazzo, F.; Honda, N.; Smith, P.; Bruce, M. ONO-7684 a novel oral FXIa inhibitor: Safety, tolerability, pharmacokinetics and pharmacodynamics in a first-in-human study. *Br. J. Clin. Pharmacol.* **2021**, *87*, 3177–3189. [CrossRef] [PubMed]
68. Wong, P.C.; Quan, M.L.; Watson, C.A.; Crain, E.J.; Harpel, M.R.; Rendina, A.R.; Luettgen, J.M.; Wexler, R.R.; Schumacher, W.A.; Seiffert, D.A. In vitro, antithrombotic and bleeding time studies of BMS-654457, a small-molecule, reversible and direct inhibitor of factor XIa. *J. Thromb. Thrombolysis* **2015**, *40*, 416–423. [CrossRef]
69. Kouyama, S.; Ono, T.; Hagio, T.; Sakimoto, S.; Miyata, H.; Tanaka, M.; Koda, T.; Tanaka, K.; Yanagida, D.; Sakai, M.; et al. Discovery of ONO-5450598, a highly orally bioavailable small molecule factor XIa inhibitor: The pharmacokinetic and pharmacological profiles. *Res. Pract. Thromb. Haemost.* **2017**, *1*, 99.
70. Wong, P.C.; Crain, E.J.; Watson, C.A.; Schumacher, W.A. A small-molecule factor XIa inhibitor produces antithrombotic efficacy with minimal bleeding time prolongation in rabbits. *J. Thromb. Thrombolysis* **2011**, *32*, 129–137. [CrossRef]
71. Büller, H.R.; Bethune, C.; Bhanot, S.; Gailani, D.; Monia, B.P.; Raskob, G.E.; Segers, A.; Verhamme, P.; Weitz, J.I. Factor XI antisense oligonucleotide for prevention of venous thrombosis. *N. Engl. J. Med.* **2015**, *372*, 232–240. [CrossRef]
72. Smiley, D.A.; Becker, R.C. Factor IXa as a target for anticoagulation in thrombotic disorders and conditions. *Drug Discov. Today* **2014**, *19*, 1445–1453. [CrossRef]
73. Vavalle, J.P.; Cohen, M.G. The REG1 anticoagulation system: A novel actively controlled factor IX inhibitor using RNA aptamer technology for treatment of acute coronary syndrome. *Future Cardiol.* **2012**, *8*, 371–382. [CrossRef]
74. Afosah, D.K.; Ofori, E.; Mottamal, M.; Al-Horani, R.A. Factor IX(a) inhibitors: An updated patent review (2003–present). *Expert Opin. Ther. Pat.* **2022**, *32*, 381–400. [CrossRef]
75. Povsic, T.J.; Vavalle, J.P.; Aberle, L.H.; Kasprzak, J.D.; Cohen, M.G.; Mehran, R.; Bode, C.; Buller, C.E.; Montalescot, G.; Cornel, J.H.; et al. A Phase 2, randomized, partially blinded, active-controlled study assessing the efficacy and safety of variable anticoagulation reversal using the REG1 system in patients with acute coronary syndromes: Results of the RADAR trial. *Eur. Heart J.* **2013**, *34*, 2481–2489. [CrossRef]
76. Lincoff, A.M.; Mehran, R.; Povsic, T.J.; Zelenkofske, S.L.; Huang, Z.; Armstrong, P.W.; Steg, P.G.; Bode, C.; Cohen, M.G.; Buller, C.; et al. Effect of the REG1 anticoagulation system versus bivalirudin on outcomes after percutaneous coronary intervention (REGULATE-PCI): A randomised clinical trial. *Lancet* **2016**, *387*, 349–356. [CrossRef]
77. Staudacher, D.L.; Putz, V.; Heger, L.; Reinöhl, J.; Hortmann, M.; Zelenkofske, S.L.; Becker, R.C.; Rusconi, C.P.; Bode, C.; Ahrens, I. Direct factor IXa inhibition with the RNA-aptamer pegnivacogin reduces platelet reactivity in vitro and residual platelet aggregation in patients with acute coronary syndromes. *Eur. Heart J. Acute Cardiovasc. Care* **2019**, *8*, 520–526. [CrossRef]
78. Chow, F.; Benincosa, L.J.; Sheth, S.B.; Wilson, D.; Davis, C.; Minthorn, E.A.; Jusko, W.J. Pharmacokinetic and pharmacodynamic modeling of humanized anti-factor IX antibody (SB 249417) in humans. *Clin. Pharmacol. Ther.* **2002**, *71*, 235–245. [CrossRef]
79. Eriksson, B.I.; Dahl, O.E.; Lassen, M.R.; Ward, D.P.; Rothlein, R.; Davis, G.; Turpie, A.G.G.; Fixit Study Group. Partial factor IXa inhibition with TTP889 for prevention of venous thromboembolism: An exploratory study. *J. Thromb. Haemost.* **2008**, *6*, 457–463. [CrossRef]
80. Badimon, L.; Vilahur, G.; Rocca, B.; Patrono, C. The Key Contribution of Platelet and Vascular Arachidonic Acid Metabolism to the Pathophysiology of Atherothrombosis. *Cardiovasc. Res.* **2021**, *117*, 2001–2015. Available online: https://pubmed.ncbi.nlm.nih.gov/33484117/ (accessed on 15 September 2022). [CrossRef]
81. Ibanez, B.; James, S.; Agewall, S.; Antunes, M.J.; Bucciarelli-Ducci, C.; Bueno, H.; Caforio, A.L.P.; Crea, F.; Goudevenos, J.A.; Halvorsen, S.; et al. 2017 ESC Guidelines for the management of acute myocardial infarction in patients presenting with ST-segment elevation. *Eur. Heart J.* **2018**, *39*, 119–177. [CrossRef]
82. Knuuti, J.; Wijns, W.; Saraste, A.; Capodanno, D.; Barbato, E.; Funck-Brentano, C.; Prescott, E.; Storey, R.F.; Deaton, C.; Cuisset, T.; et al. 2019 ESC Guidelines for the diagnosis and management of chronic coronary syndromes. *Eur. Heart J.* **2020**, *41*, 407–477. [CrossRef]

83. Collet, J.P.; Thiele, H.; Barbato, E.; Barthélémy, O.; Bauersachs, J.; Bhatt, D.L.; Dendale, P.; Dorobantu, M.; Edvardsen, T.; Folliguet, T.; et al. 2020 ESC Guidelines for the management of acute coronary syndromes in patients presenting without persistent ST-segment elevationThe Task Force for the management of acute coronary syndromes in patients presenting without persistent ST-segment elevation of the European Society of Cardiology (ESC). *Eur. Heart J.* **2021**, *42*, 1289–1367.
84. Walker, J.; Cattaneo, M.; Badimon, L.; Agnelli, G.; Chan, A.T.; Lanas, A.; Rocca, B.; Rothwell, P.; Patrignani, P.; Langley, R.; et al. Highlights from the 2019 International Aspirin Foundation Scientific Conference, Rome, 28 June 2019: Benefits and risks of antithrombotic therapy for cardiovascular disease prevention. *Ecancermedicalscience* **2020**, *14*, 998. [CrossRef] [PubMed]
85. Mehta, S.R.; Yusuf, S.; Peters, R.J.; Bertrand, M.E.; Lewis, B.S.; Natarajan, M.K.; Malmberg, K.; Rupprecht, H.; Zhao, F.; Chrolavicius, S.; et al. Effects of pretreatment with clopidogrel and aspirin followed by long-term therapy in patients undergoing percutaneous coronary intervention: The PCI-CURE study. *Lancet* **2001**, *358*, 527–533. [CrossRef]
86. Wiviott, S.D.; Braunwald, E.; McCabe, C.H.; Montalescot, G.; Ruzyllo, W.; Gottlieb, S.; Neumann, F.J.; Ardissino, D.; De Servi, S.; Murphy, S.A.; et al. Prasugrel versus Clopidogrel in Patients with Acute Coronary Syndromes. *N. Engl. J. Med.* **2007**, *357*, 2001–2015. [CrossRef] [PubMed]
87. Wallentin, L.; Becker, R.C.; Budaj, A.; Cannon, C.P.; Emanuelsson, H.; Held, C.; Horrow, J.; Husted, S.; James, S.; Katus, H.; et al. Ticagrelor versus clopidogrel in patients with acute coronary syndromes. *N. Engl. J. Med.* **2009**, *361*, 1045–1057. [CrossRef] [PubMed]
88. Schrör, K.; Siller-Matula, J.M.; Huber, K. Pharmacokinetic basis of the antiplatelet action of prasugrel. *Fundam. Clin. Pharmacol.* **2012**, *26*, 39–46. [CrossRef]
89. Li, J.; Vootukuri, S.; Shang, Y.; Negri, A.; Jiang, J.-K.; Nedelman, M.; Diacovo, T.G.; Filizola, M.; Thomas, C.J.; Coller, B.S. A novel αIIbβ3 antagonist for prehospital therapy of myocardial infarction. *Arterioscler. Thromb. Vasc. Biol.* **2014**, *34*, 2321–2329. [CrossRef]
90. Xu, Q.; Yin, J.; Si, L.Y. Efficacy and safety of early versus late glycoprotein IIb/IIIa inhibitors for PCI. *Int. J. Cardiol.* **2013**, *162*, 210–219. [CrossRef]
91. Kleindorfer, D.O.; Towfighi, A.; Chaturvedi, S.; Cockroft, K.M.; Gutierrez, J.; Lombardi-Hill, D.; Kamel, H.; Kernan, W.N.; Kittner, S.J.; Leira, E.C.; et al. 2021 Guideline for the Prevention of Stroke in Patients with Stroke and Transient Ischemic Attack: A Guideline From the American Heart Association/American Stroke Association. *Stroke* **2021**, *52*, E364–E467. [CrossRef]
92. Amarenco, P.; Denison, H.; Evans, S.R.; Himmelmann, A.; James, S.; Knutsson, M.; Ladenvall, P.; Molina, C.A.; Wang, Y.; Johnston, S.C.; et al. Ticagrelor Added to Aspirin in Acute Ischemic Stroke or Transient Ischemic Attack in Prevention of Disabling Stroke: A Randomized Clinical Trial. *JAMA Neurol.* **2021**, *78*, 177–185. [CrossRef]
93. Sahara, N.; Kuwashiro, T.; Okada, Y. Cerebral infarction and transient ischemic attack. *Nihon Rinsho* **2016**, *74*, 666–670.
94. Cave, B.; Rawal, A.; Ardeshna, D.; Ibebuogu, U.N.; Sai-Sudhakar, C.B.; Khouzam, R.N. Targeting ticagrelor: A novel therapy for emergency reversal. *Ann. Transl. Med.* **2019**, *7*, 410. [CrossRef]
95. Sang, Y.; Roest, M.; de Laat, B.; de Groot, P.G.; Huskens, D. Interplay between platelets and coagulation. *Blood Rev.* **2021**, *46*, 100733. [CrossRef]
96. Zheng, B.; Li, J.; Jiang, J.; Xiang, D.; Chen, Y.; Yu, Z.; Zeng, H.; Ge, J.; Dai, X.; Liu, J.; et al. Safety and efficacy of a platelet glycoprotein Ib inhibitor for patients with non-ST segment elevation myocardial infarction: A phase Ib/IIa study. *Pharmacotherapy* **2021**, *41*, 828–836. [CrossRef]
97. Li, T.-T.; Fan, M.-L.; Hou, S.-X.; Li, X.-Y.; Barry, D.M.; Jin, H.; Luo, S.-Y.; Kong, F.; Lau, L.-F.; Dai, X.-R.; et al. A novel snake venom-derived GPIb antagonist, anfibatide, protects mice from acute experimental ischaemic stroke and reperfusion injury. *Br. J. Pharmacol.* **2015**, *172*, 3904–3916. [CrossRef]
98. Gong, P.; Li, R.; Jia, H.-Y.; Ma, Z.; Li, X.Y.; Dai, X.-R.; Luo, S.-Y. Anfibatide Preserves Blood-Brain Barrier Integrity by Inhibiting TLR4/RhoA/ROCK Pathway After Cerebral Ischemia/Reperfusion Injury in Rat. *J. Mol. Neurosci.* **2020**, *70*, 71–83. [CrossRef]
99. Sun, Y.; Langer, H.F. Platelets, Thromboinflammation and Neurovascular Disease. *Front. Immunol.* **2022**, *13*, 843404. [CrossRef]
100. Bartunek, J.; Barbato, E.; Heyndrickx, G.; Vanderheyden, M.; Wijns, W.; Holz, J.B. Novel antiplatelet agents: ALX-0081, a Nanobody directed towards von Willebrand factor. *J. Cardiovasc. Transl. Res.* **2013**, *6*, 355–363. [CrossRef]
101. Kovacevic, K.D.; Jilma, B.; Zhu, S.; Gilbert, J.C.; Winter, M.-P.; Toma, A.; Hengstenberg, C.; Lang, I.; Kubica, J.; Siller-Matula, J.M. von Willebrand Factor Predicts Mortality in ACS Patients Treated with Potent P2Y12 Antagonists and is Inhibited by Aptamer BT200 Ex Vivo. *Thromb. Haemost.* **2020**, *120*, 1282–1290. [CrossRef]
102. Kovacevic, K.D.; Greisenegger, S.; Langer, A.; Gelbenegger, G.; Buchtele, N.; Pabinger, I.; Petroczi, K.; Zhu, S.; Gilbert, J.C.; Jilma, B. The aptamer BT200 blocks von Willebrand factor and platelet function in blood of stroke patients. *Sci. Rep.* **2021**, *11*, 3092. [CrossRef]
103. Nayak, M.K.; Son, D.J.; Fuentes, E. Modulation of Glycoprotein VI and Its Downstream Signaling Pathways as an Antiplatelet Target. *Int. J. Mol. Sci.* **2022**, *23*, 9882.
104. Vilahur, G.; Gutiérrez, M.; Arzanauskaite, M.; Mendieta, G.; Ben-Aicha, S.; Badimon, L. Intracellular platelet signalling as a target for drug development. *Vasc. Pharmacol.* **2018**, *111*, 22–25. [CrossRef] [PubMed]
105. Borst, O.; Gawaz, M. Glycoprotein VI—Novel target in antiplatelet medication. *Pharmacol. Ther.* **2021**, *217*, 107630. [CrossRef]
106. Nurden, A.T. Clinical significance of altered collagen-receptor functioning in platelets with emphasis on glycoprotein VI. *Blood Rev.* **2019**, *38*, 100592. [CrossRef] [PubMed]
107. Arthur, J.F.; Dunkley, S.; Andrews, R.K. Platelet glycoprotein VI-related clinical defects. *Br. J. Haematol.* **2007**, *139*, 363–372. [CrossRef] [PubMed]

108. Dalby, A.; Mezzano, D.; Rivera, J.; Watson, S.P.; Morgan, N.V. Introduction of an ancient founder glycoprotein VI (GP6) mutation into the Chilean population. *Blood Adv.* **2022**, *6*, 5866–5869. [CrossRef]
109. Arai, M.; Yamamoto, N.; Moroi, M.; Akamatsu, N.; Fukutake, K.; Tanoue, K. Platelets with 10% of the normal amount of glycoprotein VI have an impaired response to collagen that results in a mild bleeding tendency. *Br. J. Haematol.* **1995**, *89*, 124–130. [CrossRef]
110. Lockyer, S.; Okuyama, K.; Begum, S.; Le, S.; Sun, B.; Watanabe, T.; Matsumoto, Y.; Yoshitake, M.; Kambayashi, J.; Tandon, N.N. GPVI-deficient mice lack collagen responses and are protected against experimentally induced pulmonary thromboembolism. *Thromb. Res.* **2006**, *118*, 371–380. [CrossRef]
111. Ungerer, M.; Rosport, K.; Bültmann, A.; Piechatzek, R.; Uhland, K.; Schlieper, P.; Gawaz, M.; Münch, G. Novel antiplatelet drug revacept (Dimeric Glycoprotein VI-Fc) specifically and efficiently inhibited collagen-induced platelet aggregation without affecting general hemostasis in humans. *Circulation* **2011**, *123*, 1891–1899. [CrossRef]
112. Mayer, K.; Hein-Rothweiler, R.; Schüpke, S.; Janisch, M.; Bernlochner, I.; Ndrepepa, G.; Sibbing, D.; Gori, T.; Borst, O.; Holdenrieder, S.; et al. Efficacy and Safety of Revacept, a Novel Lesion-Directed Competitive Antagonist to Platelet Glycoprotein VI, in Patients Undergoing Elective Percutaneous Coronary Intervention for Stable Ischemic Heart Disease: The Randomized, Double-blind, Placebo-Controlled ISAR-PLASTER Phase 2 Trial. *JAMA Cardiol.* **2021**, *6*, 753–761.
113. Lebozec, K.; Jandrot-Perrus, M.; Avenard, G.; Favre-Bulle, O.; Billiald, P. Design, development and characterization of ACT017, a humanized Fab that blocks platelet's glycoprotein VI function without causing bleeding risks. *MAbs* **2017**, *9*, 945–958. [CrossRef]
114. Renaud, L.; Lebozec, K.; Voors-Pette, C.; Dogterom, P.; Billiald, P.; Perrus, M.J.; Pletan, Y.; Machacek, M. Population Pharmacokinetic/Pharmacodynamic Modeling of Glenzocimab (ACT017) a Glycoprotein VI Inhibitor of Collagen-Induced Platelet Aggregation. *Pharmacomet. J. Clin. Pharmacol.* **2020**, *2020*, 1198–1208. [CrossRef]
115. Voors-Pette, C.; Lebozec, K.; Dogterom, P.; Jullien, L.; Billiald, P.; Ferlan, P.; Renaud, L.; Favre-Bulle, O.; Avenard, G.; Machacek, M.; et al. Safety and Tolerability, Pharmacokinetics, and Pharmacodynamics of ACT017, an Antiplatelet GPVI (Glycoprotein VI) Fab. *Arterioscler. Thromb. Vasc. Biol.* **2019**, *39*, 956–964. [CrossRef]
116. Sakai, K.; Someya, T.; Harada, K.; Yagi, H.; Matsui, T.; Matsumoto, M. Novel aptamer to von Willebrand factor A1 domain (TAGX-0004) shows total inhibition of thrombus formation superior to ARC1779 and comparable to caplacizumab. *Haematologica* **2020**, *105*, 2631–2638. [CrossRef]
117. Markus, H.S.; McCollum, C.; Imray, C.; Goulder, M.A.; Gilbert, J.; King, A. The von Willebrand inhibitor ARC1779 reduces cerebral embolization after carotid endarterectomy: A randomized trial. *Stroke* **2011**, *42*, 2149–2153. [CrossRef]
118. Lapchak, P.A.; Doyan, S.; Fan, X.; Woods, C.M. Synergistic Effect of AJW200, a von Willebrand Factor Neutralizing Antibody with Low Dose (0.9 mg/mg) Thrombolytic Therapy Following Embolic Stroke in Rabbits. *J. Neurol. Neurophysiol.* **2013**, *4*, 10.4172/2155-9562.10001466. [CrossRef]
119. Kageyama, S.; Yamamoto, H.; Nakazawa, H.; Matsushita, J.; Kouyama, T.; Gonsho, A.; Ikeda, Y.; Yoshimoto, R. Pharmacokinetics and pharmacodynamics of AJW200, a humanized monoclonal antibody to von Willebrand factor, in monkeys. *Arterioscler. Thromb. Vasc. Biol.* **2002**, *22*, 187–192. [CrossRef]
120. Wu, D.; Vanhoorelbeke, K.; Cauwenberghs, N.; Meiring, M.; Depraetere, H.; Kotze, H.F.; Deckmyn, H. Inhibition of the von Willebrand (VWF)-collagen interaction by an antihuman VWF monoclonal antibody results in abolition of in vivo arterial platelet thrombus formation in baboons. *Blood* **2002**, *99*, 3623–3628. [CrossRef]
121. Scully, M.; Cataland, S.R.; Peyvandi, F.; Coppo, P.; Knöbl, P.; Kremer Hovinga, J.A.; Metjian, A.; de la Rubia, J.; Pavenski, K.; Callewaert, F.; et al. Caplacizumab Treatment for Acquired Thrombotic Thrombocytopenic Purpura. *N. Engl. J. Med.* **2019**, *380*, 335–346. [CrossRef]
122. Fontayne, A.; Meiring, M.; Lamprecht, S.; Roodt, J.; Demarsin, E.; Barbeaux, P.; Deckmyn, H. The humanized anti-glycoprotein Ib monoclonal antibody h6B4-Fab is a potent and safe antithrombotic in a high shear arterial thrombosis model in baboons. *Thromb. Haemost.* **2008**, *100*, 670–677. [CrossRef]
123. Yang, J.; Ji, S.; Dong, N.; Zhao, Y.; Ruan, C. Engineering and characterization of a chimeric anti-platelet glycoprotein Ibα monoclonal antibody and preparation of its Fab fragment. *Hybridoma* **2010**, *29*, 125–132. [CrossRef]
124. Schulte, V.; Reusch, H.P.; Pozgajová, M.; Varga-Szabó, D.; Gachet, C.; Nieswandt, B. Two-phase antithrombotic protection after anti-glycoprotein VI treatment in mice. *Arterioscler. Thromb. Vasc. Biol.* **2006**, *26*, 1640–1647. [CrossRef] [PubMed]
125. Nieswandt, B.; Schulte, V.; Bergmeier, W.; Mokhtari-Nejad, R.; Rackebrandt, K.; Cazenave, J.-P.; Ohlmann, P.; Gachet, C.; Zirngibl, H. Long-term antithrombotic protection by in vivo depletion of platelet glycoprotein VI in mice. *J. Exp. Med.* **2001**, *193*, 459–469. [CrossRef] [PubMed]
126. Lin, Y.-C.; Ko, Y.-C.; Hung, S.-C.; Lin, Y.-T.; Lee, J.-H.; Tsai, J.-Y.; Kung, P.-H.; Tsai, M.-C.; Chen, Y.-F.; Wu, C.-C. Selective Inhibition of PAR4 (Protease-Activated Receptor 4)-Mediated Platelet Activation by a Synthetic Nonanticoagulant Heparin Analog. *Arterioscler. Thromb. Vasc. Biol.* **2019**, *39*, 694–703. [CrossRef]
127. Kung, P.H.; Hsieh, P.W.; Lin, Y.T.; Lee, J.H.; Chen, I.H.; Wu, C.C. HPW-RX40 prevents human platelet activation by attenuating cell surface protein disulfide isomerases. *Redox Biol.* **2017**, *13*, 266–277. [CrossRef] [PubMed]
128. Khodier, C.; VerPlank, L.; Nag, P.P.; Pu, J.; Wurst, J.; Pilyugina, T.; Dockendorff, C.; Galinski, C.N.; Scalise, A.A.; Passam, F.; et al. Identification of ML359 as a small molecule inhibitor of protein disulfide isomerase. In *Probe Reports from the NIH Molecular Libraries Program*; National Center for Biotechnology Information (US): Bethesda, MD, USA, 2010.

129. Adili, R.; Tourdot, B.E.; Mast, K.; Yeung, J.; Freedman, J.C.; Green, A.; Luci, D.K.; Jadhav, A.; Simeonov, A.; Maloney, D.J.; et al. First Selective 12-LOX Inhibitor, ML355, Impairs Thrombus Formation and Vessel Occlusion In Vivo With Minimal Effects on Hemostasis. *Arterioscler. Thromb. Vasc. Biol.* **2017**, *37*, 1828–1839. [CrossRef] [PubMed]
130. Zheng, Z.; Pinson, J.-A.; Mountford, S.J.; Orive, S.; Schoenwaelder, S.M.; Shackleford, D.; Powell, A.; Nelson, E.M.; Hamilton, J.R.; Jackson, S.P.; et al. Discovery and antiplatelet activity of a selective PI3Kβ inhibitor (MIPS-9922). *Eur. J. Med. Chem.* **2016**, *122*, 339–351. [CrossRef]
131. Schwarz, M.; Meade, G.; Stoll, P.; Ylanne, J.; Bassler, N.; Chen, Y.C.; Hagemeyer, C.E.; Ahrens, I.; Moran, N.; Kenny, D.; et al. Conformation-specific blockade of the integrin GPIIb/IIIa: A novel antiplatelet strategy that selectively targets activated platelets. *Circ. Res.* **2006**, *99*, 25–33. [CrossRef]
132. Pang, A.; Cheng, N.; Cui, Y.; Bai, Y.; Hong, Z.; Delaney, M.K.; Zhang, Y.; Chang, C.; Wang, C.; Liu, C.; et al. High-loading Gα 13-binding EXE peptide nanoparticles prevent thrombosis and protect mice from cardiac ischemia/reperfusion injury. *Sci. Transl. Med.* **2020**, *12*, eaaz7287. [CrossRef]
133. Boldron, C.; Besse, A.; Bordes, M.F.; Tissandié, S.; Yvon, X.; Gau, B.; Badorc, A.; Rousseaux, T.; Barré, G.; Meneyrol, J.; et al. N-[6-(4-butanoyl-5-methyl-1H-pyrazol-1-yl)pyridazin-3-yl]-5-chloro-1-[2-(4-methylpiperazin-1-yl)-2-oxoethyl]-1H -indole-3-carboxamide (SAR216471), a novel intravenous and oral, reversible, and directly acting P2Y12 antagonist. *J. Med. Chem.* **2014**, *57*, 7293–7316. [CrossRef]
134. Kong, D.; Xue, T.; Guo, B.; Cheng, J.; Liu, S.; Wei, J.; Lu, Z.; Liu, H.; Gong, G.; Lan, T.; et al. Optimization of P2Y 12 Antagonist Ethyl 6-(4-((Benzylsulfonyl)carbamoyl)piperidin-1-yl)-5-cyano-2-methylnicotinate (AZD1283) Led to the Discovery of an Oral Antiplatelet Agent with Improved Druglike Properties. *J. Med. Chem.* **2019**, *62*, 3088–3106. [CrossRef]
135. Yang, W.; Wang, Y.; Lai, A.; Qiao, J.X.; Wang, T.C.; Hua, J.; Price, L.A.; Shen, H.; Chen, X.-Q.; Wong, P.; et al. Discovery of 4-aryl-7-hydroxyindoline based P2Y1 antagonists as novel antiplatelet agents. *J. Med. Chem.* **2014**, *57*, 6150–6164. [CrossRef]
136. Wong, P.C.; Watson, C.; Crain, E.J. The P2Y1 receptor antagonist MRS2500 prevents carotid artery thrombosis in cynomolgus monkeys. *J. Thromb. Thrombolysis* **2016**, *41*, 514–521. [CrossRef]
137. Gremmel, T.; Yanachkov, I.B.; Yanachkova, M.I.; Wright, G.E.; Wider, J.; Undyala, V.V.; Michelson, A.D.; Frelinger, A.L., III; Przyklenk, K. Synergistic inhibition of both P2Y1 and P2Y12 adenosine diphosphate receptors as novel approach to rapidly attenuate platelet-mediated thrombosis. *Arterioscler. Thromb. Vasc. Biol.* **2016**, *36*, 501–509. [CrossRef] [PubMed]
138. Chang, C.-H.; Chung, C.-H.; Tu, Y.-S.; Tsai, C.-C.; Hsu, C.-C.; Peng, H.-C.; Tseng, Y.J.; Huang, T.-F. Trowaglerix Venom Polypeptides as a Novel Antithrombotic Agent by Targeting Immunoglobulin-Like Domains of Glycoprotein VI in Platelet. *Arterioscler. Thromb. Vasc. Biol.* **2017**, *37*, 1307–1314. [CrossRef]
139. van Eeuwijk, J.M.; Stegner, D.; Lamb, D.J.; Kraft, P.; Beck, S.; Thielmann, I.; Kiefer, F.; Walzog, B.; Stoll, G.; Nieswandt, B. The Novel Oral Syk Inhibitor, Bl1002494, Protects Mice from Arterial Thrombosis and Thromboinflammatory Brain Infarction. *Arterioscler. Thromb. Vasc. Biol.* **2016**, *36*, 1247–1253. [CrossRef]
140. Tullemans, B.M.E.; Nagy, M.; Sabrkhany, S.; Griffioen, A.W.; Oude Egbrink, M.G.A.; Aarts, M.; Heemskerk, J.W.M.; Kuijpers, M.J.E. Tyrosine Kinase Inhibitor Pazopanib Inhibits Platelet Procoagulant Activity in Renal Cell Carcinoma Patients. *Front. Cardiovasc. Med.* **2018**, *5*, 142. [CrossRef] [PubMed]
141. Rigg, R.A.; Aslan, J.E.; Healy, L.D.; Wallisch, M.; Thierheimer, M.L.; Loren, C.P.; Pang, J.; Hinds, M.T.; Gruber, A.; McCarty, O.J. Oral administration of Bruton's tyrosine kinase inhibitors impairs GPVI-mediated platelet function. *Am. J. Physiol. Cell Physiol.* **2016**, *310*, C373–C380. [CrossRef]
142. Perrella, G.; Montague, S.J.; Brown, H.C.; Garcia Quintanilla, L.; Slater, A.; Stegner, D.; Thomas, M.; Heemskerk, J.W.M.; Watson, S.P. Role of Tyrosine Kinase Syk in Thrombus Stabilisation at High Shear. *Int. J. Mol. Sci.* **2022**, *23*, 493. [CrossRef]
143. Harbi, M.H.; Smith, C.W.; Nicolson, P.L.R.; Watson, S.P.; Thomas, M.R. Novel antiplatelet strategies targeting GPVI, CLEC-2 and tyrosine kinases. *Platelets* **2021**, *32*, 29–41. [CrossRef]
144. Nicolson, P.L.R.; Nock, S.H.; Hinds, J.; Garcia-Quintanilla, L.; Smith, C.W.; Campos, J.; Brill, A.; Pike, J.A.; Khan, A.O.; Poulter, N.S.; et al. Low-dose Btk inhibitors selectively block platelet activation by CLEC-2. *Haematologica* **2021**, *106*, 208–219. [CrossRef]
145. De Candia, E. Mechanisms of platelet activation by thrombin: A short history. *Thromb. Res.* **2012**, *129*, 250–256. [CrossRef] [PubMed]
146. Bohula, E.A.; Aylward, P.E.; Bonaca, M.P.; Corbalan, R.L.; Kiss, R.G.; Murphy, S.A.; Scirica, B.M.; White, H.; Braunwald, E.; Morrow, D.A. Efficacy and Safety of Vorapaxar with and without a Thienopyridine for Secondary Prevention in Patients with Previous Myocardial Infarction and No History of Stroke or Transient Ischemic Attack: Results from TRA 2°P-TIMI 50. *Circulation* **2015**, *132*, 1871–1879. [CrossRef] [PubMed]
147. Jones, W.S.; Tricoci, P.; Huang, Z.; Moliterno, D.J.; Harrington, R.A.; Sinnaeve, P.R.; Strony, J.; Van de Werf, F.; White, H.D.; Held, C.; et al. Vorapaxar in patients with peripheral artery disease and acute coronary syndrome: Insights from Thrombin Receptor Antagonist for Clinical Event Reduction in Acute Coronary Syndrome (TRACER). *Am. Heart J.* **2014**, *168*, 588–596. [CrossRef] [PubMed]
148. Kosova, E.C.; Bonaca, M.P.; Dellborg, M.; He, P.; Morais, J.; Ophuis, T.O.; Scirica, B.M.; Tendera, M.; Theroux, P.; Braunwald, E.; et al. Vorapaxar in patients with coronary artery bypass grafting: Findings from the TRA 2°P-TIMI 50 trial. *Eur. Heart J. Acute Cardiovasc. Care* **2017**, *6*, 164–172. [CrossRef] [PubMed]

149. Cavender, M.A.; Scirica, B.M.; Bonaca, M.P.; Angiolillo, D.J.; Dalby, A.J.; Dellborg, M.; Morais, J.; Murphy, S.A.; Ophuis, T.O.; Tendera, M.; et al. Vorapaxar in patients with diabetes mellitus and previous myocardial infarction: Findings from the thrombin receptor antagonist in secondary prevention of atherothrombotic ischemic events-TIMI 50 trial. *Circulation* **2015**, *131*, 1047–1053. [CrossRef]
150. Franchi, F.; Rollini, F.; Faz, G.; Rivas, J.R.; Rivas, A.; Agarwal, M.; Briceno, M.; Wali, M.; Nawaz, A.; Silva, G.; et al. Pharmacodynamic Effects of Vorapaxar in Prior Myocardial Infarction Patients Treated with Potent Oral P2Y 12 Receptor Inhibitors with and Without Aspirin: Results of the VORA-PRATIC Study. *J. Am. Heart Assoc.* **2020**, *9*, e015865. [CrossRef]
151. Gurbel, P.A.; Bliden, K.P.; Turner, S.E.; Tantry, U.S.; Gesheff, M.G.; Barr, T.P.; Covic, L.; Kuliopulos, A. Cell-Penetrating Pepducin Therapy Targeting PAR1 in Subjects with Coronary Artery Disease. *Arterioscler. Thromb. Vasc. Biol.* **2016**, *36*, 189–197. [CrossRef]
152. Kuliopulos, A.; Gurbel, P.A.; Rade, J.J.; Kimmelstiel, C.D.; Turner, S.E.; Bliden, K.P.; Fletcher, E.K.; Cox, D.H.; Covic, L.; TRIP-PCI Investigators. PAR1 (Protease-Activated Receptor 1) Pepducin Therapy Targeting Myocardial Necrosis in Coronary Artery Disease and Acute Coronary Syndrome Patients Undergoing Cardiac Catheterization: A Randomized, Placebo-Controlled, Phase 2 Study. *Arterioscler. Thromb. Vasc. Biol.* **2020**, *40*, 2990–3003. [CrossRef]
153. Meah, M.N.; Raftis, J.; Wilson, S.J.; Perera, V.; Garonzik, S.M.; Murthy, B.; Everlof, J.G.; Aronson, R.; Luettgen, J.; Newby, D.E. Antithrombotic Effects of Combined PAR (Protease-Activated Receptor)-4 Antagonism and Factor Xa Inhibition. *Arterioscler. Thromb. Vasc. Biol.* **2020**, *40*, 2678–2685. [CrossRef]
154. Merali, S.; Wang, Z.; Frost, C.; Callejo, M.; Hedrick, M.; Hui, L.; Shropshire, S.M.; Xu, K.; Bouvier, M.; DeSouza, M.M.; et al. New oral protease-activated receptor 4 antagonist BMS-986120: Tolerability, pharmacokinetics, pharmacodynamics, and gene variant effects in humans. *Platelets* **2022**, *33*, 969–978. [CrossRef]
155. Wong, P.C.; Seiffert, D.; Bird, J.E.; Watson, C.A.; Bostwick, J.S.; Giancarli, M.; Allegretto, N.; Hua, J.; Harden, D.; Guay, J.; et al. Blockade of protease-activated receptor- 4(PAR4) provides robust antithrombotic activity with low bleeding. *Sci. Transl. Med.* **2017**, *9*, eaaf5294. [CrossRef]
156. Wilson, S.J.; Ismat, F.A.; Wang, Z.; Cerra, M.; Narayan, H.; Raftis, J.; Gray, T.J.; Connell, S.; Garonzik, S.; Ma, X.; et al. PAR4 (Protease-Activated Receptor 4) Antagonism With BMS-986120 Inhibits Human Ex Vivo Thrombus Formation. *Arterioscler. Thromb. Vasc. Biol.* **2018**, *38*, 448–456. [CrossRef]
157. Nylander, S.; Kull, B.; Björkman, J.A.; Ulvinge, J.C.; Oakes, N.; Emanuelsson, B.M.; Andersson, M.; Skärby, T.; Inghardt, T.; Fjellström, O.; et al. Human target validation of phosphoinositide 3-kinase (PI3K)β: Effects on platelets and insulin sensitivity, using AZD6482 a novel PI3Kβ inhibitor. *J. Thromb. Haemost.* **2012**, *10*, 2127–2136. [CrossRef]
158. Nylander, S.; Wågberg, F.; Andersson, M.; Skärby, T.; Gustafsson, D. Exploration of efficacy and bleeding with combined phosphoinositide 3-kinase β inhibition and aspirin in man. *J. Thromb. Haemost.* **2015**, *13*, 1494–1502. [CrossRef]
159. Milluzzo, R.P.; Franchina, G.A.; Capodanno, D.; Angiolillo, D.J. Selatogrel, a novel P2Y 12 inhibitor: A review of the pharmacology and clinical development. *Expert Opin. Investig. Drugs* **2020**, *29*, 537–546. [CrossRef]
160. Crescence, L.; Darbousset, R.; Caroff, E.; Hubler, F.; Riederer, M.A.; Panicot-Dubois, L.; Dubois, C. Selatogrel, a reversible P2Y12 receptor antagonist, has reduced off-target interference with haemostatic factors in a mouse thrombosis model. *Thromb. Res.* **2021**, *200*, 133–140. [CrossRef]
161. Hardy, A.R.; Jones, M.L.; Mundell, S.J.; Poole, A.W. Reciprocal cross-talk between P2Y1 and P2Y12 receptors at the level of calcium signaling in human platelets. *Blood* **2004**, *104*, 1745–1752. [CrossRef]
162. Tscharre, M.; Michelson, A.D.; Gremmel, T. Novel Antiplatelet Agents in Cardiovascular Disease. *J. Cardiovasc. Pharmacol. Ther.* **2020**, *25*, 191–200. [CrossRef]
163. Kereiakes, D.J.; Henry, T.D.; DeMaria, A.N.; Bentur, O.; Carlson, M.; Seng Yue, C.; Martin, L.H.; Midkiff, J.; Mueller, M.; Meek, T.; et al. First Human Use of RUC-4: A Nonactivating Second-Generation Small-Molecule Platelet Glycoprotein IIb/IIIa (Integrin αIIbβ3) Inhibitor Designed for Subcutaneous Point-of-Care Treatment of ST-Segment-Elevation Myocardial Infarction. *J. Am. Heart Assoc.* **2020**, *9*, e016552. [CrossRef]
164. Bor, W.L.; Zheng, K.L.; Tavenier, A.H.; Gibson, C.M.; Granger, C.B.; Bentur, O.; Lobatto, R.; Postma, S.; Coller, B.S.; van't Hof, A.W.J.; et al. Pharmacokinetics, pharmacodynamics, and tolerability of subcutaneous administration of a novel glycoprotein IIb/IIIa inhibitor, RUC-4, in patients with ST-segment elevation myocardial infarction. *EuroIntervention* **2021**, *17*, 401–410. [CrossRef]
165. Kim, K.; Hahm, E.; Li, J.; Holbrook, L.-M.; Sasikumar, P.; Stanley, R.G.; Ushio-Fukai, M.; Gibbins, J.; Cho, J. Platelet protein disulfide isomerase is required for thrombus formation but not for hemostasis in mice. *Blood* **2013**, *122*, 1052–1061. [CrossRef] [PubMed]
166. Flaumenhaft, R.; Furie, B.; Zwicker, J.I. Therapeutic implications of protein disulfide isomerase inhibition in thrombotic disease. *Arterioscler. Thromb. Vasc. Biol.* **2015**, *35*, 16–23. [CrossRef] [PubMed]
167. Alenazy, F.O.; Thomas, M.R. Novel antiplatelet targets in the treatment of acute coronary syndromes. *Platelets* **2021**, *32*, 15–28. [CrossRef]
168. Zwicker, J.I.; Schlechter, B.L.; Stopa, J.D.; Liebman, H.A.; Aggarwal, A.; Puligandla, M.; Caughey, T.; Bauer, K.A.; Kuemmerle, N.; Wong, E.; et al. Targeting protein disulfide isomerase with the flavonoid isoquercetin to improve hypercoagulability in advanced cancer. *JCI Insight* **2019**, *4*, e125851. [CrossRef] [PubMed]
169. Stopa, J.D.; Neuberg, D.; Puligandla, M.; Furie, B.; Flaumenhaft, R.; Zwicker, J.I. Protein disulfide isomerase inhibition blocks thrombin generation in humans by interfering with platelet factor V activation. *JCI Insight* **2017**, *2*, e89373. [CrossRef]

170. Stainer, A.R.; Sasikumar, P.; Bye, A.P.; Unsworth, A.J.; Holbrook, L.M.; Tindall, M.; Lovegrove, J.A.; Gibbins, J.M. The Metabolites of the Dietary Flavonoid Quercetin Possess Potent Antithrombotic Activity, and Interact with Aspirin to Enhance Antiplatelet Effects. *TH Open* **2019**, *3*, e244–e258. [CrossRef]
171. Ibanez, B.; Vilahur, G.; Badimon, J.J. Plaque progression and regression in atherothrombosis. *J. Thromb. Haemost.* **2007**, *5*, 292–299. [CrossRef]
172. Badimon, L.; Bugiardini, R.; Cenko, E.; Cubedo, J.; Dorobantu, M.; Duncker, D.J.; Estruch, R.; Milicic, D.; Tousoulis, D.; Vasiljevic, Z.; et al. Position paper of the European Society of Cardiology-working group of coronary pathophysiology and microcirculation: Obesity and heart disease. *Eur. Heart J.* **2017**, *38*, 1951–1958. [CrossRef]
173. Choi, B.; Vilahur, G.; Yadegar, D.; Viles-Gonzalez, J.; Badimon, J. The role of high-density lipoprotein cholesterol in the prevention and possible treatment of cardiovascular diseases. *Curr. Mol. Med.* **2006**, *6*, 571–587. [CrossRef]

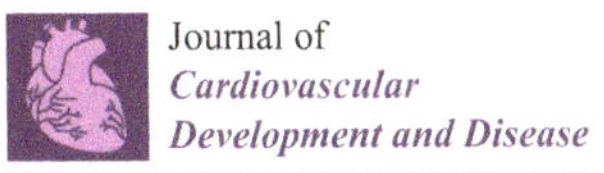
Journal of
Cardiovascular Development and Disease

Review

Factor XI/XIa Inhibition: The Arsenal in Development for a New Therapeutic Target in Cardio- and Cerebrovascular Disease

Juan J. Badimon [1], Gines Escolar [2] and M. Urooj Zafar [1,*]

[1] Cardiovascular Institute, Icahn School of Medicine at Mount Sinai, New York, NY 10029, USA
[2] Department of Hematopathology, Hospital Clinic, 08036 Barcelona, Spain
* Correspondence: urooj.zafar@mssm.edu; Tel.: +1-(212)-241-8484

Abstract: Despite major advancements in the development of safer and more effective anticoagulant agents, bleeding complications remain a significant concern in the treatment of thromboembolic diseases. Improvements in our understanding of the coagulation pathways highlights the notion that the contact pathway—specifically factor XI (FXI)—has a greater role in the etiopathogenesis of thrombosis than in physiological hemostasis. As a result, a number of drugs targeting FXI are currently in different stages of testing and development. This article aims to review the different strategies directed towards FXI-inhibition with a brief summation of the agents in clinical development, and to comment on the therapeutic areas that could be explored for potential indications. Therapeutics targeting FXI/FXIa inhibition have the potential to usher in a new era of anticoagulation therapy.

Keywords: factor XI; factor XI inhibitor; thrombosis; new drugs

Citation: Badimon, J.J.; Escolar, G.; Zafar, M.U. Factor XI/XIa Inhibition: The Arsenal in Development for a New Therapeutic Target in Cardio- and Cerebrovascular Disease. *J. Cardiovasc. Dev. Dis.* **2022**, *9*, 437. https://doi.org/10.3390/jcdd9120437

Academic Editor: Aleš Blinc

Received: 14 October 2022
Accepted: 3 December 2022
Published: 6 December 2022

Publisher's Note: MDPI stays neutral with regard to jurisdictional claims in published maps and institutional affiliations.

1. Introduction

Thromboembolism and its associated complications remain a huge healthcare burden worldwide. The central role of thrombosis is observable in a variety of cardiovascular disorders, most notably in coronary artery disease (CAD), atrial fibrillation and stroke, peripheral arterial disease (PAD), and venous thromboembolism (VTE). Ischemic heart disease and stroke collectively are responsible for nearly 25% of all deaths worldwide [1], whereas estimates for the incidence rate for VTE, comprising deep vein thrombosis (DVT) and pulmonary embolism (PE), range from 115 to 269 per 100,000 people worldwide [2]. The impact of thrombosis extends beyond the cardiovascular arena and is increasingly being encountered in pathologies as diverse as cancer, immunological diseases, and even psychiatric disorders. Even among patients with human immunodeficiency virus (HIV) infection, who are living longer thanks to improvements in antiretroviral treatment, there is evidence of increased thrombosis [3], which contributes to their increasing morbidity and mortality from cardiovascular causes (6–15% of total mortality) [4,5].

The high human and financial cost of thromboembolic events underscore the need for newer and better therapeutic options for the management of thrombotic disorders. The challenge is in developing an agent that has potent antithrombotic effects but minimal bleeding risk, as it requires a very fine balancing act in modulating the hemostatic processes. Current anticoagulant options for clinical treatment of thrombotic disorders include antithrombin activators (unfractionated heparin), low molecular weight heparins (LMWHs and fondaparinux), vitamin K antagonists (VKA—warfarin), direct inhibitors of activated factor X (rivaroxaban, apixaban, edoxaban, betrixaban), and direct inhibitors of thrombin (hirudins, argatroban, and dabigatran). Although heparins and warfarin are low-cost options with a high degree of efficacy, both are associated with drawbacks that limit their clinical use. Heparin-induced thrombocytopenia, although infrequent, can be potentially lethal, and the development of osteoporosis and risk of contamination are additional factors to consider when using unfractionated heparin [6]. Some of these shortcomings have been reduced by LMWH and fondaparinux [6]. Warfarin presents the limitation of a narrow therapeutic window and major food- and drug-interactions [7]. The significant intra- and inter-patient variability of response of VKA makes frequent blood testing for dose-adjustment a cumbersome necessity. The last few years have provided a much-improved treatment option in direct oral anticoagulants (DOACs) that are convenient in administration while being potent and equally effective to VKA, often with a lower risk of bleeding [8–11]. Even so, the annual rate of major bleeding in patients on DOAC treatment remains significant [12], approximately 5% in elderly patients with atrial fibrillation (AF) [13]. This is partly why an unacceptably high proportion of AF patients—nearly one-third—do not receive the prophylactic anticoagulation they require. Even among those that do receive anticoagulation therapy, nearly half do not receive the proper doses [14]. The need for newer, safer anticoagulants is therefore high, and novel targets for therapeutic intervention are constantly under investigation.

2. Distinguishing Physiological Hemostasis from Pathological Thrombosis

Hemostasis is the normal, physiological process by which the clotting cascade seals up vascular damage to limit blood loss following injury. Thrombosis, on the other hand, encompasses various pathological conditions where the normally physiological clotting processes end up generating blood clot(s) inside the vascular lumen that are disruptive to the normal flow of blood. Thrombin generation and fibrin formation are the culminating steps in both hemostasis and thrombosis, but with important differences in the pathways involved.

Hemostasis is commonly triggered when tissue factor (TF) within the adventitial layer of blood vessels gets exposed to blood. Injury to vasculature that can lead to bleeding activates a series of soluble plasma proteins that act together in a cascade of enzyme activation events and culminate in the formation of platelet-fibrin clot(s). Because of the relatively high concentration of TF in such scenarios, the generation of thrombin is rapid and intense, quickly forming a hemostatic plug that seals the inciting TF away from blood. This disrupts the amplification of the coagulation processes through feedback mechanisms to the point of becoming pathological.

The concentration of TF in thrombosis is lower relative to hemostasis, but its duration of contact with blood components often lasts longer. Whether triggered by TF from disruption of an atherosclerotic plaque or activated monocytes/macrophages recruited to the site of injury or inflammation, or by implanted medical devices or neutrophil extracellular traps (NETs), these scenarios depend on the feedback mechanisms of the coagulation cascade for the growth and stabilization of the thrombus. This clot or thrombus can impede the flow of blood to the distal tissues and organs, leading to ischemia and necrosis, manifesting as clinical events including acute coronary syndrome, stroke, or deep vein thrombosis.

3. A Brief Review of the Classic Coagulation Cascade

The two major pathways for triggering blood clotting cascade are well known; (1) the tissue factor pathway and (2) the contact pathway. Both pathways trigger a series of cascading events that generate a blood clot (Figure 1) with the purpose to separate and seal the triggering agent from blood, thereby preventing its further contact with plasma components and arresting the thrombotic process.

3.1. Tissue Factor Pathway

This pathway is also known as the 'Extrinsic' pathway, as it is triggered by plasma components coming into contact with an agent that is extrinsic to blood (i.e., TF). The contact may happen when TF, normally embedded in the vascular wall, is exposed to blood due to rupture of a plaque, or when TF is expressed on the surface of cells active in inflammatory and immunological processes (e.g., monocytes and macrophages).

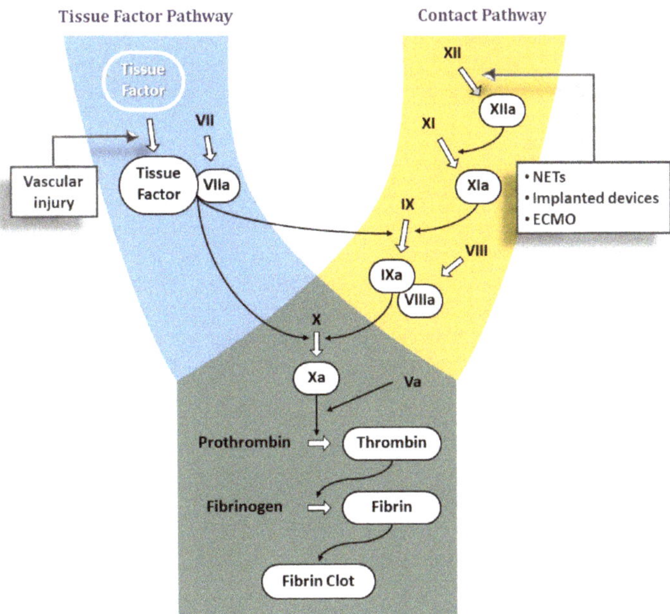

Figure 1. The classic model of the coagulation cascade with the Tissue Factor/extrinsic, the Contact/intrinsic and the common pathways. Triggering factors for both pathways are shown in square boxes. NETs: neutrophil extracellular traps, ECMO: extracorporeal membrane oxygenation.

Tissue Factor is an integral cell-membrane protein that forms a complex with the coagulation factor VIIa (FVIIa), normally present in plasma in the inactive zymogen form FVII. The TF:FVIIa complex is a potent activator of coagulation and converts factors IX (FIX) and X (FX) to the active forms FIXa and FXa, respectively (Figure 1). Each of these active enzymes assembles with its protein cofactor (FVIIIa and FVa, respectively) on suitable membrane surfaces to further propagate the coagulation cascade. The end result is a large burst of thrombin, the last serine protease in the clotting cascade. Thrombin not only converts fibrinogen into fibrin via limited proteolysis—which in turn assembles into a fibrin clot—but is also one of the most potent activators of platelets. The activation and aggregation of platelets contributes to the formation of a hemostatic plug. Additionally, thrombin also activates FV, FVIII, and FXI, the latter two of which are part of the contact pathway. Thus, the initial thrombin generated by the TF pathway can lead to activation of the contact pathway.

3.2. Contact Pathway

Also known as the 'Intrinsic' pathway, this pathway is triggered when blood comes into contact with anionic surfaces, such as extracellular DNA, RNA from activated or dying cells including neutrophil extracellular traps (NETs) released by activated neutrophils [15], or polyphosphates from the dense granules of activated platelets or microorganisms [16], or those on artificial surfaces [17]. This leads to a change in the conformation of plasma factor XII (FXII) into the active factor XII (FXIIa) [18,19]. FXIIa activates Prekallikrein to Kallikrein, which in turn reciprocally activates FXII to FXIIa in a positive feedback loop [20]. Downstream, FXIIa activates FXI to FXIa, which in turn leads to proteolysis of factor IX (FIX) to the active form (FIXa). The complex of FIXa and FVIIIa then activates FX to FXa at the point where the TF and contact pathways converge to form the final common pathway (Figure 1). The end result of all these interactions again is thrombin generation and formation of a blood clot.

The hemostatic process is kept in check by various inhibitory mechanisms that shut down the coagulation pathways, thereby localizing the hemostatic plug. These inhibitory mechanisms include proteins such as the tissue factor pathway inhibitor (TFPI) that inhibits FXa [21], activated protein C (APC), which degrades FVa and FVIIIa [22], and antithrombin (AT), which, in addition to factors IIa, Xa, and IXa [23], can also inhibit FVIIa and FXIa [24,25].

Significance of Contact Pathway in Thrombosis

Our understanding of the coagulation system in thrombotic pathophysiology has improved significantly in recent years, with the classical cascade being superseded by the cell-based model of coagulation (Figure 2). The TF pathway is understood to play a larger role in the 'initiation' and 'propagation' phases of coagulation, functioning more in normal hemostasis than in thrombosis. The contact pathway is more important in 'amplifying' the coagulation response, and despite its important role in clot formation in vitro, may contribute minimally to hemostasis in vivo, as supported by the lack of bleeding tendencies in patients deficient in FXII [26]. The contact pathway does appear, however, to have an important role in thrombotic disorders. Increased activity of plasma FXII, FXI, or kallikrein has been associated with atherosclerosis [27] and myocardial infarction [28,29], whereas severe FXI deficiency has been associated with reduced risk of stroke and deep vein thrombosis [30,31]. Deficiency in FXII in animal models has been reported to be protective against arterial thrombosis [32] and ischemic brain injury [33].

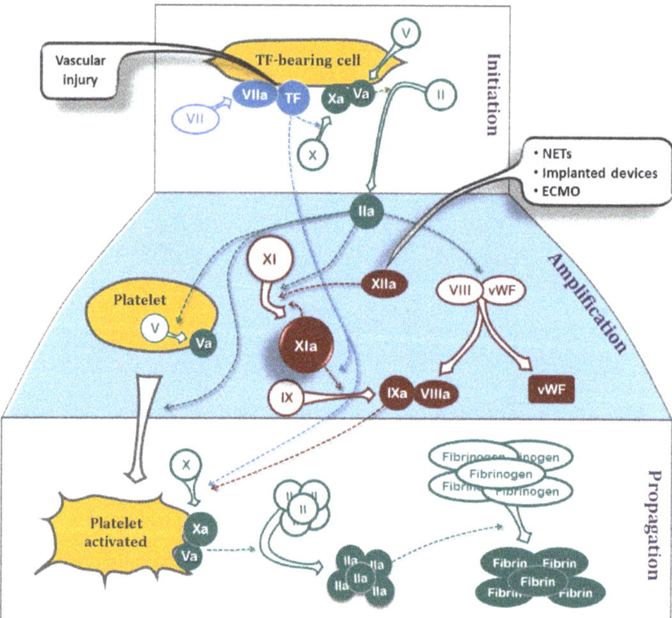

Figure 2. Cell-based coagulation model illustrating the Initiation, Amplification, and Propagation of the coagulation process. In the initiation phase, a small amount of thrombin and FIXa is generated on the surface of the tissue factor (TF)-bearing cell that then diffuse away towards platelets. In the amplification phase, this thrombin activates platelets (releasing factor Va from α granules), acts on vWF-VIII to release vWF and activated factor VIIIa, and generates activated factor XIa. The role of factor XI/XIa is primarily in the amplification stage where it activates factor IX and, in a feedback loop, promotes further activation of zymogen factor XI to active factor XIa. The propagation phase involves assembly of the various enzymes generated earlier to advance the process towards fibrin generation and clot formation. Components of the classic TF-, contact- and common-pathway are shown in blue, dark red, and green, respectively, along with triggers (black boxes) for each pathway.

Selective modulation of the contact pathway theoretically should lower the risk of thrombosis without increasing bleeding. Development of drugs that act by inhibiting components of the contact pathway is currently in high gear, with factor XI (FXI), and to a lesser degree factor XII (FXII), being the most prominent targets [34–36]. Evidence from epidemiological studies supporting their role in thrombosis is stronger for FXI than it is for FXII [37].

4. Factor XI as a Therapeutic Target

Factor XI is a blood coagulation zymogen produced by the liver that is part of the early phase of the contact pathway [38]. It is converted to the active serine protease FXIa by thrombin, FXIIa, and by FXIa itself and in turn activates FIX to further advance the coagulation process [38]. FXI plays an important part in blood coagulation because its feedback activation amplifies in vivo thrombin generation and fibrin formation [39]. The additional thrombin formed via the FXI feedback loop also promotes the activation of Thrombin Activatable Fibrinolysis Inhibitor (TAFI), which increases the clot's resistance to fibrinolysis, thereby helping to stabilize the formed clot.

The greater role of FXIa in thrombosis compared to hemostasis is evident from several epidemiological and genetic studies. Higher levels of circulating FXI levels are associated with increased risk for venous and arterial thrombosis, including stroke [40,41]. Deficiency of FXI (Hemophilia C, Plasma Thromboplastin Antecedent Deficiency, Rosenthal Syndrome) is rare and characterized by little to no bleeding tendency. Bleeding risk with factor XI deficiency selectively increases in tissues with high fibrinolytic activity (e.g., following dental surgery, tonsillectomy, and prostate surgery) [42]. Most frequent presentations involve nosebleeds or bleeding after tooth extractions. In fact, patients suffering from congenital FXI deficiency appear to have some degree of protection from thrombotic events, with lower rates of ischemic stroke and venous thromboembolism [30,43]. Moreover, hemorrhaging does not correlate with the levels of FXI in blood, i.e., bleeding is not restricted to patients with severe deficiency, and individuals with similar levels of FXI can experience different degrees of bleeding.

5. Pharmacologic Strategies for Factor XI Inhibition

Given the larger role FXI is thought to play in thrombosis than in hemostasis, novel approaches to inhibit its generation and activity are being explored as new therapeutic strategies (Figure 3). These include: (a) Antisense Oligonucleotides (ASOs) that act on the liver to knockdown hepatic synthesis of FXI, (b) small molecules that target the FXI active site or the heparin allosteric site on FXIa, (c) monoclonal antibodies that act by blocking the activation or inhibiting the activity, and (d) Aptamers.

In addition to their varying mechanisms of action, these strategies also differ in their routes of administration (oral vs. parenteral), the onset of action, and the duration of effect. Parenteral administration is a requirement for ASOs, aptamers and monoclonal antibodies, whereas small molecule agents offer the option of either parenteral or oral administration. The varied onset and duration of action may present a broad set of treatment options depending on the pathology at hand; acute thrombotic events requiring quick-acting agents whereas longer-acting options, such as antibodies, would be more suitable for chronic prophylactic and preventative measures. Similarly, for conditions presenting a high risk of bleeding complications such as trauma or surgery, shorter-acting agents would be preferable.

Inhibition of FXIa as a therapeutic option may also allow the possibility to easily reverse the effects of treatment, as has been tested successfully in animal models. In a rabbit AV-shunt model of thrombosis, the antithrombotic effects of a small molecule FXIa inhibitor (71.3 $\pm$ 5.2% lowering of thrombus weight vs. vehicle) were completely abolished by non-specific reversal agents (222% and 64% increase in thrombus weight vs. vehicle with FEIBA and NovoSeven, respectively) [44]. In another rabbit study, the addition of a specific reversal agent fully normalized the 210% prolongation in APTT produced by an anti-FXIa antibody [45]. This availability of reversing strategies for FXIa inhibition would be a significant advantage for this class of drugs, similar to the one available for some DOACs. Inability to reverse treatment effects can magnify the concerns about bleeding risks associated with any antithrombotic agent, thereby hampering its proper clinical utilization. As an example, although it is still possible to reverse the effects of antiplatelet drugs [46–48], the lack of a convenient and simple reversal strategy mandates that bleeding risk be always at the forefront of any discussion involving antiplatelet drugs.

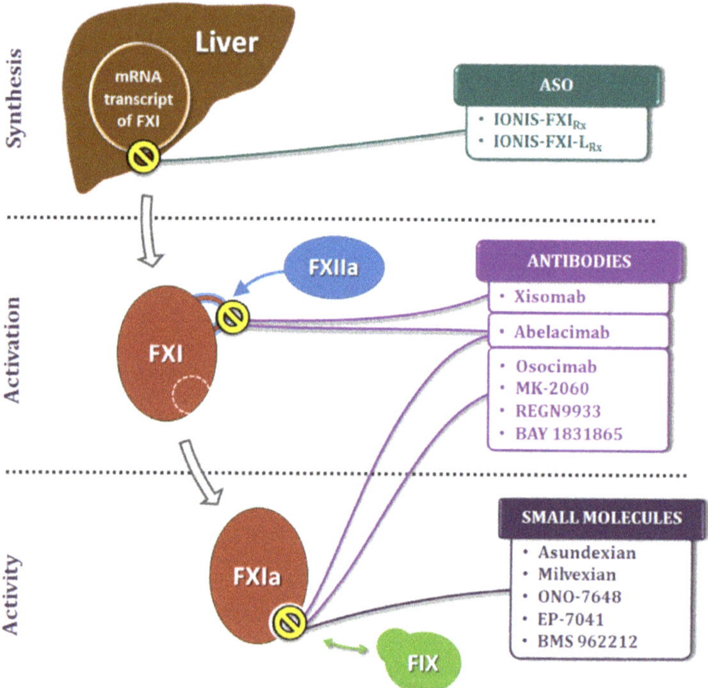

Figure 3. Sites of action (represented by the yellow circles) of the factor XI/XIa inhibitory drugs currently at different stages of clinical development. The anti-FXI ASOs (Antisense Oligonucleotides) block mRNA transcription of FXI in the hepatocytes, thus inhibiting its synthesis. Some monoclonal antibodies attach to the catalytic domain of FXI and block its FXIIa-mediated conversion to the active FXIa form, thus locking it in the inactive zymogen state. Most of the currently in development anti-FXIa antibodies act similar to the small molecule FXIa inhibitors and bind to the active site(s) on FXIa, thereby blocking its activity.

A summary of FXIa inhibitors in more advanced stages of clinical development is presented in Table 1.

5.1. Antisense Oligonucleotides (ASOs)

Antisense Oligonucleotides are short, single-stranded nucleic acid sequences that pair with specific regions of mRNA and regulate its gene expression [49,50], thereby downgrading the hepatic synthesis of FXI. Their benefits include high specificity, predictable pharmacokinetics (PK), and long half-life. Furthermore, ASOs lack the drug–drug interactions commonly seen with conventional therapeutic agents. However, as nucleic acids in general are susceptible to degradation by nucleases, ASOs require some sort of chemical modification to confer nuclease resistance and enhance intracellular stability.

1. IONIS-FXIRx.

This agent (formerly known as ISIS 416858) is the ASO furthest in clinical development. Administered subcutaneously, IONIS-FXIRx has been shown to produce a concentration-dependent reduction in FXI antigen and activity levels [51]. In a phase II study of 315 patients undergoing total knee replacement, IONIS-FXIRx reduced the risk of postoperative VTE more than enoxaparin, without increasing the risk of bleeding. Rates of VTE were 27% and 4% among patients treated with 200 and 300 mg doses of IONIS-FXIRx, respectively, versus 30% in patients who received enoxaparin 40 mg once-daily [52]. Additionally, rates of major or clinically relevant non-major bleeding were also lower with IONIS-FXIRx (3% with both doses versus 8% with enoxaparin) [52].

IONIS-FXIRx has also been tested in 49 patients with end-stage renal disease (ESRD) requiring hemodialysis, where it produced a dose-dependent reduction in FXI antigen level and activity, and in aPTT without changing INR [53]. No drug-related serious adverse events or accumulation of the drug were observed after 12 weeks of treatment. A larger phase II study of the safety, PK, and pharmacodynamics (PD) was completed in 2019 in patients with ESRD, but no results are available to-date (NCT03358030).

2. IONIS-FXI-LRx.

A second-generation, ligand-conjugated antisense (LICA) agent named IONIS-FXI-LRx is also under clinical development. Its increased potency allows for once-monthly administration at lower doses, which helps to reduce the potential for injection-site reactions seen with IONIS-FXIRx. A phase II study of the safety, PK, and PD of IONIS-FXI-LRx in patients with ESRD was recently completed, but results are not yet available (NCT03582462).

Table 1. Inhibitors of FXI/FXIa in various stages of clinical development.

Compounds	Route	Stage	Indication	N	Status
ASO [1]					
IONIS-FXI_{Rx}	S.C. [2]	Phase II	Total knee arthroplasty	315	Published [52]
		Phase II	ESRD [4]	49	Published [53]
		Phase II	ESRD [4]	213	Completed (NCT03358030)
IONIS-FXI-L_{Rx}	S.C. [2]	Phase II	ESRD [4]	307	Completed (NCT04534114)
Small molecule					
Asundexian	Oral	Phase II	Myocardial infarction	1601	Published [54]
		Phase II	Ischemic stroke	1808	Published [55]
		Phase II	AF [5]	753	Published [56]
		Phase III	*AF [5]; Stroke and TIA [6]*	*30,000*	*Announced [57]*
Milvexian	Oral	Phase II	Total knee arthroplasty	1242	Published [58]
		Phase II	Stroke and brain MRI [7]		NCT03766581 [59]
ONO-7684	Oral	Phase I	PK [8] & PD [9] in healthy	48 + 24	Published [60]
EP-7041	I.V. [3]	Phase II	Thrombocytopenia, COVID-19	90	Not recruiting (NCT05040776)
BMS-962212	I.V. [3]	Phase I	PK [8] & PD [9] in healthy	691	Completed (NCT03197779)
Antibodies					
Abelacimab	S.C. [2]	Phase II	Total knee arthroplasty	412	Published [61]
		Phase II	AF [5]	1200	Not recruiting (NCT04755283)
		Phase III	Cancer-associated VTE [10]	1655	Recruiting (NCT05171049)
		Phase III	GI/GU-associated VTE [10]	1020	Recruiting (NCT05171075)
Osocimab	I.V. [3]	Phase II	Total knee arthroplasty	813	Published [62]
Xisomab 3G3	I.V. [3]	Phase II	ESRD [4]	27	Published [63]
		Phase II	Thrombosis in chemotherapy	50	Recruiting (NCT04465760)
MK-2060	I.V. [3]	Phase II	ESRD [4]	489	Recruiting (NCT05027074)
REGN9933	I.V. [3]	Phase I	PK [8] & PD [9] in healthy	72	Recruiting (NCT05102136)

[1] Antisense Oligonucleotides; [2] subcutaneous; [3] intravenous; [4] end stage renal disease; [5] atrial fibrillation; [6] transient ischemic attack; [7] magnetic resonance imaging; [8] pharmacokinetic; [9] pharmacodynamic; [10] venous thrombo embolism.

5.2. Small Molecules

5.2.1. Small Molecules Targeting the Active Site on FXIa

A number of small molecules that inhibit FXIa activity by binding to the active site are in existence. These agents act by attaching to either S1, S2, or both pockets of FXIa [64].

1. Asundexian (BAY 2433334).

This small molecule is in the most advance stages of development among FXIa inhibitors, with results from three phase 2 studies published recently. The first to be published was a dose-finding trial that compared asundexian with placebo for the prevention of major adverse cardiac events in patients with recent acute MI on dual-antiplatelet therapy [54]. Patients (n = 1601) were randomized within 5 days of an MI to oral asundexian 10, 20, or 50 mg or placebo, given once-daily for 6–12 months in addition to aspirin plus a $P2Y_{12}$ inhibitor. Over a year of follow-up, asundexian produced dose-dependent inhibition of FXIa activity without significant increase in bleeding (Bleeding Academic Research Consortium (BARC) bleeding type 2, 3, or 5: 7.6%, 8.1%, and 10.5%, respectively, with asundexian doses, vs. 9.0% with placebo) and had low rates of ischemic events (composite of cardiovascular death, MI, stroke, or stent thrombosis: 6.8%, 6.0%, and 5.5%, respectively, vs. 5.5% with placebo).

The second phase II trial, called PACIFIC-Stroke, compared asundexian 10, 20, or 50 mg vs. placebo for the secondary prevention of recurrent stroke. The study included 1808 patients with acute (<48 h) non-cardioembolic ischemic stroke, treated with single or dual antiplatelet therapy [55]. In this trial, asundexian did not reduce the primary efficacy endpoint—a composite of recurrent symptomatic ischemic stroke and MRI-detected covert brain infarcts at 26 weeks (19%, 22%, and 20% with asundexian 10, 20, and 50 mg, respectively, vs. 19% with placebo). Rates of major or clinically relevant non-major bleeding were 4%, 3%, and 4%, respectively, with asundexian vs. 2% with placebo.

The third phase II trial called the PACIFIC-AF trial, compared treatment with asundexian with apixaban in patients with nonvalvular atrial fibrillation (n = 753). Patients were randomly assigned to asundexian 20 or 50 mg once-daily or the standard 5 mg twice-daily dose of apixaban [56]. Both doses of asundexian had lower rates of major or clinically relevant non-major bleeding at 12 weeks vs. apixaban, with ratios of incidence proportions of the primary composite endpoint being 0.50 and 0.16 for asundexian 20 and 50 mg versus apixaban, respectively. Rates of thrombotic and cardiovascular events were reported to be comparable with the two treatments, but the thrombotic endpoints were exploratory due to sample size considerations.

The reporting of the asundexian phase II results has been followed by a recent announcement of a phase III development program. The OCEANIC program is expected to enroll up to 30,000 patients in two large multinational studies, OCEANIC-AF and OCEANIC-Stroke, involving atrial fibrillation and non-cardioembolic ischemic stroke or high-risk transient ischemic attack, respectively [57].

2. Milvexian (JNJ-70033093/BMS-986177).

This orally active agent is also in the advance stage of clinical development among FXIa inhibitors. In a phase II dose-finding trial in patients undergoing elective knee arthroplasty (AXIOMATIC-TKR; n = 1242), postoperative FXIa inhibition with milvexian was effective for the prevention of venous thromboembolism, with a dose-related response in both once-daily and twice-daily administrations [58]. With twice-daily administration of milvexian 25, 50, 100, and 200 mg, the dose-response relationship was statistically significant and the incidence of VTE significantly lower than the prespecified benchmark of 30% (21%, 11%, 9%, and 8%, respectively). In the same trial, rate of VTE with subcutaneous enoxaparin was 21%. Milvexian also showed promising results on the safety side, with low rates of any bleeding, major bleeding, or clinically relevant non-major bleeding relative to enoxaparin.

The findings from the second phase II trial (AXIOMATIC-SSP; n = 2366) for the prevention of new ischemic stroke in patients following acute ischemic stroke or transient ischemic attack were recently presented at the 2022 European Society of Cardiology Congress in

Barcelona [59]. All patients received aspirin plus clopidogrel for 21 days, followed by aspirin alone thereafter. While the rate of the primary efficacy endpoint—a composite of ischemic overt stroke or covert stroke detected by brain MRI at 90 days—was numerically lower at the 50 mg and 100 mg twice-daily doses, there was no apparent dose-response. For clinical ischemic strokes (i.e., excluding covert brain infarction), milvexian doses from 25 to 100 mg twice-daily showed an ~30% relative risk reduction versus placebo. The rate of major bleeding was moderately increased with milvexian 50 mg twice-daily and above, but with no apparent dose-relation.

Based on the overall findings from the phase II studies, milvexian is moving towards further studies in phase III trials. Interestingly, investigation of antidotes to milvexian has also moved on to clinical stages (NCT04543383) after animal testing [44].

3. Other Small-Molecules in Early Development

A number of other oral and parenteral inhibitors of FXIa are in earlier phases of development. These include ONO-7684, an orally active agent that was well-tolerated in a phase I study with healthy volunteers. This study reported low overall incidence of adverse events with no evidence to suggest bleeding risk [60].

The parenteral small molecules under development include EP-7041. A placebo-controlled study to evaluate its safety, PK, and PD was conducted in healthy volunteers [65]. The drug was well-tolerated except for some cases of mild headache (23%) and infusion site bruising (7%). EP-7041 exhibited rapid onset–offset and dose-related increases of aPTT without affecting PT. Despite these positive results, there was no further development with this agent until 2021, when an IND application for its use as an investigational treatment for COVID-19 patients in ICU was accepted by the FDA (NCT05040776).

BMS-962212 is another parenterally administered, FXIa-inhibiting small molecule investigated in healthy participants. In testing of multiple doses, the drug was well-tolerated, with no bleeding events and mild adverse events in 17.6% participants [66]. Dose-dependent changes in aPTT and FXI were observed with maximal effects by approximately 2 h, and no changes in PT or INR.

5.2.2. Small Molecules Targeting Heparin Allosteric Site on FXIa

This group of FXIa-directed agents exert their inhibitory effects by attaching to the heparin-binding site on the catalytic domain of FXIa. Given the structural similarities between the active sites of various serine proteases, it is believed that allosteric inhibition would have the advantage of being more specific. Some of the sulfated glycosaminoglycan (SPGG) mimetic compounds under development in this group not only exhibit a highly selective inhibition of FXIa than any other target in the coagulation cascade, but also display a reversal of their anticoagulant effects with FXI and serum albumin [67]. Protamine could also reverse the anticoagulant effects of SPGG, providing potential ways for the development of antidotes.

5.3. Monoclonal Antibodies

A number of monoclonal antibodies that block either FXI activation or FXIa protease activity are currently under development for the treatment of thrombotic disorders. The antithrombotic effects and bleeding risk of these antibodies are at various stages of testing.

1. Abelacimab (MAA868).

A monoclonal antibody that binds the procoagulant enzymatic site of both FXI (zymogen) and the active form FXIa [68]. By binding to the catalytic domain, abelacimab locks both the FXI and activated FXIa in an inactive, zymogen-like conformation, thereby taking them out of the coagulation system. In a phase I testing, the pharmacodynamic effects of a single subcutaneous administration lasted up to 4 weeks or longer, suggesting the possibility of a once-monthly dosing [68].

Abelacimab has been compared with enoxaparin for the prevention of VTE in patients undergoing elective knee arthroplasty (n = 412) in a phase 2 study with promising results.

Patients were randomized to single intravenous administration of abelacimab 30, 75, or 150 mg, or to subcutaneous enoxaparin 40 mg [61]. Rates of VTE were 13%, 5%, and 4% with abelacimab doses, respectively, vs. 22% with enoxaparin, assessed by venography or objective confirmation of symptomatic events 8–12 days after the operation. Bleeding risk was low, with occurrence in 2% of cases with the lower two doses of abelacimab and none of the patients in the highest dose abelacimab or the enoxaparin groups.

A larger phase II study to compare the bleeding risk of abelacimab vs. rivaroxaban in patients with AF at moderate-to-high risk of stroke (AZALEA-TIMI 71) is currently listed as 'Active, not recruiting' and plans to enroll 1200 patients (NCT04755283). Interestingly, the more advanced studies with this agent are in the prevention of cancer-associated VTE, with two active phase III trials comparing the efficacy of abelacimab vs. dalteparin (MAGNOLIA; NCT05171075) and vs. apixaban (ASTER; NCT05171049).

2. Osocimab (BAY 1213790).

This is a fully human IgG1 antibody. Its crystal structure analysis has shown a novel allosteric mechanism of action, with the antibody binding to a region adjacent to the FXIa active site, leading to structural rearrangements and blocking of activity.

Osocimab has been compared with enoxaparin and apixaban for thromboprophylaxis in patients undergoing elective knee arthroplasty in the phase II FOXTROT trial (n = 813). A single intravenous administration of osocimab (given postoperatively at 0.3, 0.6, 1.2, or 1.8 mg/kg, or preoperatively at 0.3 or 1.8 mg/kg) was tested vs. once-daily enoxaparin (40 mg subcutaneous) and twice-daily apixaban (2.5 mg oral) [62] to prevent the incidence of VTE (assessed between 10 and 13 days postoperatively with bilateral venography or confirmed symptomatic deep vein thrombosis or pulmonary embolism). Postoperatively osocimab administration was noninferior at all, but the lowest dose vs. enoxaparin (VTE rates of 18%, 8%, 13%, and 14% vs. 20%, respectively). Preoperative osocimab dosing at 1.8 mg/kg was in fact superior to enoxaparin in preventing VTE (9% vs. 20% VTE, respectively), but also had the highest rate of major or clinically relevant nonmajor bleeding (4.7%) among all osocimab doses, although it was still lower than enoxaparin (5.9%). Comparisons with apixaban (12% VTE and 2% bleeding) were exploratory, and no statistical hypothesis was defined. Given the combination of efficacy and safety, the 0.6- and 1.2 mg/kg doses of osocimab appear to be most promising for future development. A phase II study in end stage renal disease patients undergoing dialysis (n = 686) was recently completed with these two doses, but findings are not yet available (NCT04523220).

3. Xisomab 3G3 (AB023).

This is a human IgG2b monoclonal antibody that binds to the apple 2 domain of FXI and FXIa and inhibits the activation of FXI by FXIIa. Despite suppressing the FXIIa-mediated activation of FXI, it leaves intact the ability of thrombin to reciprocally activate FXI, as well as the enzymatic active site of the formed FXIa itself. In a small (n = 24) study with ESRD patients on chronic hemodialysis, there were fewer occlusive events requiring hemodialysis circuit exchange and lower levels of thrombin-antithrombin complexes and C-reactive protein after xisomab administration compared with data collected prior to dosing [63]. Another phase II trial to assess the efficacy of xisomab in preventing catheter-associated thrombosis in cancer patients receiving chemotherapy is currently underway (NCT04465760).

4. Other Antibodies in Clinical Testing.

Other agents from different manufacturers are also in early stages of clinical development. These include MK-2060, which has a placebo-controlled phase II study actively recruiting to evaluate the efficacy and safety in 489 patients with ESRD on hemodialysis (NCT05027074), and REGN9933, with a phase I, placebo-controlled PK and PD study recruiting healthy participants (NCT05102136).

5.4. Aptamers

Aptamers are single-stranded oligonucleotides that act as potent antagonists by binding to their target protein. A number of specific aptamers have been developed that serve as strong anticoagulants by disrupting complex interactions on their target proteins [69].

To date, aptamers targeting FXI directly or indirectly are in very early stages, with none reaching clinical development. In laboratory testing, an agent designated Factor ELeven Inhibitory APtamer (FELIAP) was shown to competitively inhibit FXIa-catalyzed FIX activation and complex formation with antithrombin, without affecting FXI activation itself. Plasma clotting and thrombin generation assays were also inhibited by this aptamer [70]. Similarly, two aptamers, designated 11.16 and 12.7, that bind to sites on the FXIa catalytic domain were shown to non-competitively inhibit FXIa activation of FIX in laboratory testing [71]. In human plasma, aPTT clotting time was also significantly prolonged by aptamer 12.7.

One of the advantages of this class of agents is that it allows the possibility of developing specific antidotes that bind to its target aptamer and disrupt its aptamer–protein interaction [72]. Furthermore, a universal antidote can also be developed that blocks the action of any aptamer [73]. Despite some promising early data, the potential of aptamers seems outmatched by that of the more direct inhibitors, including monoclonal antibodies and small-molecule inhibitors. Research with these agents has thus lagged behind and even declined, but their further development as an additional therapeutic tool remains relevant.

6. Fields for Therapeutic Investigation

6.1. Active Areas of Investigations

Inhibitors of FXI/FXIa are being investigated as alternatives to standard anticoagulation therapy with heparins, VKAs, and DOACs. As such, the active areas of investigations include the usual indications for anticoagulants.

6.1.1. Atrial Fibrillation

It is the most common clinically significant arrhythmia [74], with an age-related risk of occurrence, and cardiac thrombus formation and systemic embolization are its most significant clinical complications, raising the risk of stroke by 4–5 fold [75,76]. The DOACs have shown better results than warfarin in preventing stroke in non-valvular AF patients, with lower or equivalent rates of bleeding complications [77]. However, the need for safer agents still persists and is even more pressing in AF patients requiring hemodialysis. There is uncertainty as to whether the benefits of VKA actually outweigh their harm in AF patients requiring hemodialysis, and trials investigating the role of DOACs in this population are mostly in the early stages. Even in the absence of AF, hemodialysis on its own is a major problem, with cardiovascular events accounting for nearly half of the mortality in these patients. The availability of a newer antithrombotic agent with a better safety profile than existing strategies could significantly improve clinical outcomes in AF patients with or without the need for hemodialysis and in those who require dialysis, with or without AF. An FXI-inhibiting strategy could be an improved therapeutic option in these patients and warrants investigation in clinical trials.

6.1.2. Venous Thromboembolism

Anticoagulant therapy is the mainstay for the prevention and treatment of VTE diseases. The development of DOACs has improved the management of VTE compared to where it was with LHMH/VKA [78]. As a result, rates of idiopathic VTE appear to be on the decline, but the incidence of non-idiopathic DVT and PE seem to be steady or increasing [79], highlighting the need for newer treatment options. Even when used at reduced doses, there is a risk of bleeding with DOAC therapy in these patients [80]. Strategies with longer-acting FXI inhibitors such as ASO and monoclonal antibodies could prove to be better treatment alternatives given their greater effect in reducing thrombosis versus impeding hemostasis.

6.2. Potential Areas for Therapeutic Investigations

Inhibitors of FXI/FXIa are currently in the early stages of clinical development, and over time the spectrum of their clinical application will evolve into specific, focused indications. The areas for the investigation of their therapeutic applications potentially include any pathology where thromboembolism plays an important role. Given the wide-ranging times of their onset and duration of action, FXIa inhibitors have the potential to develop into therapeutic strategies for the treatment and prevention of both acute and chronic, venous, and arterial thromboembolic disorders.

6.2.1. Antiphospholipid Syndrome

Antiphospholipid antibody syndrome (APS) often manifests with symptoms of arterial and venous thrombosis, with DVT being the most common venous presentation. Current management of APS-related VTE is the same as any VTE and involves anticoagulation with heparin, followed by warfarin. Among the DOACS, rivaroxaban has been compared against warfarin to treat patients with thrombotic APS (RAPS study) but did not reach the non-inferiority threshold for the study's primary outcome (endogenous thrombin potential—ETP) [81]. A larger trial of rivaroxaban versus warfarin was terminated early due to "unbalance in the composite endpoint between arms" without further information (TRAPS study; NCT02157272). Another trial for secondary prevention of thrombosis with apixaban in APS patients (ASTRO-APS; NCT02295475) is currently 'Active, not recruiting' [82]. Prevention of the thrombotic complications in APS may be a potential therapeutic area to explore using the new anti-FXI agents.

6.2.2. Sickle Cell Disease (SCD)

Sickle cell disease is an autosomal recessive disorder of hemoglobin β-chain, often manifesting as chronic anemia or acute vaso-occlusive crises. Stroke is a major complication of SCD, with a prevalence rate of at least 11% in SCD patients by the age of 20 years [83]. Although the pathophysiology of stroke in SCD is not fully understood, the association is well established [84,85]. A number of variables are thought to play a role, including inflammation and TF derived from endothelial cells and monocytes, that increase the propensity for thrombosis in these patients [86–89]. Periodic red cell transfusion is the only intervention proven to prevent stroke in SCD patients in randomized trials [90]. Although it may be premature to test the benefits of FXI-inhibiting strategies in SCD patients in large-scale clinical trials, pre-clinical studies to explore treatment effect on the elevated thrombotic tendency of this population may be warranted.

6.2.3. Implantable Devices/Blood Contact with Artificial Surfaces

Implantable devices that come into contact with blood, such as stents and mechanical heart valves, left ventricular assist devices, and indwelling central venous lines and ports used in chemotherapy, are frequently associated with thrombosis. Contact of blood with artificial surfaces in extracorporeal membrane oxygenation (ECMO) also causes frequent thrombotic complications. Interestingly, thromboembolism is the second major complication reported with ECMO, surpassed only by bleeding [91]. Systemic anticoagulation is recommended in ECMO, though this may be undesirable in patients at high risk of bleeding [92]. This in turn can lead to the failure of these devices and life-threatening consequences. The success rate of DOACs in preventing thrombotic events in patients with implanted devices has so far been disappointing. Not only are the DOACs non-viable treatment options in patients with devices, but are also in fact contraindicated in patients with mechanical heart valves, where warfarin is still the anticoagulant of choice. Some DOAC trials in patients with devices were terminated due to higher thrombotic and bleeding events in treated patients [93], while others were stopped due to safety reasons (NCT02872649).

Mechanical devices initiate coagulation through the contact pathway by activating FXII, leading to the local generation of TF [94]. Depletion of FXI in in vitro experiments have been shown to abolish this thrombin generation [95]. Dabigatran has been less successful in this application than warfarin in both basic and clinical testing, and given their mechanism of action, FXa inhibitors are unlikely to fare any better. None of the DOACs are approved for preventing thrombotic complications in patients with mechanical valves. FXI-directed strategies theoretically may be the most suitable for device-related treatment scenarios as they may present comparable efficacy to warfarin with a better safety profile. Prevention/treatment of thrombosis related to implantable devices appears to be one area perfectly suited for FXI-inhibiting agents and needs clinical investigation.

6.2.4. Myocardial Infarction

The mainstay of CAD treatment is dual antiplatelet therapy with aspirin and one of the $P2Y_{12}$–receptor inhibitors (clopidogrel, ticagrelor, or prasugrel). Among the DOACs, rivaroxaban is the only one to successfully undergo phase III evaluation in ACS patients in combination with dual antiplatelet therapy. It reduced the risk of death from cardiovascular causes, myocardial infarction, and stroke, but increased the risk of major bleeding and intracranial hemorrhage [96]. FXI-directed strategies could prove to be safer than rivaroxaban in ACS patients. Not only could they block contact activation on stents, but could also prevent FXI-mediated thrombus stabilization and growth.

7. Conclusions

Recent advances in the understanding of the contact pathway, especially of its significant role in thrombus stabilization and growth vs. in the initiation of clot formation, have opened up new targets for therapeutic intervention. FXI is one such promising target. Existing DOACs have improved treatment options compared to the classic heparins and VKA, but the bleeding risks associated with their use are substantial enough to expand the focus onto the development of their antidotes. Early indications are that FXI-directed strategies could offer similar protection against thrombotic events as DOACs, but with the added benefit of lower bleeding risk. Furthermore, the spectrum of modalities for FXI inhibition presents a range of options in both types of administration and duration of effect. With the possibility of once- or twice-monthly injections, some FXI-directed agents could also improve treatment compliance compared to current therapies. Altogether, FXIa inhibitors could be a therapeutic option in a broad spectrum of clinical scenarios that should be investigated in human trials.

8. Future Directions

The broad spectrum of strategies available to modulate FXI/FXIa, including ASOs, small molecules, antibodies, and aptamers, present opportunities to explore therapeutic indications applicable in a wide variety of clinical scenarios. Several of the FXI-directed agents discussed in this review are currently undergoing clinical evaluations in phase II and phase III trials.

In addition to investigating the effectiveness of FXI-directed strategies versus anticoagulants (i.e., heparins, warfarin, and DOACs), their safety and efficacy should also be assessed in combination with anti-platelet agents because a large swath of the population is on chronic aspirin therapy with or without a $P2Y_{12}$-receptor inhibitor.

Author Contributions: Conceptualization, M.U.Z., G.E. and J.J.B.; writing—original draft preparation, M.U.Z.; writing—review and editing, J.J.B., G.E. and M.U.Z. All authors have read and agreed to the published version of the manuscript.

Funding: This research received no external funding.

Institutional Review Board Statement: Not applicable.

Informed Consent Statement: Not applicable.

Conflicts of Interest: The authors declare no conflict of interest.

References

1. Lozano, R.; Naghavi, M.; Foreman, K.; Lim, S.; Shibuya, K.; Aboyans, V.; Abraham, J.; Adair, T.; Aggarwal, R.; Ahn, S.Y.; et al. Global and regional mortality from 235 causes of death for 20 age groups in 1990 and 2010: A systematic analysis for the Global Burden of Disease Study 2010. *Lancet* **2012**, *380*, 2095–2128. [CrossRef] [PubMed]
2. Day, I.S.C.f.W.T. Thrombosis: A major contributor to the global disease burden. *J. Thromb. Haemost.* **2014**, *12*, 1580–1590. [CrossRef]
3. O'Brien, M.P.; Zafar, M.U.; Rodriguez, J.C.; Okoroafor, I.; Heyison, A.; Cavanagh, K.; Rodriguez-Caprio, G.; Weinberg, A.; Escolar, G.; Aberg, J.A.; et al. Targeting thrombogenicity and inflammation in chronic HIV infection. *Sci. Adv.* **2019**, *5*, eaav5463. [CrossRef]
4. Hemkens, L.G.; Bucher, H.C. HIV infection and cardiovascular disease. *Eur. Heart J.* **2014**, *35*, 1373–1381. [CrossRef] [PubMed]
5. Alonso, A.; Barnes, A.E.; Guest, J.L.; Shah, A.; Shao, I.Y.; Marconi, V. HIV Infection and Incidence of Cardiovascular Diseases: An Analysis of a Large Healthcare Database. *J. Am. Heart Assoc.* **2019**, *8*, e012241. [CrossRef] [PubMed]
6. Garcia, D.A.; Baglin, T.P.; Weitz, J.I.; Samama, M.M. Parenteral anticoagulants: Antithrombotic Therapy and Prevention of Thrombosis, 9th ed: American College of Chest Physicians Evidence-Based Clinical Practice Guidelines. *Chest* **2012**, *141*, e24S–e43S. [CrossRef] [PubMed]
7. Ageno, W.; Gallus, A.S.; Wittkowsky, A.; Crowther, M.; Hylek, E.M.; Palareti, G. Oral anticoagulant therapy: Antithrombotic Therapy and Prevention of Thrombosis, 9th ed: American College of Chest Physicians Evidence-Based Clinical Practice Guidelines. *Chest* **2012**, *141*, e44S–e88S. [CrossRef] [PubMed]
8. Zafar, M.U.; Vorchheimer, D.A.; Gaztanaga, J.; Velez, M.; Yadegar, D.; Moreno, P.R.; Kunitada, S.; Pagan, J.; Fuster, V.; Badimon, J.J. Antithrombotic effects of factor Xa inhibition with DU-176b: Phase-I study of an oral, direct factor Xa inhibitor using an ex-vivo flow chamber. *Thromb. Haemost.* **2007**, *98*, 883–888. [CrossRef]
9. Zafar, M.U.; Farkouh, M.E.; Osende, J.; Shimbo, D.; Palencia, S.; Crook, J.; Leadley, R.; Fuster, V.; Chesebro, J.H. Potent arterial antithrombotic effect of direct factor-Xa inhibition with ZK-807834 administered to coronary artery disease patients. *Thromb. Haemost.* **2007**, *97*, 487–492. [CrossRef] [PubMed]
10. Viles-Gonzalez, J.F.; Gaztanaga, J.; Zafar, U.M.; Fuster, V.; Badimon, J.J. Clinical and experimental experience with factor Xa inhibitors. *Am. J. Cardiovasc. Drugs* **2004**, *4*, 379–384. [CrossRef] [PubMed]
11. Sardar, P.; Chatterjee, S.; Lavie, C.J.; Giri, J.S.; Ghosh, J.; Mukherjee, D.; Lip, G.Y. Risk of major bleeding in different indications for new oral anticoagulants: Insights from a meta-analysis of approved dosages from 50 randomized trials. *Int. J. Cardiol.* **2015**, *179*, 279–287. [CrossRef]
12. Hellenbart, E.L.; Faulkenberg, K.D.; Finks, S.W. Evaluation of bleeding in patients receiving direct oral anticoagulants. *Vasc. Health Risk Manag.* **2017**, *13*, 325–342. [CrossRef]
13. Ruff, C.T.; Giugliano, R.P.; Braunwald, E.; Hoffman, E.B.; Deenadayalu, N.; Ezekowitz, M.D.; Camm, A.J.; Weitz, J.I.; Lewis, B.S.; Parkhomenko, A.; et al. Comparison of the efficacy and safety of new oral anticoagulants with warfarin in patients with atrial fibrillation: A meta-analysis of randomised trials. *Lancet* **2014**, *383*, 955–962. [CrossRef]
14. Alamneh, E.A.; Chalmers, L.; Bereznicki, L.R. Suboptimal Use of Oral Anticoagulants in Atrial Fibrillation: Has the Introduction of Direct Oral Anticoagulants Improved Prescribing Practices? *Am. J. Cardiovasc. Drugs* **2016**, *16*, 183–200. [CrossRef]
15. Sorvillo, N.; Cherpokova, D.; Martinod, K.; Wagner, D.D. Extracellular DNA NET-Works With Dire Consequences for Health. *Circ. Res.* **2019**, *125*, 470–488. [CrossRef]
16. Baker, C.J.; Smith, S.A.; Morrissey, J.H. Polyphosphate in thrombosis, hemostasis, and inflammation. *Res. Pract. Thromb. Haemost.* **2019**, *3*, 18–25. [CrossRef]
17. Nossel, H.L. Differential consumption of coagulation factors resulting from activation of the extrinsic (tissue thromboplastin) or the intrinsic (foreign surface contact) pathways. *Blood* **1967**, *29*, 331–340. [CrossRef]
18. Silverberg, M.; Dunn, J.T.; Garen, L.; Kaplan, A.P. Autoactivation of human Hageman factor. Demonstration utilizing a synthetic substrate. *J. Biol. Chem.* **1980**, *255*, 7281–7286. [CrossRef] [PubMed]
19. Tankersley, D.L.; Finlayson, J.S. Kinetics of activation and autoactivation of human factor XII. *Biochemistry* **1984**, *23*, 273–279. [CrossRef] [PubMed]
20. Franke, L.; Schewe, H.J.; Muller, B.; Campman, V.; Kitzrow, W.; Uebelhack, R.; Berghofer, A.; Muller-Oerlinghausen, B. Serotonergic platelet variables in unmedicated patients suffering from major depression and healthy subjects: Relationship between 5HT content and 5HT uptake. *Life Sci.* **2000**, *67*, 301–315. [CrossRef]
21. Wood, J.P.; Ellery, P.E.; Maroney, S.A.; Mast, A.E. Biology of tissue factor pathway inhibitor. *Blood* **2014**, *123*, 2934–2943. [CrossRef] [PubMed]
22. Griffin, J.H.; Fernandez, J.A.; Gale, A.J.; Mosnier, L.O. Activated protein C. *J. Thromb. Haemost.* **2007**, *5* (Suppl. S1), 73–80. [CrossRef] [PubMed]
23. Olson, S.T.; Swanson, R.; Raub-Segall, E.; Bedsted, T.; Sadri, M.; Petitou, M.; Herault, J.P.; Herbert, J.M.; Bjork, I. Accelerating ability of synthetic oligosaccharides on antithrombin inhibition of proteinases of the clotting and fibrinolytic systems. Comparison with heparin and low-molecular-weight heparin. *Thromb. Haemost.* **2004**, *92*, 929–939. [CrossRef]
24. Larsen, K.S.; Ostergaard, H.; Bjelke, J.R.; Olsen, O.H.; Rasmussen, H.B.; Christensen, L.; Kragelund, B.B.; Stennicke, H.R. Engineering the substrate and inhibitor specificities of human coagulation Factor VIIa. *Biochem. J.* **2007**, *405*, 429–438. [CrossRef]
25. Zhao, M.; Abdel-Razek, T.; Sun, M.F.; Gailani, D. Characterization of a heparin binding site on the heavy chain of factor XI. *J. Biol. Chem.* **1998**, *273*, 31153–31159. [CrossRef]

26. Renne, T.; Schmaier, A.H.; Nickel, K.F.; Blomback, M.; Maas, C. In vivo roles of factor XII. *Blood* **2012**, *120*, 4296–4303. [CrossRef] [PubMed]
27. Colhoun, H.M.; Zito, F.; Norman Chan, N.; Rubens, M.B.; Fuller, J.H.; Humphries, S.E. Activated factor XII levels and factor XII 46C>T genotype in relation to coronary artery calcification in patients with type 1 diabetes and healthy subjects. *Atherosclerosis* **2002**, *163*, 363–369. [CrossRef] [PubMed]
28. Grundt, H.; Nilsen, D.W.; Hetland, O.; Valente, E.; Fagertun, H.E. Activated factor 12 (FXIIa) predicts recurrent coronary events after an acute myocardial infarction. *Am. Heart J.* **2004**, *147*, 260–266. [CrossRef]
29. Doggen, C.J.; Rosendaal, F.R.; Meijers, J.C. Levels of intrinsic coagulation factors and the risk of myocardial infarction among men: Opposite and synergistic effects of factors XI and XII. *Blood* **2006**, *108*, 4045–4051. [CrossRef] [PubMed]
30. Salomon, O.; Steinberg, D.M.; Koren-Morag, N.; Tanne, D.; Seligsohn, U. Reduced incidence of ischemic stroke in patients with severe factor XI deficiency. *Blood* **2008**, *111*, 4113–4117. [CrossRef] [PubMed]
31. Salomon, O.; Steinberg, D.M.; Zucker, M.; Varon, D.; Zivelin, A.; Seligsohn, U. Patients with severe factor XI deficiency have a reduced incidence of deep-vein thrombosis. *Thromb. Haemost.* **2011**, *105*, 269–273. [CrossRef] [PubMed]
32. Renne, T.; Pozgajova, M.; Gruner, S.; Schuh, K.; Pauer, H.U.; Burfeind, P.; Gailani, D.; Nieswandt, B. Defective thrombus formation in mice lacking coagulation factor XII. *J. Exp. Med.* **2005**, *202*, 271–281. [CrossRef] [PubMed]
33. Kleinschnitz, C.; Stoll, G.; Bendszus, M.; Schuh, K.; Pauer, H.U.; Burfeind, P.; Renne, C.; Gailani, D.; Nieswandt, B.; Renne, T. Targeting coagulation factor XII provides protection from pathological thrombosis in cerebral ischemia without interfering with hemostasis. *J. Exp. Med.* **2006**, *203*, 513–518. [CrossRef] [PubMed]
34. Gailani, D.; Bane, C.E.; Gruber, A. Factor XI and contact activation as targets for antithrombotic therapy. *J. Thromb. Haemost.* **2015**, *13*, 1383–1395. [CrossRef] [PubMed]
35. Muller, F.; Gailani, D.; Renne, T. Factor XI and XII as antithrombotic targets. *Curr. Opin. Hematol.* **2011**, *18*, 349–355. [CrossRef] [PubMed]
36. Schumacher, W.A.; Luettgen, J.M.; Quan, M.L.; Seiffert, D.A. Inhibition of factor XIa as a new approach to anticoagulation. *Arter. Thromb. Vasc. Biol.* **2010**, *30*, 388–392. [CrossRef]
37. Key, N.S. Epidemiologic and clinical data linking factors XI and XII to thrombosis. *Hematol. Am. Soc. Hematol. Educ. Prog.* **2014**, *2014*, 66–70. [CrossRef]
38. Davie, E.W.; Fujikawa, K.; Kisiel, W. The coagulation cascade: Initiation, maintenance, and regulation. *Biochemistry* **1991**, *30*, 10363–10370. [CrossRef]
39. Vondemborne, P.A.K.; Meijers, J.C.M.; Bouma, B.N. Feedback Activation of Factor-Xi by Thrombin in Plasma Results in Additional Formation of Thrombin That Protects Fibrin Clots from Fibrinolysis. *Blood* **1995**, *86*, 3035–3042. [CrossRef]
40. Meijers, J.C.; Tekelenburg, W.L.; Bouma, B.N.; Bertina, R.M.; Rosendaal, F.R. High levels of coagulation factor XI as a risk factor for venous thrombosis. *N. Engl. J. Med.* **2000**, *342*, 696–701. [CrossRef] [PubMed]
41. Gill, D.; Georgakis, M.K.; Laffan, M.; Sabater-Lleal, M.; Malik, R.; Tzoulaki, I.; Veltkamp, R.; Dehghan, A. Genetically Determined FXI (Factor XI) Levels and Risk of Stroke. *Stroke J. Cereb. Circ.* **2018**, *49*, 2761–2763. [CrossRef] [PubMed]
42. Salomon, O.; Steinberg, D.M.; Seligshon, U. Variable bleeding manifestations characterize different types of surgery in patients with severe factor XI deficiency enabling parsimonious use of replacement therapy. *Haemophilia* **2006**, *12*, 490–493. [CrossRef] [PubMed]
43. Preis, M.; Hirsch, J.; Kotler, A.; Zoabi, A.; Stein, N.; Rennert, G.; Saliba, W. Factor XI deficiency is associated with lower risk for cardiovascular and venous thromboembolism events. *Blood* **2017**, *129*, 1210–1215. [CrossRef]
44. Wang, X.; Li, Q.; Du, F.; Caruso, J.; Nawrocki, A.; Chintala, M. Reversal of Antithrombotic Effects of FXIa Inhibitor Milvexian (BMS-986177/JNJ-70033093) by Non-specific Agents in a Rabbit AV-shunt Model of Thrombosis. In Proceedings of the ISTH 2021 Congress, Philadelphia, PA, USA, 17–21 July 2021.
45. David, T.; Kim, Y.C.; Ely, L.K.; Rondon, I.; Gao, H.; O'Brien, P.; Bolt, M.W.; Coyle, A.J.; Garcia, J.L.; Flounders, E.A.; et al. Factor XIa-specific IgG and a reversal agent to probe factor XI function in thrombosis and hemostasis. *Sci. Transl. Med.* **2016**, *8*, 353ra112. [CrossRef]
46. Vilahur, G.; Choi, B.G.; Zafar, M.U.; Viles-Gonzalez, J.F.; Vorchheimer, D.A.; Fuster, V.; Badimon, J.J. Normalization of platelet reactivity in clopidogrel-treated subjects. *J. Thromb. Haemost.* **2007**, *5*, 82–90. [CrossRef] [PubMed]
47. Zafar, M.U.; Santos-Gallego, C.; Vorchheimer, D.A.; Viles-Gonzalez, J.F.; Elmariah, S.; Giannarelli, C.; Sartori, S.; Small, D.S.; Jakubowski, J.A.; Fuster, V.; et al. Platelet function normalization after a prasugrel loading-dose: Time-dependent effect of platelet supplementation. *J. Thromb. Haemost.* **2013**, *11*, 100–106. [CrossRef]
48. Zafar, M.U.; Smith, D.A.; Baber, U.; Sartori, S.; Chen, K.; Lam, D.W.; Linares-Koloffon, C.A.; Rey-Mendoza, J.; Jimenez Britez, G.; Escolar, G.; et al. Impact of Timing on the Functional Recovery Achieved With Platelet Supplementation after Treatment With Ticagrelor. *Circ. Cardiovasc. Interv.* **2017**, *10*, 8. [CrossRef]
49. Kole, R.; Krainer, A.R.; Altman, S. RNA therapeutics: Beyond RNA interference and antisense oligonucleotides. *Nat. Rev. Drug Discov.* **2012**, *11*, 125–140. [CrossRef]
50. Shen, X.; Corey, D.R. Chemistry, mechanism and clinical status of antisense oligonucleotides and duplex RNAs. *Nucleic Acids Res.* **2018**, *46*, 1584–1600. [CrossRef]
51. Younis, H.S.; Crosby, J.; Huh, J.I.; Lee, H.S.; Rime, S.; Monia, B.; Henry, S.P. Antisense inhibition of coagulation factor XI prolongs APTT without increased bleeding risk in cynomolgus monkeys. *Blood* **2012**, *119*, 2401–2408. [CrossRef]

52. Buller, H.R.; Bethune, C.; Bhanot, S.; Gailani, D.; Monia, B.P.; Raskob, G.E.; Segers, A.; Verhamme, P.; Weitz, J.I.; Investigators, F.-A.T. Factor XI antisense oligonucleotide for prevention of venous thrombosis. *N. Engl. J. Med.* **2015**, *372*, 232–240. [CrossRef] [PubMed]
53. Walsh, M.; Bethune, C.; Smyth, A.; Tyrwhitt, J.; Jung, S.W.; Yu, R.Z.; Wang, Y.; Geary, R.S.; Weitz, J.; Bhanot, S.; et al. Phase 2 Study of the Factor XI Antisense Inhibitor IONIS-FXIRx in Patients With ESRD. *Kidney Int. Rep.* **2022**, *7*, 200–209. [CrossRef] [PubMed]
54. Rao, S.V.; Kirsch, B.; Bhatt, D.L.; Budaj, A.; Coppolecchia, R.; Eikelboom, J.; James, S.K.; Jones, W.S.; Merkely, B.; Keller, L.; et al. A Multicenter, Phase 2, Randomized, Placebo-Controlled, Double-Blind, Parallel-Group, Dose-Finding Trial of the Oral Factor XIa Inhibitor Asundexian to Prevent Adverse Cardiovascular Outcomes Following Acute Myocardial Infarction. *Circulation* **2022**, *146*, 1196–1206. [CrossRef] [PubMed]
55. Shoamanesh, A.; Mundl, H.; Smith, E.E.; Masjuan, J.; Milanov, I.; Hirano, T.; Agafina, A.; Campbell, B.; Caso, V.; Mas, J.L.; et al. Factor XIa inhibition with asundexian after acute non-cardioembolic ischaemic stroke (PACIFIC-Stroke): An international, randomised, double-blind, placebo-controlled, phase 2b trial. *Lancet* **2022**, *400*, 997–1007. [CrossRef] [PubMed]
56. Piccini, J.P.; Caso, V.; Connolly, S.J.; Fox, K.A.A.; Oldgren, J.; Jones, W.S.; Gorog, D.A.; Durdil, V.; Viethen, T.; Neumann, C.; et al. Safety of the oral factor XIa inhibitor asundexian compared with apixaban in patients with atrial fibrillation (PACIFIC-AF): A multicentre, randomised, double-blind, double-dummy, dose-finding phase 2 study. *Lancet* **2022**, *399*, 1383–1390. [CrossRef]
57. BusinessWire. *Bayer Initiates Phase III Study Program for Investigational Oral FXIa Inhibitor Asundexian*; BusinessWire: San Francisco, CA, USA, 2022.
58. Weitz, J.I.; Strony, J.; Ageno, W.; Gailani, D.; Hylek, E.M.; Lassen, M.R.; Mahaffey, K.W.; Notani, R.S.; Roberts, R.; Segers, A.; et al. Milvexian for the Prevention of Venous Thromboembolism. *N. Engl. J. Med.* **2021**, *385*, 2161–2172. [CrossRef]
59. ESCPressOffice. *Milvexian Shows Potential to Reduce the Risk of Ischaemic Stroke*; ESCPressOffice: Barcelona, Spain, 2022.
60. Beale, D.; Dennison, J.; Boyce, M.; Mazzo, F.; Honda, N.; Smith, P.; Bruce, M. ONO-7684 a novel oral FXIa inhibitor: Safety, tolerability, pharmacokinetics and pharmacodynamics in a first-in-human study. *Br. J. Clin. Pharm.* **2021**, *87*, 3177–3189. [CrossRef]
61. Verhamme, P.; Yi, B.A.; Segers, A.; Salter, J.; Bloomfield, D.; Buller, H.R.; Raskob, G.E.; Weitz, J.I.; Investigators, A.-T. Abelacimab for Prevention of Venous Thromboembolism. *N. Engl. J. Med.* **2021**, *385*, 609–617. [CrossRef]
62. Weitz, J.I.; Bauersachs, R.; Becker, B.; Berkowitz, S.D.; Freitas, M.C.S.; Lassen, M.R.; Metzig, C.; Raskob, G.E. Effect of Osocimab in Preventing Venous Thromboembolism Among Patients Undergoing Knee Arthroplasty: The FOXTROT Randomized Clinical Trial. *JAMA* **2020**, *323*, 130–139. [CrossRef]
63. Lorentz, C.U.; Tucker, E.I.; Verbout, N.G.; Shatzel, J.J.; Olson, S.R.; Markway, B.D.; Wallisch, M.; Ralle, M.; Hinds, M.T.; McCarty, O.J.T.; et al. The contact activation inhibitor AB023 in heparin-free hemodialysis: Results of a randomized phase 2 clinical trial. *Blood* **2021**, *138*, 2173–2184. [CrossRef]
64. Al-Horani, R.A.; Desai, U.R. Factor XIa inhibitors: A review of the patent literature. *Expert Opin. Ther. Pat.* **2016**, *26*, 323–345. [CrossRef] [PubMed]
65. Hayward, N.J.; Goldberg, D.I.; Morrel, E.M.; Friden, P.M.; Bokesch, P.M. Phase 1a/1b Study of EP-7041: A Novel, Potent, Selective, Small Molecule FXIa Inhibitor. *Circulation* **2017**, *136*, 13747.
66. Perera, V.; Luettgen, J.M.; Wang, Z.; Frost, C.E.; Yones, C.; Russo, C.; Lee, J.; Zhao, Y.; LaCreta, F.P.; Ma, X.; et al. First-in-human study to assess the safety, pharmacokinetics and pharmacodynamics of BMS-962212, a direct, reversible, small molecule factor XIa inhibitor in non-Japanese and Japanese healthy subjects. *Br. J. Clin. Pharm.* **2018**, *84*, 876–887. [CrossRef] [PubMed]
67. Al-Horani, R.A.; Gailani, D.; Desai, U.R. Allosteric inhibition of factor XIa. Sulfated non-saccharide glycosaminoglycan mimetics as promising anticoagulants. *Thromb. Res.* **2015**, *136*, 379–387. [CrossRef] [PubMed]
68. Koch, A.W.; Schiering, N.; Melkko, S.; Ewert, S.; Salter, J.; Zhang, Y.; McCormack, P.; Yu, J.; Huang, X.; Chiu, Y.H.; et al. MAA868, a novel FXI antibody with a unique binding mode, shows durable effects on markers of anticoagulation in humans. *Blood* **2019**, *133*, 1507–1516. [CrossRef] [PubMed]
69. White, R.R.; Sullenger, B.A.; Rusconi, C.P. Developing aptamers into therapeutics. *J. Clin. Investig.* **2000**, *106*, 929–934. [CrossRef] [PubMed]
70. Donkor, D.A.; Bhakta, V.; Eltringham-Smith, L.J.; Stafford, A.R.; Weitz, J.I.; Sheffield, W.P. Selection and characterization of a DNA aptamer inhibiting coagulation factor XIa. *Sci. Rep.* **2017**, *7*, 2102. [CrossRef]
71. Woodruff, R.S.; Ivanov, I.; Verhamme, I.M.; Sun, M.F.; Gailani, D.; Sullenger, B.A. Generation and characterization of aptamers targeting factor XIa. *Thromb. Res.* **2017**, *156*, 134–141. [CrossRef]
72. Rusconi, C.P.; Roberts, J.D.; Pitoc, G.A.; Nimjee, S.M.; White, R.R.; Quick, G., Jr.; Scardino, E.; Fay, W.P.; Sullenger, B.A. Antidote-mediated control of an anticoagulant aptamer in vivo. *Nat. Biotechnol.* **2004**, *22*, 1423–1428. [CrossRef]
73. Oney, S.; Lam, R.T.; Bompiani, K.M.; Blake, C.M.; Quick, G.; Heidel, J.D.; Liu, J.Y.; Mack, B.C.; Davis, M.E.; Leong, K.W.; et al. Development of universal antidotes to control aptamer activity. *Nat. Med.* **2009**, *15*, 1224–1228. [CrossRef]
74. Go, A.S.; Hylek, E.M.; Phillips, K.A.; Chang, Y.; Henault, L.E.; Selby, J.V.; Singer, D.E. Prevalence of diagnosed atrial fibrillation in adults: National implications for rhythm management and stroke prevention: The AnTicoagulation and Risk Factors in Atrial Fibrillation (ATRIA) Study. *JAMA* **2001**, *285*, 2370–2375. [CrossRef]
75. Wolf, P.A.; Abbott, R.D.; Kannel, W.B. Atrial fibrillation as an independent risk factor for stroke: The Framingham Study. *Stroke J. Cereb. Circ.* **1991**, *22*, 983–988. [CrossRef]
76. Kannel, W.B.; Benjamin, E.J. Status of the epidemiology of atrial fibrillation. *Med. Clin. N. Am.* **2008**, *92*, 17–40. [CrossRef] [PubMed]

77. Lopez-Lopez, J.A.; Sterne, J.A.C.; Thom, H.H.Z.; Higgins, J.P.T.; Hingorani, A.D.; Okoli, G.N.; Davies, P.A.; Bodalia, P.N.; Bryden, P.A.; Welton, N.J.; et al. Oral anticoagulants for prevention of stroke in atrial fibrillation: Systematic review, network meta-analysis, and cost effectiveness analysis. *BMJ Clin. Res. Ed* **2017**, *359*, j5058. [CrossRef] [PubMed]
78. van Es, N.; Coppens, M.; Schulman, S.; Middeldorp, S.; Buller, H.R. Direct oral anticoagulants compared with vitamin K antagonists for acute venous thromboembolism: Evidence from phase 3 trials. *Blood* **2014**, *124*, 1968–1975. [CrossRef] [PubMed]
79. Heit, J.A.; Petterson, T.M.; Farmer, S.A.; Bailey, K.R.; Joseph Melton, L. Trends in the incidence of deep vein thrombosis and pulmonary embolism: A 35-year population-based study. *Blood* **2006**, *108*, 1488. [CrossRef]
80. van der Hulle, T.; Kooiman, J.; den Exter, P.L.; Dekkers, O.M.; Klok, F.A.; Huisman, M.V. Effectiveness and safety of novel oral anticoagulants as compared with vitamin K antagonists in the treatment of acute symptomatic venous thromboembolism: A systematic review and meta-analysis. *J. Thromb. Haemost.* **2014**, *12*, 320–328. [CrossRef]
81. Cohen, H.; Hunt, B.J.; Efthymiou, M.; Arachchillage, D.R.; Mackie, I.J.; Clawson, S.; Sylvestre, Y.; Machin, S.J.; Bertolaccini, M.L.; Ruiz-Castellano, M.; et al. Rivaroxaban versus warfarin to treat patients with thrombotic antiphospholipid syndrome, with or without systemic lupus erythematosus (RAPS): A randomised, controlled, open-label, phase 2/3, non-inferiority trial. *Lancet. Haematol.* **2016**, *3*, e426–e436. [CrossRef]
82. Woller, S.C.; Stevens, S.M.; Kaplan, D.A.; Branch, D.W.; Aston, V.T.; Wilson, E.L.; Gallo, H.M.; Johnson, E.G.; Rondina, M.T.; Lloyd, J.F.; et al. Apixaban for the Secondary Prevention of Thrombosis Among Patients With Antiphospholipid Syndrome: Study Rationale and Design (ASTRO-APS). *Clin. Appl. Thromb./Hemost. Off. J. Int. Acad. Clin. Appl. Thromb./Hemost.* **2016**, *22*, 239–247. [CrossRef]
83. Ohene-Frempong, K.; Weiner, S.J.; Sleeper, L.A.; Miller, S.T.; Embury, S.; Moohr, J.W.; Wethers, D.L.; Pegelow, C.H.; Gill, F.M. Cerebrovascular accidents in sickle cell disease: Rates and risk factors. *Blood* **1998**, *91*, 288–294.
84. Adams, R.; McKie, V.; Nichols, F.; Carl, E.; Zhang, D.L.; McKie, K.; Figueroa, R.; Litaker, M.; Thompson, W.; Hess, D. The use of transcranial ultrasonography to predict stroke in sickle cell disease. *N. Engl. J. Med.* **1992**, *326*, 605–610. [CrossRef] [PubMed]
85. Adams, R.J.; McKie, V.C.; Carl, E.M.; Nichols, F.T.; Perry, R.; Brock, K.; McKie, K.; Figueroa, R.; Litaker, M.; Weiner, S.; et al. Long-term stroke risk in children with sickle cell disease screened with transcranial Doppler. *Ann. Neurol.* **1997**, *42*, 699–704. [CrossRef] [PubMed]
86. Switzer, J.A.; Hess, D.C.; Nichols, F.T.; Adams, R.J. Pathophysiology and treatment of stroke in sickle-cell disease: Present and future. *Lancet. Neurol.* **2006**, *5*, 501–512. [CrossRef]
87. Hillery, C.A.; Panepinto, J.A. Pathophysiology of stroke in sickle cell disease. *Microcirculation* **2004**, *11*, 195–208. [CrossRef]
88. Shet, A.S.; Aras, O.; Gupta, K.; Hass, M.J.; Rausch, D.J.; Saba, N.; Koopmeiners, L.; Key, N.S.; Hebbel, R.P. Sickle blood contains tissue factor-positive microparticles derived from endothelial cells and monocytes. *Blood* **2003**, *102*, 2678–2683. [CrossRef] [PubMed]
89. Glassberg, J.; Rahman, A.H.; Zafar, M.; Cromwell, C.; Punzalan, A.; Badimon, J.J.; Aledort, L. Application of phospho-CyTOF to characterize immune activation in patients with sickle cell disease in an ex vivo model of thrombosis. *J. Immunol. Methods* **2018**, *453*, 11–19. [CrossRef] [PubMed]
90. Adams, R.J.; McKie, V.C.; Hsu, L.; Files, B.; Vichinsky, E.; Pegelow, C.; Abboud, M.; Gallagher, D.; Kutlar, A.; Nichols, F.T.; et al. Prevention of a first stroke by transfusions in children with sickle cell anemia and abnormal results on transcranial Doppler ultrasonography. *N. Engl. J. Med.* **1998**, *339*, 5–11. [CrossRef]
91. Sy, E.; Sklar, M.C.; Lequier, L.; Fan, E.; Kanji, H.D. Anticoagulation practices and the prevalence of major bleeding, thromboembolic events, and mortality in venoarterial extracorporeal membrane oxygenation: A systematic review and meta-analysis. *J. Crit. Care* **2017**, *39*, 87–96. [CrossRef]
92. McMichael, A.B.V.; Ryerson, L.M.; Ratano, D.; Fan, E.; Faraoni, D.; Annich, G.M. 2021 ELSO Adult and Pediatric Anticoagulation Guidelines. *ASAIO J.* **2022**, *68*, 303–310. [CrossRef]
93. Eikelboom, J.W.; Connolly, S.J.; Brueckmann, M.; Granger, C.B.; Kappetein, A.P.; Mack, M.J.; Blatchford, J.; Devenny, K.; Friedman, J.; Guiver, K.; et al. Dabigatran versus warfarin in patients with mechanical heart valves. *N. Engl. J. Med.* **2013**, *369*, 1206–1214. [CrossRef]
94. Jaffer, I.H.; Fredenburgh, J.C.; Hirsh, J.; Weitz, J.I. Medical device-induced thrombosis: What causes it and how can we prevent it? *J. Thromb. Haemost.* **2015**, *13* (Suppl. S1), S72–S81. [CrossRef] [PubMed]
95. Jaffer, I.H.; Stafford, A.R.; Fredenburgh, J.C.; Whitlock, R.P.; Chan, N.C.; Weitz, J.I. Dabigatran is Less Effective Than Warfarin at Attenuating Mechanical Heart Valve-Induced Thrombin Generation. *J. Am. Heart Assoc.* **2015**, *4*, e002322. [CrossRef] [PubMed]
96. Mega, J.L.; Braunwald, E.; Wiviott, S.D.; Bassand, J.P.; Bhatt, D.L.; Bode, C.; Burton, P.; Cohen, M.; Cook-Bruns, N.; Fox, K.A.; et al. Rivaroxaban in patients with a recent acute coronary syndrome. *N. Engl. J. Med.* **2012**, *366*, 9–19. [CrossRef] [PubMed]

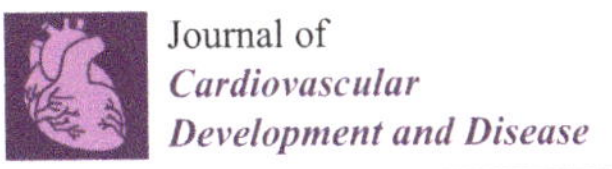
Journal of
Cardiovascular
Development and Disease

Review

Anticoagulation for Left Ventricle Thrombus—Case Series and Literature Review for Use of Direct Oral Anticoagulants

Akshyaya Pradhan [1], Monika Bhandari [1], Pravesh Vishwakarma [1], Chiara Salimei [2], Ferdinando Iellamo [2], Rishi Sethi [1] and Marco Alfonso Perrone [2,*]

1 Department of Cardiology, King George's Medical University, Lucknow 226003, India
2 Department of Cardiology and CardioLab, University of Rome Tor Vergata, 00133 Rome, Italy
* Correspondence: marco.perrone@uniroma2.it

Abstract: Left ventricular thrombus is a known complication following acute myocardial infarction that can lead to systemic thromboembolism. To obviate the risk of thromboembolism, the patient needs anticoagulation in addition to dual antiplatelet therapy. However, combining antiplatelets with anticoagulants substantially increases the bleeding risk. Traditionally, vitamin K antagonists (VKAs) have been the sheet anchor for anticoagulation in this scenario. The use of direct oral anticoagulants has significantly attenuated the bleeding risk associated with anticoagulation for atrial fibrillation and venous thromboembolism. Furthermore, in patients with atrial fibrillation undergoing percutaneous coronary intervention, the use of direct oral anticoagulants (DOACs) in conjunction with antiplatelets has been found to be noninferior in reducing ischemic events while significantly attenuating the bleeding compared with VKA. After initial case reports, multiple observational and nonrandomized studies have now safely and effectively utilized direct oral anticoagulants for anticoagulation in left ventricular thrombus. Here, we report a series of two cases presenting with left ventricular thrombus following acute myocardial infarction. In this case series, we try to address the issues concerning the choice and duration of anticoagulation in the case of postinfarct left ventricular thrombus. Pending the results of large randomized control trials, the judicious use of direct oral anticoagulant is warranted when taking into consideration the ischemic and bleeding profile in an individualized approach.

Keywords: left ventricular thrombus; anticoagulation; anterior wall myocardial infarction; dual therapy; direct oral anticoagulants

Citation: Pradhan, A.; Bhandari, M.; Vishwakarma, P.; Salimei, C.; Iellamo, F.; Sethi, R.; Perrone, M.A. Anticoagulation for Left Ventricle Thrombus—Case Series and Literature Review for Use of Direct Oral Anticoagulants. *J. Cardiovasc. Dev. Dis.* **2023**, *10*, 41. https://doi.org/10.3390/jcdd10020041

Academic Editors: Giovanni Cimmino and Plinio Cirillo

Received: 3 December 2022
Revised: 18 January 2023
Accepted: 19 January 2023
Published: 23 January 2023

1. Introduction

A left ventricular (LV) thrombus is a known complication following acute myocardial infarction (AMI) that can lead to systemic thromboembolism. With the increasing use of timely thrombolysis and primary percutaneous interventions (PCIs), along with the unabated use of secondary prevention medications, the complications following AMI are decreasing and survival is improving [1]. After myocardial infarction (MI), LV thrombus still remains as high as 15% in the PCI era [2,3]. An LV thrombus usually occurs within 1 month post ST elevation MI, mostly occurs in the setting of acute anterior wall MI, and is associated with poor outcomes. The consideration of optimal anticoagulation, along with the decision of revascularization, makes decision-making a challenge. Echocardiography is the standard screening tool for detecting a thrombus, but sometimes contrast echocardiography might be required for confirming the diagnosis. The American College of Cardiology (ACC)/American Heart Association (AHA) guidelines for the management of AMI recommend oral anticoagulants (OAC) in addition to dual antiplatelet (DAPT) agents for the treatment and prevention of LV thrombi in acute MI [4]. However, the use of triple therapy comes at the cost of increased bleeding complications [5]. Bleeding following anticoagulation is also associated with an increase in mortality. Hence, balancing the ischemic benefits against bleeding events is a common clinical dilemma. The introduction of direct oral anticoagulants (DOACs) has revolutionized the scenario of the anticoagulation of vascular thromboembolism, including atrial fibrillation (AF). Studies conducted to assess the efficacy of dual therapy (single antiplatelet with OAC) in patients with acute coronary syndrome (ACS) and atrial fibrillation (AF) undergoing PCI have shown encouraging results with respect to attenuated bleeding and preserved efficacy. Additionally, DOACs have been found to be comparable with vitamin K antagonists (VKAs) [6–9]. Although these studies did not involve patients with an LV thrombus per se, a large body of affirmative data in the form of case reports, case series, observational studies and small randomized studies has emerged regarding the safety and efficacy of DOACs in treating LV thrombus. In this case series, we try to address this fairly common yet underestimated and underrepresented situation.

2. Case Summary

2.1. Case 1

A 46-year-old man with conventional cardiovascular risk factors presented with complaints of severe sudden onset chest pain of a 4-day duration. On examination, he had a dyskinetic apex with an LV third heart sound. His electrocardiogram was suggestive of anterior wall ST elevation MI, and his echocardiography showed a 1.8 cm × 1.5 cm clot at the apex (Figure 1a) and attendant severe LV dysfunction. The patient underwent coronary angiography, which revealed the 95% stenosis of the proximal left anterior descending artery with poor contractility of LV. In view of his severe LV dysfunction, late presentation, and pain-free status, he was subjected to myocardial perfusion imaging. Anticoagulation with VKA was initiated and was targeted to an INR 2.0–3.0. DOACs were not used, because the patient refused owing to financial constraints. Stress imaging (Technetium-99 single-photon emission computerized tomography) did not reveal any evidence of viability, and dual therapy was continued for 1 month. The LV thrombus resolved by the end of 1 month, but we still continued dual therapy (clopidogrel and oral warfarin), along with optimal medical treatment. He is planned for repeat echocardiography after 3 months and is under follow-up.

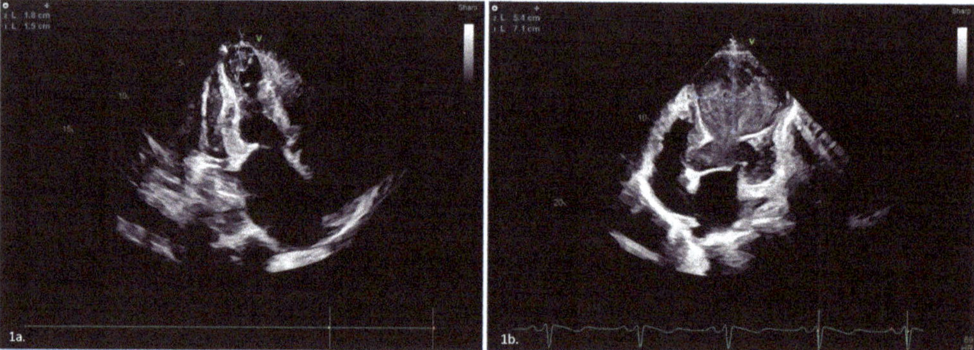

Figure 1. Echocardiographic demonstration of thrombus in two cases managed by different anticoagulation regimens. (**1a**)—two-dimensional echocardiography in apical view showing homogenous echo dense mass (1.8 × 1.5 cm) at apex of left ventricle, suggestive of thrombus. (**1b**)—two-dimensional echocardiography in apical view showing large echo dense mass (5.4 × 7.1 cm) at apex of a dilated and akinetic left ventricle, suggestive of thrombus.

2.2. Case 2

A 66-year-old man presented to the emergency department on seventh-day post anterior wall ST elevation myocardial infarction. He was a known hypertensive and had an episode of ischemic stroke 3 years back. His echocardiography revealed a 5.4 cm × 7.1 cm LV thrombus at the apex with a cavity (Figure 1b). Because of the very late presentation, he was offered an option of ischemia-guided or symptom-guided revascularization. The patient opted for medical management and dual therapy with rivaroxaban, and clopidogrel was initiated at discharge. Interestingly, this thrombus revealed early partial resolution with dual therapy at the end of 1 month, and dual therapy was continued till 3 months. The repeat echocardiographic evaluation at 3 months failed to demonstrate any evidence of residual thrombus. He is presently on dual antiplatelet therapy and is doing well on follow-up.

3. LV Thrombus—An Overview

LV thrombus usually forms in the akinetic or dyskinetic segments of the ventricle post myocardial infarction (MI). The stasis of blood is maximal in these areas. Additionally, MI causes damage to the endocardium, and post MI, there is hypercoagulability; hence, according to the Virchow's triad, these areas are prone to developing an LV thrombus. The incidence of LV thrombus post MI has been estimated to be between 15% and 25% in anterior wall MI by using cardiac MRI (magnetic resonant imaging) [3]. The incidence of LV thrombi is decreasing thanks to the increasing use of primary PCI, neurohormonal antagonists, adverse LV remodeling preventing agents and potent antithrombotic regimens. A cardiac MRI (CMRI) is the most specific and sensitive modality, making it the gold standard for detection of LV thrombus. However, in view of its limited availability, an echocardiogram remains the diagnostic tool of choice.

A recent study by Maniwa et al. has shown the incidence of systemic embolization to be 16.3% overall and 2.9% in patients maintaining an adequate therapeutic range of anticoagulation [10]. In another study, acute ischemic stroke occurred in 11.8% of patients with an LV thrombus who received anticoagulation as compared with 44.1% in those who were not on anticoagulants [11]. The protrusion of thrombi into the cavity, non resolving thrombi, and recurrent thrombi were predictors of stroke.

Once an LV thrombus is detected, patients should be immediately started on anticoagulation. In this scenario, anticoagulation provides benefits with respect to systemic thromboembolism, whereas antiplatelets provide benefits regarding ischemic events. According to the 2013 ACC/AHA ST elevation MI (STEMI) guidelines, it is reasonable to add OACs to DAPTs for patients with a STEMI and an asymptomatic LV thrombus, for 3 months, targeting a lower international normalized ratio (INR) goal of 2.0 to 2.5 [4]. On the other hand, the 2014 AHA/American Stroke Association (ASA) stroke prevention guidelines recommend anticoagulation for a similar duration, but with an INR target of 2.5 [12]. The 2017 European Society of Cardiology (ESC)'s STEMI guidelines recommend OACs for at least 6 months if there is an LV thrombus [13]. After 6 months, OACs are to be guided by repeated echocardiography and balancing bleeding risk and the need for concomitant antiplatelet therapy. However, these guidelines are not based on any randomized prospective studies in this scenario (AMI with an LV thrombus).

The guidelines recommend VKAs in the setting of LV thrombus because of more clinical experience. However, the need for frequent monitoring using the international normalized ratio (INR), food–drug interactions, and an inability to achieve the target therapeutic rate (TTR) are the major limitations of VKAs. Several observational studies and case reports have been conducted in this regard.

4. Clinical Experience of Combination of OACs with Antiplatelets

It is well known that combining OACs with DAPTs substantially increases bleeding risk [5,14]. Triple therapy, however, may be initially considered in patients with high ischemic risk (recurrent MI, a suboptimal stent placement, or a history of stent thrombosis) [12]. While there are no studies that have compared dual therapy with triple therapy in the setting of MI and an LV thrombus, indirect evidence for the safety and efficacy of DOACs plus dual antiplatelets comes from trials of AF patients undergoing PCI (Figure 2 and Table 1) [15]. The WOEST and the ISAR-triple were the initial trials that included patients of AMI with AF requiring PCI that compared triple therapy against dual therapy with VKA [16,17]. In the WOEST trial, there was a significant reduction in serious bleeding (44% vs. 19.1%), and in the ISAR-triple trial, the shortening of clopidogrel therapy's duration from 6 months to 6 weeks was found to be noninferior for both ischemic and bleeding events. Interestingly, the WOEST trial included patients not only with AF but also with other indications for anticoagulation. The four pivotal randomized trials of DOACs in patients of AMI with AF who underwent PCI have also shown the benefits of dual therapy (SAPT with DOACs) in reducing bleeding events primarily compared with VKA-based dual or triple therapy [6–9]. There were no differences in the ischemic events with DOAC-based therapy, though most of these studies were not powered enough for the detection of ischemic end points.

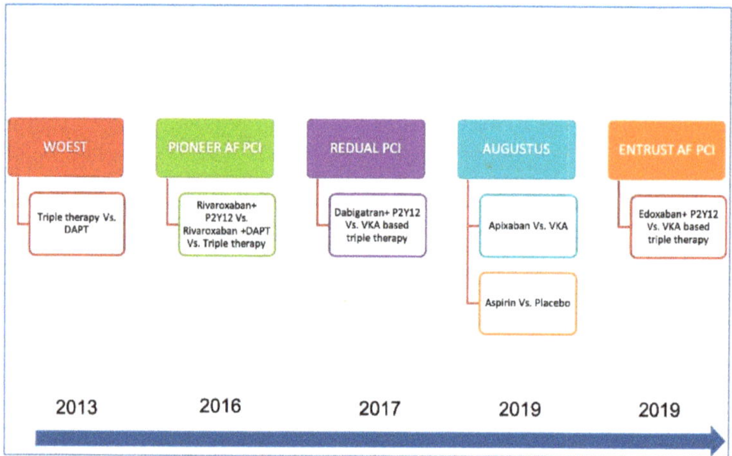

Figure 2. Timeline of pivotal trials comparing conventional triple therapy with dual therapy (either VKA or NOAC based). In the PIONEER AF PCI study, two doses of rivaroxaban were tried—2.5 mg and 5 mg. Similarly, in REDUAL PCI, both 110 mg and 150 mg doses were studied. AUGUSTUS PCI was a 2 × 2 factorial evaluating apixaban versus warfarin and aspirin versus no aspirin. (DAPT—dual antiplatelet therapy; P2Y12—clopidogrel; VKA—vitamin K antagonist).

Meta-analyses of DOAC-based dual therapy have clearly shown that, compared with triple therapy (OACs with DAPT), their utilization leads to a marked decline in bleeding episodes without any increase in ischemic events. A meta-analysis of pivotal RCTs, by Lopes et al., revealed that the odds ratios (ORs) for TIMI major bleeding were 0.58 (95% CI, 0.31–1.08) for VKAs plus a P2Y12 regime, 0.49 (95% CI, 0.30–0.82) for DOACs plus a P2Y12 inhibitor regime, and 0.70 (95% CI, 0.38–1.23) for DOACs plus a DAPT regime, respectively, using VKAs plus a DAPT regime for comparison. Concurrently, the ORs for MACE were 0.96 (95% CI, 0.60–1.46) for VKAs plus a P2Y12 inhibitor, 1.02 (95% CI, 0.71–1.47) for DOACs plus a P2Y12 inhibitor, and 0.94 (95% CI, 0.60–1.45) for DOACs plus a DAPT, respectively [18]. The positive data from these studies set the stage for exploring novel oral anticoagulants (NOACs) in an LV thrombus in conjunction with antiplatelets.

Table 1. Trials of oral anticoagulation therapy comparing dual therapy with triple therapy in patients with atrial fibrillation with acute coronary syndrome. Notes: DAPT—dual antiplatelet; ISTH-International Society of Thrombosis and Hemostasis; VKA—vitamin K antagonist; C—clopidogrel; A—aspirin; TIMI—thrombolysis in myocardial infarction.

Trial	Year	Drugs Compared	Number	Follow-Up	Primary End Points
WOEST	2013	VKA + C, VKA + DAPT	563	12 months	Total number of TIMI bleeding events
ISAR-TRIPLE	2015	VKA + A, VKA + DAPT	614	9 months	Composite of death, MI, definite stent thrombosis, stroke, and TIMI major bleeding
PIONEER- AF PCI	2016	Rivaroxaban (2.5/5) + C, VKA + DAPT	1415	12 months	A composite of major bleeding or minor bleeding event according to the TIMI or bleeding requiring medical attention.
REDUAL PCI	2017	Dabigatran (110/150) + C, VKA + DAPT	2725	24 months	A composite of major or clinically relevant nonmajor bleeding event according to ISTH
AUGUSTUS PCI	2019	VKA + C, Apixaban + C, VKA + DAPT	4614	6 months	A composite of major or clinically relevant nonmajor bleeding event according to ISTH
ENTRUST AF PCI	2019	Edoxaban + C, VKA + DAPT	1506	12 months	Major or clinically relevant nonmajor bleeding event according to ISTH

5. DOACs in LV Thrombus—The Clinical Experience

5.1. Case Reports

The initial data emerged with multiple case reports that demonstrated a resolution of an LV thrombus with use of DOACs [19–25]. Most of these patients had an LV thrombus in the setting of acute MI. One of these cases was of hypertrophic cardiomyopathy, while two had nonischemic heart failure. The majority of patients had a resolution of the thrombus by the end of 1 month with DOAC use, while one case demonstrated thrombus resolution by as early as 7 days [24]. None of these reported any bleeding or systemic embolism with DOACs.

5.2. Observational Studies and Case Series

Iqbal et al. performed a retrospective observational cohort study comparing DOAC therapy with VKAs in patients with an LV thrombus [26]. In this study, 74% patients received warfarin, and 26% patients received DOACs. There was no significant difference in the rate of stroke or that of other thromboembolic events between the groups (2% vs. 0%, respectively, $p = 0.55$). There were six episodes of clinically significant bleeding in the study, all of which were seen with warfarin-based triple therapy (10% vs. 0%, $p = 0.13$). The indication of fewer bleeding events with DOAC-based triple therapy as compared with VKA-based triple therapy was clearly apparent. Subsequently, multiple retrospective and prospective observational studies and case series have evaluated DOACs for anticoagulation in the context of an LV thrombus, as summarized in Table 2 [27–35]. Thrombus resolution on follow-up was the major end point in most of the studies, and DOACs were similar or superior to VKAs in all of them (Figure 3). Both the rate of resolution and the time of resolution were equal or better with DOACs. Bleeding events were similar or lower with DOACs in all the studies described in the Table when compared with VKAs. In one study, bleeding events necessitating transfusion were noted with DOACs, but all these patients had concomitant antiplatelets [29]. Additionally, systemic embolism and stroke rates were evaluated by many, and DOACs were again as efficacious as VKAs in the majority. Although inherently limited by their nonrandomized nature, the variability of the types, the doses of DOACs utilized, and the nonuniform end points evaluated, the plethora of studies do herald the era of DOAC anticoagulation for LV thrombi.

Table 2. Data from observational studies comparing anticoagulation with DOACs vs. VKAs in patients with LV thrombus. Notes: DOAC—direct oral anticoagulant; OAC—oral anticoagulant; VKA—vitamin K antagonist; LV—left ventricle; HR—hazard ratio.

Study	Number of Patients	Anticoagulant Profile	End Points	Follow-Up	Outcome
Robinson et al. (2018) [27]	84	No OAC: 16 patients Warfarin: 40 patients NOACs: 35 patients Other OACs: 7 patients	Survival free of stroke and systemic embolism	1 year	No difference 88% vs. 77.9%, $p = 0.719$.
Jaidka et al. (2018) [28]	49	Warfarin: 37 patients NOACS: 12 patients	Thrombus resolution, embolic events, bleeding events	6 months	No difference in bleeding or embolic events. Thrombus resolution also not different between VKAs and NOACs (69.2% vs. 88.9%; $p = 0.245$).
Fleddermann et al. (2019) [29]	52	Only NOAC—apixaban = 26 Rivaroxaban = 24 Dabigatran = 2	Rate of LV thrombus resolution; bleeding	264 days	83% had resolution of LV thrombus on follow-up echocardiogram. 1 cardioembolic event and 4 bleeding events requiring transfusion.
Daher et al. (2020) [30]	59	Warfarin: 42 patients NOACS: 17 patients	Rate of LV thrombus resolution	3 months	Thrombus resolution was similar in patients on NOACs (70.6%) and those on VKAs (71.4%; $p = 0.9$).
Jones et al. (2020) [31]	101	Warfarin: 60 patients NOACS: 41 patients	Primary—rate of LV thrombus resolution; secondary—rate of bleeding	2.2 years	Thrombus resolution earlier and greater with NOACs (82% vs. 64.4%, $p = 0.0018$). Bleeding rates lower with NOACs (0% vs. 6.7%, $p = 0.030$). No difference in rates of systemic thromboembolism (5% vs. 2.4%, $p = 0.388$).
Guddeti et al. (2020) [32]	99	Warfarin: 80 patients NOACS: 19 patients	Occurrence of ischemic stroke, bleeding, and thrombus resolution	1 year	No difference between stroke within 1 year or bleeding between two groups (numerically higher event in warfarin group); thrombus resolution was similar between groups (80% vs. 81%, $p = 0.9$).
Alcalai et al. (2020) [33]	25	Warfarin: 12 patients Apixaban: 13 patients	Primary end point: thrombus resolution Secondary end point: systemic embolism, major bleeding, and death from any cause	3 months	Complete thrombus resolution in all patients with warfarin and 12 out of 13 patients in apixaban group. 2 major bleeding events in warfarin group and none in apixaban group.
Robinson et al. (2020) [34]	514	Warfarin: 300 patients NOACs (apixaban in majority): 185 patients No OAC: 93 patients 64 switched regimens	Stroke and systemic embolism (SSE)	~1 year (351 days)	NOAC use associated with higher SSE risk compared with VKA use (HR—2.64–2.71); prior stroke or embolism also associated with higher SSE risk.
Albabtain et al. (2021) [35]	63	Warfarin: 35 patients NOAC (rivaroxaban): 28 patients	Time to thrombus resolution, bleeding, stroke, and mortality	9.5 months	Median time to thrombus resolution faster with NOACs (9 months vs. 3 months, $p = 0.019$); no difference in embolism, bleeding, or mortality.

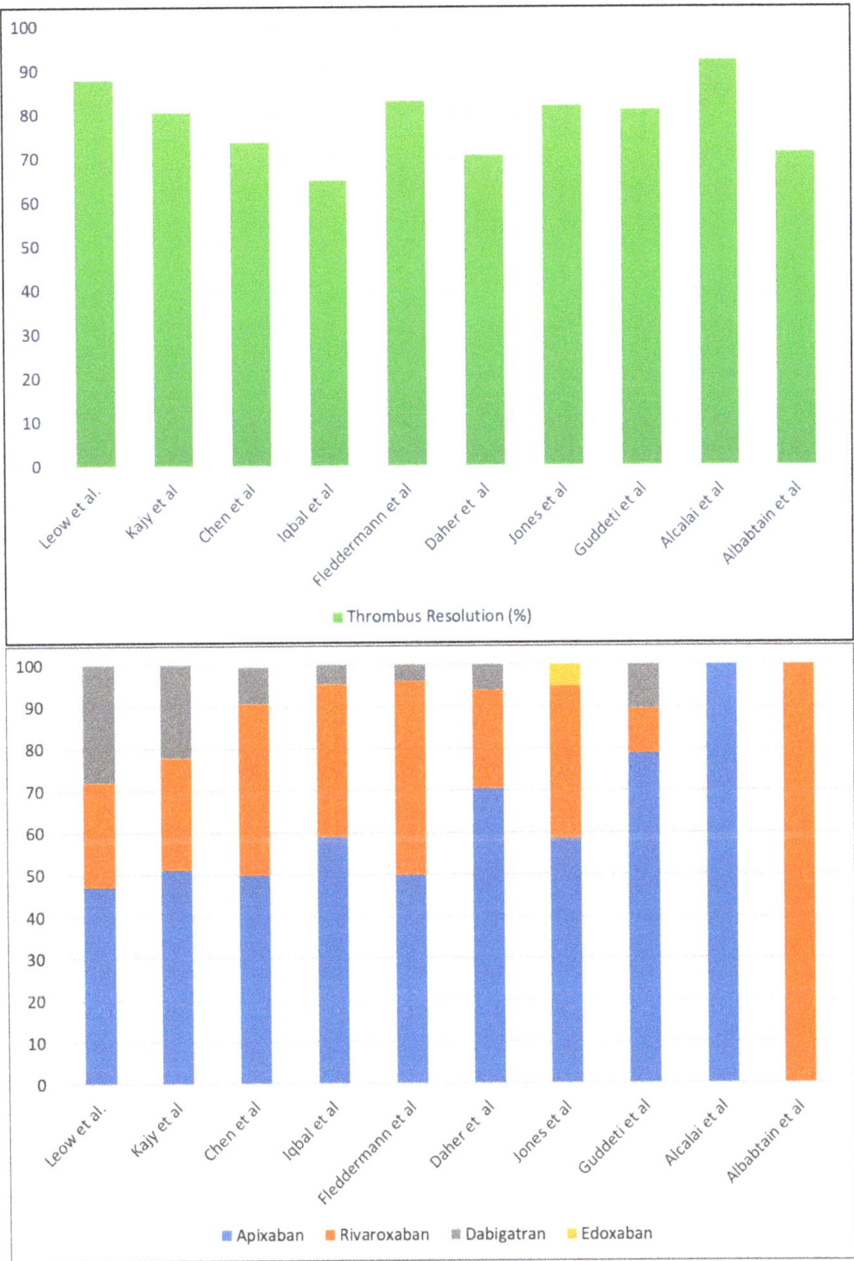

Figure 3. Rates of thrombus resolution observed in various studies/meta-analyses of studies utilizing DOACs for anticoagulation for LV thrombi (upper panel) and the proportion of various DOACs used in these studies (lower panel). Factor Xa inhibitor—apixaban has been the most extensively utilized DOAC, and edoxaban was the least favored. The first three columns represent meta-analyses, while the other seven represent individual studies.

5.3. Randomized Controlled Trial Experience

As noted above, there is paucity of RCTs for NOAC use in LV thrombi. The recently published NO-LVT study is possibly the only RCT comparing rivaroxaban (20 mg OD) with warfarin [36]. The main outcome was thrombus resolution at 1, 3, and 6 months, assessed by echocardiography, while bleeding and systemic embolism were secondary end points. In the warfarin arm, enoxaparin bridging was employed until the INR reached 2–3. Of the 79 patients randomized, complete thrombus resolution was seen in 72%, 77%, and 87% at months 1, 3, and 6, respectively, with rivaroxaban. The corresponding figures with VKAs were 48%, 67%, and 80%. At 1 month, the odds of thrombus resolution were higher with rivaroxaban compared with VKAs (187 OR—2.8; $p = 0.03$). No embolic events (stroke or systemic embolism) were seen with rivaroxaban, while bleeding was numerically lower with rivaroxaban. The major limitation was the use of TTE for assessing LV thrombi, but that is the general practice worldwide. The trial comes as a shot in the arm for DOAC use in LV thrombi.

5.4. Meta-Analysis and Systematic Reviews

Many meta-analyses, meta-summaries, and systematic reviews based on these observational studies and case reports have also shown the equivalent efficacy and better safety of DOACs in treating patients with an LV thrombus (Table 3) [37–45].

One of the largest meta-analyses was conducted by Chen et al., which included 2467 patients on anticoagulation for an LV thrombus. The common theme again was the better efficacy of DOACs for thrombus resolution, with no difference in systemic embolism or stroke. The risk of bleeding was also found to be similar between VKAs and DOACs in most of these meta-analyses.

Three recent meta-analyses of anticoagulation in LV thrombi have also shown similar results. In a systemic meta-analysis conducted by Shu Fang et al., which included 2262 patients from 12 observational studies, there was no difference in safety or efficacy. The rates of systemic embolism and stroke were 18.8% for DOACs and 22.6% for VKAs, OR = 1.01, and 8.8% vs. 11.4%, OR = 0.76, respectively. Thrombus resolution also showed similar trends in two groups (80.6% for DOACs vs. 80.2% for VKAs). However, it was noted that there were fewer bleeding and systemic embolism episodes, although these were not statistically significant. Thus, DOACs might be a safer option [46].

Similar results were reported in a meta-analysis conducted by Tetsuji Ketano et al. comprising 2612 patients. There was no difference in thrombus resolution (0.75 for VKAs vs. 0.75 for DOACs), stroke (0.06 for VKAs vs. 0.02 for DOACs), or any embolism (0.08 for VKAs vs. 0.03 for DOACs. The odds ratio for major bleeding was 0.06 for VKAs vs. 0.03 for DOACs [47].

Another meta-analysis, conducted by H. da Silva Ferraira, showed almost-equivalent efficacy and safety for both types of OACs (stroke/systemic embolism for DOACs: 109/618 vs. 386/1814 for VKAs (OR 0.86; 95% CI, 0.55–1.33; $p = 0.50$); any bleeding event: 8.7% of DOAC patients and 8.3% of VKA patients (OR 0.96, 95% CI 0.62–1.48, $p = 0.88$)) [48].

Thus, it can be concluded from these meta-analyses that DOACs are noninferior to VKAs in terms of thrombus resolution, with no difference in the risk of stroke or embolism.

Table 3. Meta-analyses, meta-summaries, and systematic reviews comparing DOAC therapy and VKA therapy in setting of LV thrombus following MI. Notes: IQR—interquartile range; DOAC—direct oral anticoagulant; OAC—oral anticoagulant; VKA—vitamin K antagonist; LV—left ventricle; OR—odds ratio.

Author (Year)	Sample Size	Study Drug	End Points	Results	Safety
Leow et al. (2018) [37]	36	Rivaroxaban (47.2%) Apixaban (25.0%) Dabigatran (27.8%)	Thrombus resolution and time to resolution.	Thrombus resolution was observed in 87.9%, and median duration of treatment to resolution was 30.0 days (IQR = 22.5–47.0).	1 nonfatal bleeding event (3.0%); no embolic events.
Kajy et al. (2020) [38]	41	Rivaroxaban (51.2%) Apixaban (26.8%) Dabigatran (22%)	Thrombus resolution and time to resolution.	Thrombus resolution—81% (R), 100% (A), and 88.9% (D). Median time of resolution—40 days (R), 36 days (A), and 24 days (D).	One nonfatal bleeding event and one stroke event were reported while on a DOAC.
Al-abcha et al. (2020) [39]	857	VKAs = 480; DOACs = 220	Primary outcome was thrombus resolution, and the secondary outcomes were major bleeding and stroke or systemic embolism (SSE).	Similar rate of thrombus resolution (odds ratio (OR) 0.97; $p = 0.90$).	Major bleeding (OR 0.62; $p = 0.27$) and systemic embolism (OR 1.86; $p = 0.05$) were not different between groups.
Chen et al. (2021) [40]	2467	Among DOAC users, apixaban (50.0%), rivaroxaban (40.8%), dabigatran (8.8%), and edoxaban (0.4%); among VKA warfarin (98.5%) was predominantly prescribed	Stroke or systemic embolism; thrombus resolution.	For prevention of stroke or systemic embolism (VKA vs. DOAC—RR: 0.96, 95% confidence interval (CI): 0.80–1.16, $p = 0.68$); for thrombus resolution (VKA vs. NOAC—RR: 0.88, 95% CI: 0.72–1.09, $p = 0.26$); for risk of stroke (VKA vs. DOAC—RR: 0.68, 95% CI: 0.47–1.00, $p = 0.048$).	For risk of any bleeding; no difference between VKAs and DOACs (RR: 0.94, 95% CI: 0.67–1.31, $p = 0.70$); for clinically relevant bleedings—lower risk with DOAC users (RR: 0.35, 95% CI: 0.13–0.92, $p = 0.03$) compared with VKA users.
Burmister et al. (2021) [41]	2153	570 on DOACs vs. 1583 on VKAs)	LV thrombus resolution, thromboembolic events, and thromboembolic stroke.	Thrombus resolution was significantly higher in DOACs compared with VKAs (RR: 1.18 (95% CI: 1.04–1.35); $p = 0.01$, I2 = 25%); no significant difference existed between DOACs and VKAs regarding overall thromboembolic events (RR: 1.10 (95% CI: 0.75–1.62); $p = 0.61$) or embolic strokes (RR: 0.63 (95% CI: 0.39–1.02); $p = 0.06$).	No difference in all-cause death (RR-0.84, $p = 0.53$) or bleeding (RR-1.00, $p = 0.9$).
Trongtorsak (2021) [42]	1771	DOACs—426 and VKAs—1345	Stroke, systemic embolism. Thrombus resolution, bleeding.	No significant differences in rates of systemic embolism or LV thrombus resolution.	Bleeding similar between two groups.
Shah et al. (2021) [43]	867		Systemic embolism and LV thrombus resolution.	Systemic embolic events (SEE)—2.7%; thrombus—86.6%.	Bleeding (composite of major and minor) and major bleeding—5.6% and 1.1%, respectively.
Abdelaziz et al. (2021) [44]	700	VKAs = 480; DOACs = 220.	Stroke or systemic embolism (SSE). Secondary outcomes were thrombus resolution, bleeding, and death.	For stroke or systemic embolism (SSE), lower rates with VKAs compared with DOACs (5.2% vs. 9%; OR = 0.54, $p = 0.05$).	Rates of thrombus resolution (OR = 1.00, $p = 0.99$) and bleeding (OR = 1.62, $p = 0.27$) and death (OR = 1.09, $p = 0.79$) were similar.

Table 3. *Cont.*

Author (Year)	Sample Size	Study Drug	End Points	Results	Safety
Tetsuji Ketano et al. (2021) [47]	2612	VKAs = 2004; DOACs = 608	Thrombus resolution, stroke, any thromboembolism, and major bleeding.	No difference in thrombus resolution (0.75 for VKAs vs. 0.75 for DOACs), stroke (0.06 for VKAs vs. 0.02 for DOACs), or any embolism (0.08 for VKAs vs. 0.03 for DOACs).	OR for major bleeding -0.06 & 0.03 for VKAs DOAC respectively.
Saleh et al. (2021) [45]	2395		Primary—thrombus resolution; secondary—occurrence of major bleeding and stroke or systemic embolization (SSE).	The rates of thrombus resolution for VKAs and DOACs were equal (71.9% vs. 71.4%; $p = 0.36$). Systemic embolism was also similar between arms (21.3% and 15.6%, respectively; $p = 0.57$).	Major bleeding rates were similar between DOACs and VKAs. (8.2% vs. 7.1%, $p = 0.57$, OR = 0.87).
Shu Fang et al. (2022) [46]	2262	VKAs = 1575; NOACs = 570	Thrombus resolution, stroke/SSE, bleeding, and mortality.	The rate for SSE (OR 1.01, $p = 0.95$) and that for thrombus resolution (OR = 1.15) were similar.	Similar bleeding risk(OR = 0.78).
H. da Silva Ferraira (2022) [48]	2432	DOACs = 618; NOACs = 1814	Stroke/SSE and bleeding events.	DOACs vs. VKAs (OR = 0.86).	8.7% for DOACs vs. 8.3% for VKAs.

5.5. Use of Anticoagulation for Prevention of LV Thrombus Formation

Previously, conventional triple therapy comprising DAPT plus VKAs was used to prevent LV thrombus formation in patients with a high-risk ST elevation MI such as large anterior wall ST elevation MI—LV ejection fraction < 30%, dyskinetic LV, or formation of LV aneurysm [46]. However, this practice was not supported by high-quality evidence. Moreover, triple therapy increased the risk of major bleeding, so interest in this area has been waning.

Zhang et al. studied the prophylactic use of rivaroxaban for LV thrombus after anterior ST elevation MI [49]. The study comprised 279 patients who underwent PCI and were randomized in a one-to-one manner to either rivaroxaban (2.5 mg twice daily for 30 days) plus DAPT or DAPT alone. The primary end point was the formation of an LV thrombus within 30 days. The net clinical adverse event included all-cause mortality, LV thrombus formation, systemic embolism, rehospitalization for cardiovascular events, and bleeding. There was a significant reduction in LV thrombus formation by rivaroxaban (0.7% vs. 8.6%). The net adverse events were also lower in the rivaroxaban group, while there were no differences in bleeding events at 30 or 180 days.

Thus, the use of the shortest possible course of triple therapy comprising DOACs with the further continuation of DAPT as required may be used for at-risk patients and is an area of research in such patients. However, further research is needed in this regard for strong validation [49].

5.6. A Note of Dissent

While the majority of case reports, case series, and nonrandomized studies favor the use of DOACs in settings involving an LV thrombus, some have produced disparate results. Robinson et al. demonstrated higher stroke and systemic embolism rates (hazard ratio: 2.6–2.7) with DOAC use for an LV thrombus [34]. The large sample size, multicenter design, and longer follow-up are strengths of the study. Similarly, Abdelaziz, in a recent meta-analysis, found that lower rates of stroke and systemic embolism were noted with VKAs vs. DOACs [43]. These results advocate caution and argue against the blanket use of DOACs for LV thrombi without additional consideration of ischemic and bleeding risks.

The genesis of LV thrombi is multifactorial, including stasis and endothelial dysfunction, whereas the left atrial (LA) thrombus is primarily stasis induced. It also noteworthy that DOACs in AF are used principally to prevent the genesis of an LA thrombus, but the thrombus is already in situ in the current scenario. In this case, the type of DOAC could be of importance: factor Xa inhibitors versus direct thrombin inhibitors. One hypothesis postulated is that dabigatran binds thrombin in a one-to-one molecular ratio, while one factor Xa leads to the generation of 1000 thrombin molecules, making factor Xa inhibition more attractive. Few cases of LV thrombi on dabigatran therapy have emerged in the literature [50,51]. The use of dabigatran for anticoagulation in mechanical heart valves was also unsuccessful in a RE-ALIGN study. However, factor Xa inhibitors have been preferentially utilized in LV thrombus studies (Figure 3).

Interestingly, Robinson et al. found no effect from oral/parenteral anticoagulation use on LV thrombus resolution during follow-up. This contrasts with previous studies and conventional wisdom. Contemporary studies have shown that prolonged anticoagulation attenuates rates of major adverse cardiovascular events and embolic events in patients with an LV thrombus [52].

An analogy can be drawn from the use of DOACs in situations including AF undergoing PCI. DOAC-based combination therapy has now shown to be noninferior in comparison to warfarin-based therapy in reducing ischemic events while showing simultaneous superiority in reducing serious bleeding in AF patients undergoing PCI [15]. However, none of the individual trials were powered enough to assess the ischemic events. In fact, some signals of numerically increased stent thrombosis have emerged in a meta-analysis, advising caution [53].

The use of an echocardiographic resolution of LV thrombi as an end point is marred by the low sensitivity of echocardiography, the varied time interval between echocardiographic acquisitions, and the differential frequency of imaging used in these studies, calling for clinical event-driven end points in future studies.

More recently, the failure of two large DOAC trials in rheumatic heart disease and prosthetic heart valves, respectively, further bolsters the role of VKAs as a first-line therapy for non-AF-based indications of OACs. Rivaroxaban failed to improve outcomes compared with VKAs in the large randomized INVICTUS study in the setting of rheumatic mitral valve disease [54]. Similarly, a trial of apixaban in the setting of prosthetic heart valves (the ON-X valve in the PROACT-Xa trial) was stopped prematurely owing to futility [55]. Though the results are not generalizable to the current context, they at least give an indication for slowing down the pace of the universal acceptance of DAOCs for LV thrombi.

6. Future Directions

Large and adequately powered RCTs comparing DOACs and VKAs with at least 6–12-month follow-ups are the need of the hour. With the prompt revascularization and institution of secondary prevention therapies attenuating the rates of LV thrombus formation following MI, this seems to be an uphill task. The **EARLYmyo-LVT (NCT03926780;** n = 280) is an ongoing study comparing rivaroxaban (15 mg OD) with warfarin (target INR: 2–2.5) as a part of triple therapy post MI. The rate of thrombus resolution at 3 months and bleeding events are the primary end points. Another ongoing study (**NCT03232398**) is comparing apixaban (5 mg BD) in LV thrombi versus warfarin (target INR: 2–3) for post MI. The primary end point again is the echocardiographic resolution of a thrombus after 3 months of therapy, and it plans to recruit 50 patients.

7. Choice for Anticoagulation—Practical Considerations and Guidelines

7.1. Utilizing Risk Scores for Decision-Making

The choice of DOACs versus VKAs for the anticoagulation regime is a matter of debate. In the absence of large RCTs, few practical considerations deserve merit. Three potential factors need to be considered: bleeding risk with VKAs (assessed by a **HAS-BLED** score), the ability to maintain therapeutic INR with VKAs (assessed by a **SAMeTT2R2** score), and financial considerations. If the patient has a high bleeding risk and/or there is difficulty in achieving therapeutic INR, DOACs should be preferred. Otherwise, VKAs should be the choice for anticoagulation. Additionally, when there are financial constraints, VKAs should be used thanks to their low cost. When combining OACs and DAPT, the duration of triple therapy should be kept to no longer than 1 month, according to the data extrapolated from trials of AF patients undergoing PCI [6–9,15].

A recently published review article on triple therapy in the setting of PCI suggested that therapy can be individualized on the basis of patients with thrombotic and bleeding risks, by taking into account the time frame post PCI. The authors suggested four time frames, 0–1 month, >1–6 months, >6–12 months, and >12 months. In the first month post PCI, all the patients can be given DOACs plus P2Y12 inhibitors, and aspirin can be added in those patients with high thrombotic but low bleeding risks. After 1 month and until 6 months, all patients are to be kept on DOACs plus P2Y12 inhibitors. In the next 6 months, patients with low bleeding risk to be kept on DOACs plus P2Y12 inhibitors, irrespective of thrombotic risk, and only on DOACs if the bleeding risk is high. Beyond 12 months, all the patients should be on DOACs only [56].

7.2. Suggested Algorithm

In medically managed patients, a dual therapy is preferred in order to curtail the bleeding risk while patients undergoing PCI will need an initial triple therapy regimen (DAPT+OAC). For one of our patients, we prescribed dual therapy with VKAs, while for another patient, we gave dual therapy with DOACs. Interestingly, both patients responded well to dual therapy, and there was a resolution of the LV thrombus at 1 month. More importantly, there were no thromboembolic events; neither were there any bleeding episodes. Certain clinical features that predict a high risk of stroke, such as a prior systemic embolism, the protrusion of a thrombus into the cavity, a recurrent thrombus, and the nonresolution of a thrombus from the initial therapy, may call for the preferential use of warfarin-based anticoagulation [10,34]. Patients with a high risk of stent thrombosis (recurrent ACS, multiple stents, complex bifurcation PCI, heavily calcified lesions, total stent length >60 mm, or bioabsorbable stents) may benefit from the extended duration of initial triple therapy [57].

A suggested algorithm regarding the choice and duration of anticoagulation use in LV thrombi that is based on the current literature is presented in Figure 4.

Nonetheless, the lack of a predictable anticoagulant response, narrow therapeutic range, and need for frequent monitoring has spurred the more widespread use of DOACs and use of DOAC-based combination therapy in AMI patients who require concomitant oral anticoagulation. Because rivaroxaban and apixaban are now off patent, the financial constraints may no longer be a valid argument in many geographical regions, leading to increased prescriptions. However, as previously detailed, there is no need to jump the queue in utilizing DOACs until their noninferiority is established in large RCTs, and they should still be alternatives to VKAs on case-by-case bases.

7.3. Guideline Track

The 2014 ASA guidelines do recommend the use of DOACs in patients who are intolerant to warfarin [12]. In patients with apical akinesis/dyskinesis, OAC use has been given a Class IIb recommendation by the 2013 ACC/AHA guidelines for the management of a STEMI, as well as by the 2014 ACC/ASA guidelines for the prevention of stroke [4,12]. The 2018 Canadian Cardiovascular Society/Canadian Association of Interventional Cardiology Guidelines recommend only VKAs for patients with an established LV thrombus undergoing PCI for acute or stable indication [57]. They suggest the discontinuation of OACs beyond 3 months if there is no echocardiographic evidence of a thrombus, similar to ACC/AHA guidelines. They do acknowledge a lack of adequate evidence in this scenario.

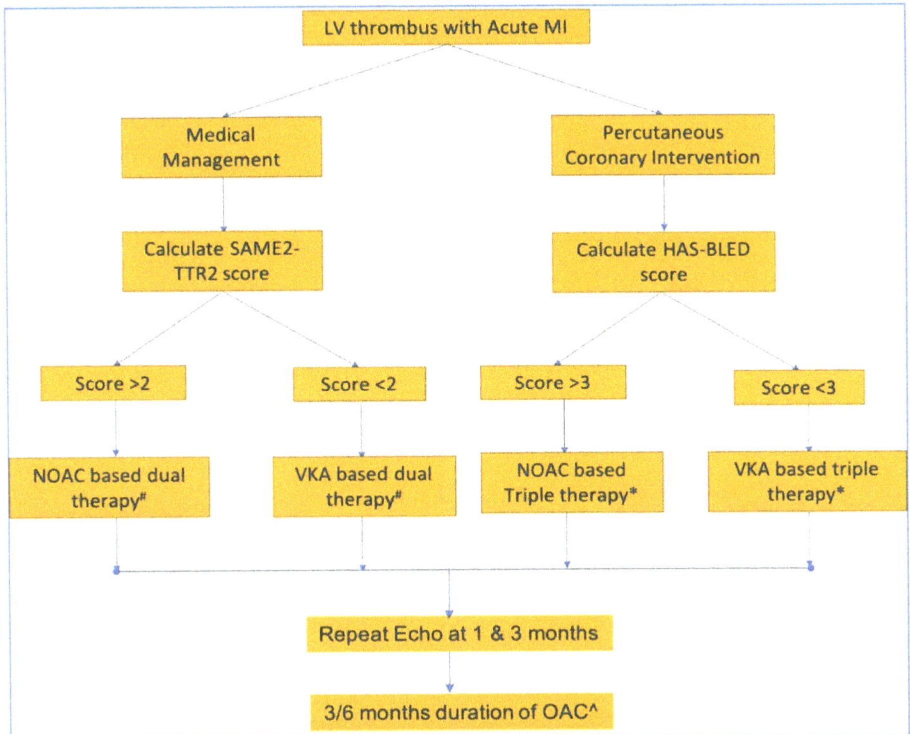

Figure 1. An approach to anticoagulation in LV thrombus complicating acute MI utilizing risk stratification scores. Dual therapy refers to a combination of single antiplatelet agent and oral anticoagulation. Triple therapy refers to a combination of dual antiplatelet therapy and oral anticoagulation. Presence of high-risk features warrants upgradation to VKAs from NOACs or triple therapy from dual. The variables used in HAS-BLED score include hypertension, abnormal renal or liver enzymes, stroke, bleeding events, labile INR, elderly age (>65 yrs), and drugs or ethanol. The SAMeT2TR2 score variables are sex (female), age (< 60 yrs), medical history, treatment strategy (rhythm control), tobacco use, and race (nonwhite race). Notes: #—prasugrel and ticagrelor should be avoided in dual or triple therapy; * generally continued for 1 month, followed by dual therapy to avoid bleeding, and in the case of embolic events, it can be continued beyond 1 month; ^- ACC/AHA and CCS/CAIS guidelines advocate a 3-month regimen, while ESC guidelines prescribe a 6-month duration.

The recent scientific statement of the AHA on the management of LV thrombi suggests using anticoagulation for 3 months, and thereafter, imaging should be performed to determine thrombus resolution. For patients with a history of a more distant MI, a longer duration of OACs up to 6 months may be considered. If there is a resolution of the LV thrombus, anticoagulation can be stopped. However, in case imaging is needed before 3 months for some other reasons and there is thrombus resolution, anticoagulation can be stopped earlier. It also suggests that if there is a clinical suggestion of a LV thrombus and the echo does not visualize a thrombus or if the echo is not confirmative, a cardiac MRI (CMR) should be conducted. It also suggests DOACs as reasonable alternatives to warfarin on the basis of supportive evidence. If the thrombus persists beyond 3 months, particularly a protruding thrombus, a trial of alternative anticoagulation should be considered: the use of DOACs with repetitive subtherapeutic INR if the patient was on warfarin or the use of warfarin if the patient was previously on DOACs [58].

Because of the relatively weak evidence, these latest guidelines suggest that the use of OACs in patients with revascularized anterior MI (usually primary PCI) may be considered. However, such a consideration should take into account the perceived risk of thrombus formation and bleeding risk and should involve shared decision-making. The treatment duration should be 1–3 months, depending on the bleeding risk [58].

Additional maneuvers that can be utilized to attenuate bleeding risk while combining antiplatelets with antithrombotic are summarized in Table 4.

Table 4. Methods to mitigate bleeding risk with a combination of antiplatelet and anticoagulant therapy.

Methods
Use of lower doses of aspirin
Proton pump inhibitor use
Avoid potent P2Y12 inhibitor—ticagrelor and prasugrel
Shorten the duration of DAPT
De-escalation of DAPT
Radial access in case of PCI
Sparing use of glycoprotein IIb/IIIa

8. Conclusions

According to the current evidence, it can be stated that if an LV thrombus is detected in a setting of AMI, (VKA-based) oral anticoagulation targeted to an INR of 2.0–3.0 has been the standard of care. DOACs have emerged as acceptable alternatives to VKAs in this scenario, owing to challenges in their use—such as their high bleeding risk, food interactions, need for repeated INR monitoring, and failure to achieve therapeutic range in many patients. A plethora of successful studies in the form of case reports, case series, observation studies, small RCT and meta-analyses have now demonstrated the utility of DOACs in better thrombus resolution and less bleeding. There have been some signals of an increased risk of stroke and systemic embolism in some studies with DOACs, but unfortunately, there have been no large randomized studies to date. Hence, if a patient is unable to achieve therapeutic INR (high SAME-TT2R2) or if they have a high bleeding risk (high HAS-BLED) with VKAs, full-dose DOACs should be prescribed instead of VKAs. This approach has the potential to attenuate bleeding risks while preserving efficacy [59]. The duration of anticoagulation is not defined but should be continued for at least 3 months, guided by a similar imaging modality to what was used earlier (or CMR if needed), to evaluate the resolution of an LV thrombus. If there is no LV thrombus on repeated echocardiographic evaluations, OACs can be stopped and DAPT should be started, which can be continued for 1 year. A repeat imaging after the cessation of OACs is prudent to detect the recurrence of a thrombus or a small nidus of a thrombus previously missed. In patients who continue to have some spontaneous echo contrast or a well-organized thrombus in the area of a wall motion abnormality, the optimal duration of OACs is not well defined. The further continuation of OACs can be made on a case-by-case basis. Large and well-designed trials comparing VKAs and DOACs in this setting are warranted.

Funding: This research received no external funding.

Conflicts of Interest: The authors declare no conflict of interest.

References

1. Sanchis-Gomar, F.; Perez-Quilis, C.; Leischik, R.; Lucia, A. Epidemiology of coronary heart disease and acute coronary syndrome. *Ann. Transl. Med.* **2016**, *4*, 256–267.
2. Solheim, S.; Seljeflot, I.; Lunde, K.; Bjørnerheim, R.; Aakhus, S.; Forfang, K.; Arnesen, H. Frequency of left ventricular thrombus in patients with anterior wall acutemyocardial infarction treated with percutaneous coronary intervention and dual antiplatelet therapy. *Am. J. Cardiol.* **2010**, *106*, 1197–1200.
3. Delewi, R.; Nijveldt, R.; Hirsch, A.; Marcu, C.B.; Robbers, L.; Hassell, M.E.; de Bruin, R.H.; Vleugels, J.; van der Laan, A.M.; Bouma, B.J.; et al. Left ventricular thrombus formation after acute myocardial infarction as assessed by cardiovascular magnetic resonance imaging. *Eur. J. Radiol.* **2012**, *81*, 3900–3904. [PubMed]

4. O'Gara, P.T.; Kushner, F.G.; Ascheim, D.D.; Casey, D.E.; Chung, M.K.; de Lemos, J.A.; Ettinger, S.M.; Fang, J.C.; Fesmire, F.M.; Franklin, B.A.; et al. 2013 ACCF/AHA guideline for the management of st-elevation myocardial infarction: A report of the American college of cardiology foundation/american heart association task force on practice guidelines. *J. Am. Coll. Cardiol.* **2013**, *61*, e78–e140. [CrossRef] [PubMed]
5. Le May, M.R.; Acharya, S.; Wells, G.A.; Burwash, I.; Chong, A.Y.; So, D.Y.; Glover, C.A.; Froeschl, M.P.; Hibbert, B.; Marquis, J.-F.; et al. Prophylactic Warfarin Therapy After Primary Percutaneous Coronary Intervention for Anterior ST-Segment Elevation Myocardial Infarction. *JACC: Cardiovasc. Interv.* **2015**, *8*, 155–162. [CrossRef]
6. Cannon, C.P.; Bhatt, D.L.; Oldgren, J.; Lip, G.Y.; Ellis, S.G.; Kimura, T.; Maeng, M.; Merkely, B.; Zeymer, U.; Gropper, S.; et al. Dual Antithrombotic Therapy with Dabigatran after PCI in Atrial Fibrillation. *N. Engl. J. Med.* **2017**, *377*, 1513–1524.
7. Gibson, C.M.; Mehran, R.; Bode, C.; Halperin, J.; Verheugt, F.W.; Wildgoose, P.; Birmingham, M.; Ianus, J.; Burton, P.; van Eickels, M.; et al. Prevention of Bleeding in Patients with Atrial Fibrillation Undergoing PCI. *N. Engl. J. Med.* **2016**, *375*, 2423–2434. [PubMed]
8. Lopes, R.D.; Heizer, G.; Aronson, R.; Vora, A.N.; Massaro, T.; Mehran, R.; Goodman, S.G.; Windecker, S.; Darius, H.; Li, J.; et al. Antithrombotic Therapy after Acute Coronary Syndrome or PCI in Atrial Fibrillation. *N. Engl. J. Med.* **2019**, *380*, 1509–1524.
9. Vranckx, P.; Valgimigli, M.; Eckardt, L.; Tijssen, J.; Lewalter, T.; Gargiulo, G.; Batushkin, V.; Campo, G.; Lysak, Z.; Vakaliuk, I.; et al. Edoxaban-based versus vitamin K antagonist-based antithrombotic regimen after successful coronary stenting in patients with atrial fibrillation (ENTRUST-AF PCI): A randomized, open-label, phase 3b trial. *Lancet* **2019**, *394*, 1335–1343.
10. Maniwa, N.; Fujino, M.; Nakai, M.; Nishimura, K.; Miyamoto, Y.; Kataoka, Y.; Asaumi, Y.; Tahara, Y.; Nakanishi, M.; Anzai, T.; et al. Anticoagulation combined with antiplatelet therapy in patients with left ventricular thrombus after first acute myocardial infarction. *Eur. Heart J.* **2017**, *39*, 201–208.
11. Leow, A.S.T.; Sia, C.H.; Tan, B.Y.Q.; Kaur, R.; Yeo, T.C.; Chan, M.Y.Y.; Tay, E.L.W.; Seet, R.C.S.; Loh, J.P.Y.; Yeo, L.L.L. Characterization of acute ischemic stroke in patients with left ventricular thrombi after myocardial infarction. *J. Thromb. Thrombolysis.* **2019**, *48*, 158–166. [PubMed]
12. Kernan, W.N.; Ovbiagele, B.; Black, H.R.; Bravata, D.M.; Chimowitz, M.I.; Ezekowitz, M.D.; Fang, M.C.; Fisher, M.; Furie, K.L.; Heck, D.V.; et al. Guidelines for the prevention of stroke in patients with stroke and transient ischemic attack: A guideline for healthcare professionals from the American Heart Association/American Stroke Association. *Stroke* **2014**, *45*, 2160–2236. [CrossRef] [PubMed]
13. Ibanez, B.; James, S.; Agewall, S.; Antunes, M.J.; Bucciarelli-Ducci, C.; Bueno, H.; Caforio, A.L.; Crea, F.; Goudevenos, J.A.; Halvorsen, S.; et al. 2017 ESC Guidelines for the management of acute myocardial infarction in patients presenting with ST-segment elevation: The task force for the management of acute myocardial infarction in patients presenting with ST-segment elevation of the European Society of Cardiology (ESC). *Eur. Heart J.* **2018**, *39*, 119–177. [PubMed]
14. Hansen, M.L.; Sørensen, R.; Clausen, M.T.; Fog-Petersen, M.L.; Raunsø, J.; Gadsbøll, N.; Gislason, G.; Folke, F.; Andersen, S.S.; Schramm, T.K.; et al. Risk of Bleeding with Single, Dual, or Triple Therapy With Warfarin, Aspirin, and Clopidogrel in Patients With Atrial Fibrillation. *Arch. Intern. Med.* **2010**, *170*, 1433–1441. [CrossRef]
15. Pradhan, A.; Bhandari, M.; Vishwakarma, P.; Sethi, R. Novel Dual Therapy: A Paradigm Shift in Anticoagulation in Patients of Atrial Fibrillation Undergoing Percutaneous Coronary Intervention. *TH Open* **2020**, *4*, e332–e343. [CrossRef]
16. Dewilde, W.J.; Oirbans, T.; Verheugt, F.W.; Kelder, J.C.; De Smet, B.J.; Herrman, J.P.; Adriaenssens, T.; Vrolix, M.; Heestermans, A.A.; Vis, M.M.; et al. Use of clopidogrel with or without aspirin in patients taking oral anticoagulant therapy and under going percutaneous coronary intervention: An open-label, randomized, controlled trial. *Lancet* **2013**, *381*, 1107–1115. [CrossRef]
17. Fiedler, K.A.; Maeng, M.; Mehilli, J.; Schulz-Schupke, S.; Byrne, R.A.; Sibbing, D.; Hoppmann, P.; Scheider, S.; Fusaro, M.; Ott, I.; et al. Duration of triple therapy in patients requiring oral anticoagulation after drug-eluting stent implantation: The ISAR-TRIPLE trial. *J. Am. Coll. Cardiol.* **2015**, *65*, 1619–1629. [CrossRef]
18. Lopes, R.D.; Hong, H.; Harskamp, R.E.; Bhatt, D.L.; Mehran, R.; Cannon, C.P.; Granger, C.B.; Verheugt, F.W.; Li, J.; Jurriën, M.; et al. Safety and Efficacy of Antithrombotic Strategies in Patients with Atrial Fibrillation Undergoing Percutaneous Coronary Intervention: A Network Meta-analysis of Randomized Controlled Trials. *JAMA Cardiol.* **2019**, *4*, 747–755.
19. Kaya, A.; Hayıroğlu, M.I.; Keskin, M.; Tekkeşin, A.I.; Alper, A.T. Resolution of left ventricular thrombus with apixaban in a patient with hypertrophic cardiomyopathy. *Turk. Kardiyol. Dern. Ars.* **2016**, *44*, 335–337.
20. Mano, Y.; Koide, K.; Sukegawa, H.; Kodaira, M.; Ohki, T. Successful resolution of a left ventricular thrombus with apixaban treatment following acute myocardial infarction. *Heart Vessels* **2016**, *31*, 118–123. [CrossRef]
21. Seecheran, R.; Seecheran, V.; Persad, S.; Seecheran, N.A. Rivaroxaban as an antithrombotic agent in a patient with ST-segment elevation myocardial infarction and left ventricular thrombus: A case report. *J. Investig. Med. High Impact Case Rep.* **2017**, *5*, 2324709617697991. [CrossRef] [PubMed]
22. Makrides, C.A. Resolution of left ventricular postinfarction thrombi in patients undergoing percutaneous coronary intervention using rivaroxaban in addition to dual antiplatelet therapy. *BMJ Case Rep.* **2016**, *2016*, bcr2016217843. [CrossRef] [PubMed]
23. Nagamoto, Y.; Shiomi, T.; Matsuura, T.; Okahara, A.; Takegami, K.; Mine, D.; Shirahama, T.; Koga, Y.; Yoshida, K.; Sadamatsu, K.; et al. Resolution of a left ventricular thrombus by the thrombolytic action of dabigatran. *Heart Vessels* **2014**, *29*, 560–562. [CrossRef]
24. Nakasuka, K.; Ito, S.; Noda, T.; Hasuo, T.; Sekimoto, S.; Ohmori, H.; Inomata, M.; Yoshida, T.; Tamai, N.; Saeki, T.; et al. Resolution of Left Ventricular Thrombus Secondary to Tachycardia-Induced Heart Failure with Rivaroxaban. *Case Rep. Med.* **2014**, *2014*, 814524. [CrossRef]

25. Pérez, M.P.; Bravo, D.S.; Trigo, J.A.G.; Castroviejo, E.V.R.d.C.; Llergo, J.T.; Cabezas, C.L.; Guerrero, J.C.F. Resolution of left ventricular thrombus by rivaroxaban. *Future Cardiol.* **2014**, *10*, 333–336. [CrossRef] [PubMed]
26. Iqbal, H.; Straw, S.; Craven, T.P.; Stirling, K.; Wheatcroft, S.B.; Witte, K.K. Direct oral anticoagulants compared to vitamin K antagonist for the management of left ventricular thrombus. *ESC Heart Fail.* **2020**, *7*, 2032–2041. [CrossRef] [PubMed]
27. Robinson, A.; Ruth, B.; Dent, J. Direct oral anticoagulants compared to warfarin for left ventricular thrombi: A single center experience. *J. Am. Coll. Cardiol.* **2018**, *71*, A981. [CrossRef]
28. Jaidka, A.; Zhu, T.; Lavi, S.; Johri, A. Treatment of left ventricular thrombus using warfarin versus direct oral anticoagulants following anterior myocardial infarction. *Can. J. Cardiol.* **2018**, *34*, S143. [CrossRef]
29. Fleddermann, A.M.; Hayes, C.H.; Magalski, A.; Main, M.L. Efficacy of Direct Acting Oral Anticoagulants in Treatment of Left Ventricular Thrombus. *Am. J. Cardiol.* **2019**, *124*, 367–372. [CrossRef] [PubMed]
30. Daher, J.; Da Costa, A.; Hilaire, C.; Ferreira, T.; Pierrard, R.; Guichard, J.B.; Romeyer, C.; Isaaz, K. Management of Left Ventricular Thrombi with Direct Oral Anticoagulants: Retrospective Comparative Study with Vitamin K Antagonists. *Clin. Drug Investig.* **2020**, *40*, 343–353. [CrossRef]
31. Jones, D.A.; Wright, P.; Alizadeh, M.A.; Fhadil, S.; Rathod, K.S.; Guttmann, O.; Knight, C.; Timmis, A.; Baumbach, A.; Wragg, A.; et al. The Use of Novel Oral Anti-Coagulant's (NOAC) compared to Vitamin K Antagonists (Warfarin) in patients with Left Ventricular thrombus after Acute Myocardial Infarction (AMI). *Eur. Heart J. Cardiovasc Pharmacother.* **2020**, *7*, 398–404. [CrossRef] [PubMed]
32. Guddeti, R.R.; Anwar, M.; Walters, R.W.; Apala, D.; Pajjuru, V.; Kousa, O.; Gujjula, N.R.; Alla, V.M. Treatment of Left Ventricular Thrombus with Direct Oral Anticoagulants: A Retrospective Observational Study. *Am. J. Med.* **2020**, *133*, 1488–1491. [CrossRef]
33. Alcalai, R.; Rashad, R.; Butnaru, A.; Moravsky, G.; Leibowitz, D. Apixaban versus Warfarin in Patients with Left Ventricular Thrombus, a prospective randomized trial. *Eur. Heart J.* **2020**, *41*, 660–667. [CrossRef]
34. Robinson, A.A.; Trankle, C.R.; Eubanks, G.; Schumann, C.; Thompson, P.; Wallace, R.L.; Gottiparthi, S.; Ruth, B.; Kramer, C.M.; Salerno, M.; et al. Off-label Use of Direct Oral Anticoagulants Compared with Warfarin for Left Ventricular Thrombi. *JAMA Cardiol.* **2020**, *5*, 685–692. [CrossRef] [PubMed]
35. Albabtain, M.A.; Alhebaishi, Y.; Al-Yafi, O.; Kheirallah, H.; Othman, A.; Alghosoon, H.; Arafat, A.A.; Alfagih, A. Rivaroxaban versus warfarin for the management of left ventricle thrombus. *Egypt. Heart J.* **2021**, *73*, 41. [CrossRef]
36. Abdelnabi, M.; Saleh, Y.; Fareed, A.; Nossikof, A.; Wang, L.; Morsi, M.; Eshak, N.; Abdelkarim, O.; Badran, H.; Almaghraby, A. Comparative Study of Oral Anticoagulation in Left Ventricular Thrombi (No-LVT Trial). *J. Am. Coll. Cardiol.* **2021**, *77*, 1590–1592. [CrossRef]
37. Leow, A.S.-T.; Sia, C.-H.; Tan, B.Y.-Q.; Loh, J.P.-Y. A meta-summary of case reports of non-vitamin K antagonist oral anticoagulant use in patients with left ventricular thrombus. *J. Thromb. Thrombolysis* **2018**, *46*, 68–73. [CrossRef]
38. Kajy, M.; Shokr, M.; Ramappa, P. Use of Direct Oral Anticoagulants in the Treatment of Left Ventricular Thrombus: Systematic Review of Current Literature. *Am. J. Ther.* **2020**, *27*, e584–e590. [CrossRef] [PubMed]
39. Al-Abcha, A.; Herzallah, K.; Saleh, Y.; Mujer, M.; Abdelkarim, O.; Abdelnabi, M.; Almaghraby, A.; Abela, G.S. The Role of Direct Oral Anticoagulants Versus Vitamin K Antagonists in the Treatment of Left Ventricular Thrombi: A Meta-Analysis and Systematic Review. *Am. J. Cardiovasc. Drugs* **2020**, *21*, 435–441. [CrossRef]
40. Chen, R.; Zhou, J.; Liu, C.; Zhou, P.; Li, J.; Wang, Y.; Zhao, X.; Chen, Y.; Song, L.; Zhao, H.; et al. Direct Oral Anticoagulants versus Vitamin K Antagonists for Patients with Left Ventricular Thrombus: A Systematic Review and Meta-Analysis. *Pol. Arch. Intern. Med.* **2021**, *131*, 429–438. [CrossRef]
41. Burmeister, C.; Beran, A.; Mhanna, M.; Ghazaleh, S.; Tomcho, J.C.; Maqsood, A.; Sajdeya, O.; Assaly, R. Efficacy and Safety of Direct Oral Anticoagulants Versus Vitamin K Antagonists in the Treatment of Left Ventricular Thrombus. *Am. J. Ther.* **2021**, *28*, e411–e419. [CrossRef]
42. Trongtorsak, A.; Thangjui, S.; Kewcharoen, J.; Polpichai, N.; Yodsuwan, R.; Kittipibul, V.; Friedman, H.J.; Estrada, A.Q. Direct oral anticoagulants *vs.* vitamin K antagonists for left ventricular thrombus: A systematic review and meta-analysis. *Acta Cardiol.* **2021**, *76*, 933–942. [CrossRef]
43. Shah, S.; Shah, K.; Turagam, M.K.; Sharma, A.; Natale, A.; Lakkireddy, D.; Garg, J. Direct oral anticoagulants to treat left ventricular thrombus—A systematic review and meta-analysis: ELECTRAM investigators. *J. Cardiovasc. Electrophysiol.* **2021**, *32*, 1764–1771. [CrossRef]
44. Abdelaziz, H.K.; Megaly, M.; Debski, M.; Abdelrahman, A.; Abdelaziz, S.; Kamal, D.; Patel, B.; More, R.; Choudhury, T. Meta-Analysis Comparing Direct Oral Anticoagulants to Vitamin K Antagonists for The Management of Left Ventricular Thrombus. *Expert Rev. Cardiovasc. Ther.* **2021**, *19*, 427–432. [CrossRef]
45. Saleh, Y.; Al-Abcha, A.; Abdelkarim, O.; Abdelnabi, M.; Almaghraby, A. Meta-Analysis Investigating the Role of Direct Oral Anticoagulants Versus Vitamin K Antagonists in the Treatment of Left Ventricular Thrombi. *Am. J. Cardiol.* **2021**, *150*, 126–128. [CrossRef]
46. Gong, Y.-J.; Fang, S.; Zhu, B.-Z.; Yang, F.; Wang, Z.; Xiang, Q. Direct oral anticoagulants compared with vitamin K antagonists for left ventricular thrombus: A systematic review and meta-analysis. *Curr. Pharm. Des.* **2022**, *28*, 1902–1910. [CrossRef] [PubMed]
47. Kitano, T.; Nabeshima, Y.; Kataoka, M.; Takeuchi, M. Therapeutic efficacy of direct oral anticoagulants and vitamin K antagonists for left ventricular thrombus: Systematic review and meta-analysis. *PLoS ONE* **2021**, *16*, e0255280. [CrossRef]
48. Ferreira, H.d.S.; Lopes, J.L.; Augusto, J.; Simões, J.; Roque, D.; Faria, D.; Ferreira, J.; Fialho, I.; Beringuilho, M.; Morais, H.; et al. Effect of direct oral anticoagulants versus vitamin K antagonists or warfarin in patients with left ventricular thrombus outcomes: A systematic review and meta-analysis. *Rev. Port. Cardiol.* **2022**, *42*, 63–70. [CrossRef]

49. Zhang, Z.; Si, D.; Zhang, Q.; Jin, L.; Zheng, H.; Qu, M.; Yu, M.; Jiang, Z.; Li, D.; Li, S.; et al. Prophylactic Rivaroxaban Therapy for Left Ventricular Thrombus After Anterior ST-Segment Elevation Myocardial Infarction. *JACC Cardiovasc. Interv.* **2022**, *15*, 861–872. [CrossRef]
50. Adar, A.; Onalan, O.; Çakan, F. Newly developed left ventricular apical thrombus under dabigatran treatment. *Blood Coagul. Fibrinolysis* **2018**, *29*, 126–128. [CrossRef]
51. Fredgart, M.; Gill, S. Left ventricular mural thrombus despite treatment with dabigatran and clopidogrel. *BMJ Case Rep.* **2018**, *2018*, bcr2017223899. [CrossRef]
52. Lattuca, B.; Bouziri, N.; Kerneis, M.; Portal, J.-J.; Zhou, J.; Hauguel-Moreau, M.; Mameri, A.; Zeitouni, M.; Guedeney, P.; Hammoudi, N.; et al. Antithrombotic Therapy for Patients with Left Ventricular Mural Thrombus. *J. Am. Coll. Cardiol.* **2020**, *75*, 1676–1685. [CrossRef]
53. Potpara, T.S.; Mujovic, N.; Proietti, M.; Dagres, N.; Hindricks, G.; Collet, J.P.; Valgimigli, M.; Heidbuchel, H.; Lip, G.Y. Revisiting the effects of omitting aspirin in combined antithrombotic therapies for atrial fibrillation and acute coronary syndromes or percutaneous coronary interventions: Meta-analysis of pooled data from the PIONEERAF-PCI, RE-DUAL PCI, and AUGUSTUS trials. *Europace* **2020**, *22*, 33–46.
54. Connolly, S.J.; Karthikeyan, G.; Ntsekhe, M.; Haileamlak, A.; El Sayed, A.; El Ghamrawy, A.; Damasceno, A.; Avezum, A.; Dans, A.M.; Gitura, B.; et al. Rivaroxaban in Rheumatic Heart Disease-Associated Atrial Fibrillation. *N. Engl. J. Med.* **2022**, *387*, 978–988. [CrossRef]
55. Brooks, M. PROACT Xa Trial of Apixaban with On-X Heart Valve Stopped. *Medscape*. 2022. Available online: https://www.medscape.com/viewarticle/981644 (accessed on 30 December 2022).
56. Hussain, A.; Minhas, A.; Sarwar, U.; Tahir, H. Triple Antithrombotic Therapy (Triple Therapy) After Percutaneous Coronary Intervention in Chronic Anticoagulation: A Literature Review. *Cureus* **2022**, *14*, e21810. [CrossRef]
57. Mehta, S.R.; Bainey, K.R.; Cantor, W.J.; Lordkipanidzé, M.; Marquis-Gravel, G.; Robinson, S.D.; Sibbald, M.; So, D.Y.; Wong, G.C.; Abunassar, J.G.; et al. 2018 Canadian Cardiovascular Society/Canadian Association of Interventional Cardiology Focused Update of the Guidelines for the Use of Antiplatelet Therapy. *Can. J. Cardiol.* **2018**, *34*, 214–233. [CrossRef]
58. Levine, G.N.; McEvoy, J.W.; Fang, J.C.; Ibeh, C.; McCarthy, C.P.; Misra, A.; Shah, Z.I.; Shenoy, C.; Spinler, S.A.; Vallurupalli, S.; et al. Management of Patients at Risk for and With Left Ventricular Thrombus: A Scientific Statement From the American Heart Association. *Circulation* **2022**, *146*, e205–e223. [CrossRef] [PubMed]
59. McCarthy, C.P.; Vaduganathan, M.; McCarthy, K.J.; Januzzi, J.L.; Bhatt, D.L.; McEvoy, J.W. Left Ventricular Thrombus After Acute Myocardial Infarction Screening, Prevention, and Treatment. *JAMA Cardiol.* **2018**, *3*, 642–649. [CrossRef]

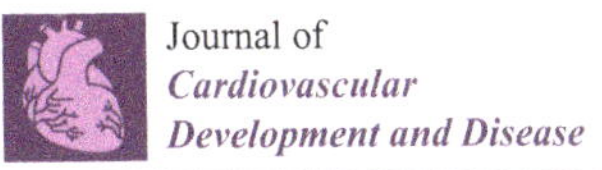
Journal of
Cardiovascular
Development and Disease

Systematic Review

Left Atrial Appendage Thrombosis and Oral Anticoagulants: A Meta-Analysis of Risk and Treatment Response

Yun-Yung Cheng [1,†], Shennie Tan [1,†], Chien-Tai Hong [1,2], Cheng-Chang Yang [1,3,*] and Lung Chan [1,2,*]

1 Department of Neurology, Shuang-Ho Hospital, Taipei Medical University, New Taipei 235, Taiwan
2 Department of Neurology, School of Medicine, College of Medicine, Taipei Medical University, Taipei 235, Taiwan
3 Brain and Consciousness Research Center, Shuang Ho Hospital, Taipei Medical University, New Taipei 235, Taiwan
* Correspondence: 19589@s.tmu.edu.tw (C.-C.Y.); cjustinmd@tmu.edu.tw (L.C.)
† These authors contributed equally to this work.

Abstract: Background: Left atrial appendage thrombus (LAAT) is the main cause of cardioembolism in patients with nonvalvular atrial fibrillation (AF). Emerging evidence indicates that direct oral anticoagulants (DOACs) may be a preferred, safer choice for patients with LAAT. However, current guidelines indicate vitamin K antagonist (VKA) as the preferred treatment for LAAT. We conducted a meta-analysis to compare the efficacy of VKA and DOAC for the treatment of LAAT. **Methods:** The search was conducted in the PubMed, Embase, Google Scholar, and Cochrane Library databases from inception to July 2022, with the language restricted to English. A first analysis was conducted to evaluate the risk of LAAT under VKA or DOAC treatment. A second analysis was conducted to compare the resolution of LAAT under VKA and DOAC treatment. **Results:** In 13 studies comparing LAAT incidence rates under VKA and DOAC treatment, significant superiority of DOAC was detected (pooled RR = 0.65, 95% CI = 0.47–0.90, $p = 0.009$) with moderate heterogeneity being identified in the pooled studies. In 13 studies comparing LAAT resolution under VKA and DOAC use, treatment with DOAC exhibited a significantly increased probability of LAAT resolution compared with VKA (pooled odds ratio = 1.52, 95% CI = 1.02–2.26, $p = 0.040$). **Conclusions:** This meta-analysis suggests a superiority of DOAC over VKA with respect to LAAT incidence in people with AF and the likelihood of LAAT resolution. Due to their established safety profile, DOAC is a preferable choice for anticoagulation, although further randomized controlled studies are warranted to provide further evidence of their suitability as a new recommended treatment.

Keywords: left atrial appendage thrombus; atrial fibrillation; stroke; oral anticoagulant

Citation: Cheng, Y.-Y.; Tan, S.; Hong, C.-T.; Yang, C.-C.; Chan, L. Left Atrial Appendage Thrombosis and Oral Anticoagulants: A Meta-Analysis of Risk and Treatment Response. *J. Cardiovasc. Dev. Dis.* **2022**, *9*, 351. https://doi.org/10.3390/jcdd9100351

Academic Editor: Giovanni Cimmino

Received: 22 August 2022
Accepted: 5 October 2022
Published: 13 October 2022

1. Introduction

Stroke is the leading cause of death and disability worldwide [1]. Ischemic stroke, which accounts for more than 70% of the overall incidence of stroke in developed countries, has various causes, such as large artery atherosclerosis in cerebral circulation, occlusion of cerebral small vessels, and cardiac embolism [2]. Of these causes, cardiac embolism contributes most to the increasing incidence of ischemic stroke [3]. Atrial fibrillation (AF) also independently contributes to the increased occurrence of ischemic stroke and is the most common sustained arrhythmia in older adults. In nonvalvular AF, the left atrial appendage (LAA) is the location most susceptible to thrombus formation, accounting for more than 90% of cases [4].

Even though LAA is the prime location of thrombus formation in AF patients, accumulative evidence shows that LAA thrombus may also occur in patients with sinus rhythm or even subclinical AF [5]. Through the advances and widespread use of medical devices, more cryptogenic strokes have been found to be related to subclinical AF [6].

Identifying the cause of stroke is vital in achieving optimal therapeutic strategies for the treatment and prevention of recurrent stroke [7]. Initiation of anticoagulation therapy with vitamin K antagonist (VKA) is the most common and conventional strategy employed for LAA thrombus (LAAT) [8]. However, these practices are slowly changing after the launch of the nonvitamin K direct oral anticoagulant (DOAC) in 2002. The introduction of the IIa inhibitor, dabigatran, and Xa inhibitors rivaroxaban, apixaban, and edoxaban in the millennium year has proved that these anticoagulants were at least as effective as VKA in AF for stroke prevention [9].

The safety profiles of DOAC have been highly recognized in many meta-analytic studies and healthcare databases [10,11]. Given that VKA requires regular coagulation monitoring and the potential effects from its interactions with drugs and food [12], DOAC's high efficacy and reliable safety profile are preferred over VKA in current clinical settings. Therefore, DOAC is now generally accepted as the treatment of choice over VKA in patients with nonvalvular AF [11,13].

This trend of switching from VKA to DOAC is not limited to the prevention of strokes from nonvalvular AF; it also extends to other forms of thromboembolism, such as deep venous thrombosis. Moreover, many emerging studies assessed the comparability of DOAC against VKA for LAAT prevention and resolution. However, the optimal treatment for LAAT is yet to be established.

To the best of our knowledge, not many large-scale randomized controlled trials have attempted to verify the differences between the roles of VKA and DOAC in the risk of LAAT formation and rate of thrombus resolution. Furthermore, the lack of large-scale cohort studies has impeded guidelines from being developed that would provide high-level recommendations for LAAT medication management.

The present study is a systematic review of the outcomes of VKA and DOAC use and was performed through an examination of real-world evidence. Further, a meta-analysis of available data was also performed to compare the effectiveness of VKA and DOAC for primary prevention and resolution of LAAT. In this meta-analysis, we included studies providing specific data on the incidence of LAAT and the LAAT resolution rate under VKA or DOAC use.

2. Methods

2.1. Research Question and Objectives

In this meta-analysis, we aimed to synthesize evidence to systematically review real-world evidence for a comparison of VKAs and DOAC with respect to their influence on the (i) risk of LAAT and (ii) resolution of LAAT.

2.2. Selection of Articles

Relevant studies, including case series and clinical trials published before July 2022 were identified from the PubMed, Embase, Google Scholar, and Cochrane databases. Only publications in English were included. We used the following sets of terms in our search: (warfarin (Title/Abstract)) OR (novel oral anticoagulant (Title/Abstract)) OR (oral anticoagulant (Title/Abstract)) OR (anticoagulant (Title/Abstract)) OR (direct oral anticoagulant (Title/Abstract)) OR (vitamin K anticoagulant (Title/Abstract)) OR (non-vitamin K oral anticoagulant (Title/Abstract)) OR (dabigatran (Title/Abstract)) OR (rivaroxaban (Title/Abstract)) OR (apixaban (Title/Abstract)) OR (edoxaban (Title/Abstract)) AND (left atrial appendage thrombus (Title/Abstract)) OR (left atrial thrombus (Title/Abstract)).

The syntax used in the database searches is detailed in the Supplementary Information (Table S1). Duplicate articles from different databases were excluded. The selection process is illustrated in Figure 1. All search records from all databases were downloaded and merged into Endnote.

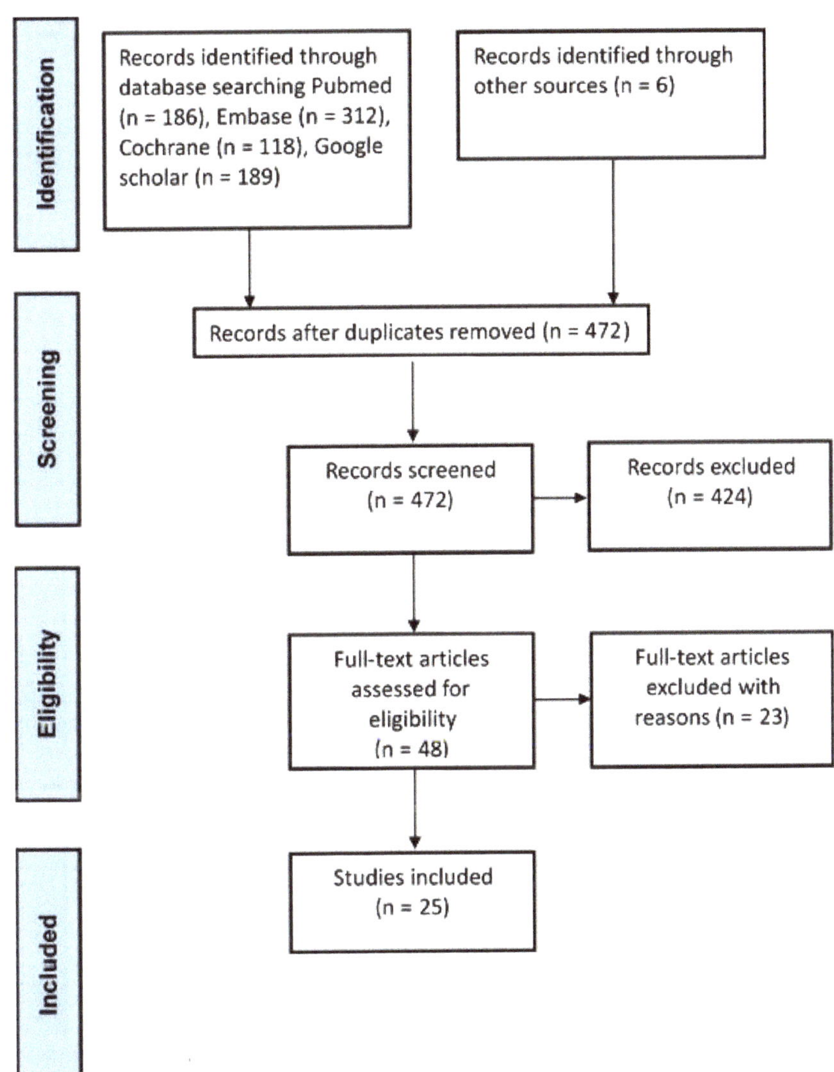

Figure 1. Flow diagram of study selection and search results.

2.3. Study Design

We included studies (1) with LAA thrombus diagnosed using transesophageal echocardiography (TEE); (2) with clear records of VKA or DOAC anticoagulant use and in which patients were appropriately anticoagulated; (3) that were cohort studies published as original articles, or case series; and (4) that were publications in English.

2.4. Data Extraction

The selected studies were independently retrieved by two reviewers (Y-Y.C. and C-C.Y.) and were further reviewed by another author (S.T.). The selected studies were reviewed to identify the type of study, year of publication, total patient population, mean patient age, percentage of patients taking VKA, percentage of patients taking DOAC, percentage of male participants, and mean duration of anticoagulant use. Any disagreements were resolved by a fourth reviewer (C-T.H.).

3. Outcomes

The efficacy of the primary prevention method was evaluated based on the risk of LAAT. A second comparison was made of the potency of VKA and DOAC in resolving LAAT.

3.1. Synthesis of Results and Measures of Inconsistency

A random-effects model was implemented to assess LAAT incidence and resolution under VKA and DOAC use. Q and I^2 were used to assess the level of heterogeneity between the included studies [14]. Q is a measure of the weighted sum of the squared deviations of the effect size of each study from the overall mean effect size and thereby serves as a test of heterogeneity significance ($p \leq 0.05$) [15]. I^2 is a measure of relative heterogeneity, estimating the percentage of the variability of effect estimates that occurs due to heterogeneity rather than due to chance. I^2 ranges from 0% to 100%, with a value of 0% indicating no observed heterogeneity and a value greater than 50% representing moderate heterogeneity [16]. Furthermore, tau-squared (T^2) measures the variance of the true effect as an estimate of absolute heterogeneity in effect sizes. When the observed variance increases or when the variance within studies decreases, T^2 increases accordingly [15].

3.2. Publication Bias

We used funnel plots [17], Egger's test [18], and the Begg and Mazumdar rank correlation test [19] to assess publication bias.

3.3. Statistical Analysis

All statistical analyses were performed using Stata 17. The meta-analysis is registered with PROSPERO (CRD42022319759) and was performed in accordance with the Preferred Reporting Items for Systematic Review and Meta-Analyses (PRISMA) guidelines [20]. Standard deviation was calculated using the provided confidence interval (CI) limits, standard errors, or interquartile ranges. The overall risk/odds ratios were pooled using a random-effects model. Publication bias was assessed using the funnel plot of each study's effect size against precision (1/SE). Publication bias was investigated using Egger's test at $p < 0.10$.

4. Results

4.1. Study Selection

After removing duplicate studies, we identified 811 articles for screening. After the exclusion of ineligible studies, 48 studies were included in the full-article assessment, and additional 23 studies were excluded because they were case reports, animal studies, or did not contain original data. Finally, 25 studies were included for qualitative synthesis. We further segregated the 25 studies into two categories: (1) cross-sectional risk analysis of developing LAAT under VKA or DOAC use and (2) analysis of LAAT resolution under VKA or DOAC use.

4.2. Study Characteristics

The characteristics of the included studies on the incidence of LAAT under VKA or DOAC use are listed in Table 1, and the characteristics of those on the LAAT resolution rate under VKA or DOAC use are listed in Table 2. 13 studies were cross-sectional analyses of the incidence of LAAT under VKA and DOAC use (involving 8609 individuals), and 13 were longitudinal analyses of the LAAT resolution rate under VKA and DOAC use (involving 922 individuals).

Table 1. Baseline characteristics of included studies for systematic review and meta-analysis of the incidence of left arterial appendage thrombus (LAAT) in patients with atrial fibrillation under vitamin K antagonist (VKA) or direct oral anticoagulant (DOAC) treatment.

Author (Year)/ Country, Study Type	Age (Years Old, Mean ± SD), Male (n/%)	CHA2DS2-VASC Score (Mean ± SD)	Anticoagulant (Duration)
Alqarawi (2019)/ Canada, prospective [21]	64 ± 11, (478/72%)	1.9 ± 1.4	VKA (INR ≥ 2) and DOAC: 258 Dab, 184 Riv, and 54 Api (>4 weeks)
Bursi (2021)/ Italy, prospective [22]	71 ± 10, (177/64%)	3.1 ± 1.4	VKA (8.1% INR < 2) and DOAC (>3 weeks)
Durmaz (2020)/Turkey, prospective [23]	69.9 ± 12.4 (LAAT) and 65.1 ± 12.1 (nLAAT), (45.6%)	3.44	VKA (INR not specified) and DOAC (>3 weeks) (61 VKA, 32 Dab, 62 Riv, 29 Api)
Frenkel (2016)/ US, retrospective [24]	65, (287/74%)	2	VKA (INR median 3.0 (IQR: 2.5 to 3.2)) and DOAC: 93 Dab, 62 Riv, and 28 Api (>4 weeks)
Kawabata (2017)/ Japan, retrospective [25]	62 ± 11, (445/79.6%)	1.9 ± 1.5	VKA (INR available in 90% of participants but not specified) and DOAC: 145 Dab, 121 Riv, 40 Api, and 5 Edo (>4 weeks)
Merino (2019)/ Europe, prospective [26]	67.3 ± 9.4 (LAAT) and 64.2 ± 10.8 (nLAAT), (733)	3.0 ± 1.4 (LAAT) 2.7 ± 1.5 (nLAAT)	VKA (INR 1.51 ± 0.61 at baseline) and DOAC (>30 days)
Uziębło-Życzkowska (2020)/Poland, retrospective [27]	63.35, (61%)	2.48 ± 1.53	VKA (INR 1.69 ± 0.86 at baseline) and DOAC (>3 weeks) (VKA 227, 240 Dab, 279 Riv, 4 Api)
Wyrembak (2017)/ US, retrospective [28]	65, (618/66%)	3.1 ± 2	VKA (INR 2.32 ± 0.59) and DOAC (>4 weeks)
Kapłon-Cie'slicka (2022)/ Poland, prospective [29]	67, (1731/63%)	3	VKA (INR not specified) and DOAC (>3 weeks) 814 Dab, 1060 Riv, 388 Api
Karwowski (2022)/ Poland, retrospective [30]	73.4 ± 10.3 (80, 50%)	3.83 ± 1.64	VKA (6 warfarin, 22 acenocoumarol; INR not reported) and DOAC (25 Dab, 83 Riv, 16 Api, and 8 Edo) (>4 weeks)
Feickert (2020)/ Germany, retrospective [31]	71.3 ± 9.0 (68, 48.2%)	4.03 ± 1.53	VKA (INR ≥2, n = 74; INR <2, n = 20) and DOAC (>3 weeks) (32 VKA, 14 Dab, 7 Api, 13 Riv, 1 Edo)
Shiraki (2022)/ Japan, retrospective [32]	65.2 ± 10.1 (193, 26.2%)	nil	VKA (INR not specified) and DOAC (>3 weeks) (120 Dab, 213 Riv, 199 Api, and 108 Edo)
Turek (2022)/ Poland, prospective [33]	65.4 (182, 61.5%)	nil	VKA (INR ≥ 2) and DOAC (>3 weeks) (145 Dab, 80 Riv, 10 Api)

nLAAT, patients without LAAT; Dab, dabigatran; Riv, rivaroxaban; Api, apixaban; Edo, edoxaban; INR, ≥ international normalized ratio.

Table 2. Baseline characteristics of included studies for systematic review and meta-analysis of LAAT resolution in patients under VKA or DOAC treatment.

Author (Year), Country/Continent, Study Type	Age, Overall (Mean ± SD)	Male, n (%)	CHA2DS2-VASc (Median/Mean)	Anticoagulant (Type, Duration)
Hao (2015), China, Retrospective [34]	57.7 ± 7.4	36 (87.8)	VKA: 1.41 ± 1.01 Dab: 1.16 ± 1.01	VKA (INR not reported) and Dab (4.2 months, median)
Hussain (2019), US, retrospective [35]	63.2	31 (69)	3.4 ± 1.7	DOAC (60 days, median) VKA (INR not reported; 116 days, median)
Kawabata (2017), Japan, retrospective [25]	64	9 (60)	3.7 ± 1.8	VKA (INR not specified) and 1 Dab (>3 weeks)
Ke (2019), China, prospective [36]	VKA: 64.2 ± 10.5 Riv: 63.7 ± 8.6	66 (82.5)	1.46	VKA (INR not reported) and Riv (12 weeks)
Lip (2015), Europe, prospective (X-TRA) and retrospective (CLOT-AF) [37]	X-TRA: 69.6 ± 11 CLOT-AF: 67.7 ± 9.6	X-TRA: 30 (50) CLOT-AF: 103 (66)	X-TRA: 4.0 CLOT-AF: 3.0	VKA (INR not reported) and DOAC: 12 Dab, 1 Riv, and 7 Api (X-TRA: 6 weeks; CLOT-AF: 3–12 weeks)
Mitamura (2015), Japan, retrospective [38]	67.3 ± 12.7	7 (87.5)	1.88	VKA (INR not reported) and Dab (21–308 days)
Nelles (2021), Germany, retrospective [39]	76.1 ± 8.3	45 (57.7)	4.3 ± 1.1	VKA (INR 2.2 ± 0.2) and DOAC: 15 Dab, 12 Api, 11 Riv and 1 Api (116 ± 79 days)
Niku (2019), Japan, retrospective [40]	71.9 ± 11.9	52 (44)	3.4	VKA (59% had INR values ≥ 2.0) and DOAC: 2 Dab, 12 Riv, and 16 Api (96 ± 72 days)
Wu (2018), US, retrospective [41]	67	33 (75)	3	VKA (INR median 2.7 (IQR 2.2, 3.2) at baseline) and DOAC: 12 Dab, 1 Riv, and 7Api (≥4 weeks)
Yang (2019), China, retrospective [42]	63.5 ± 10.9	52 (72.2)	2	VKA (INR not reported) and DOAC: 26 Dab and 29 Riv (101.5 days)
Mazur (2021) Russia, retrospective [43]	59.7± 9.8	41 (60.3)	2.22 ± 1.40	VKA (INR between 2 and 3) and DOAC (>3 weeks): 14 Dab, 14 Riv, 3 Api
Faggiano (2022) Italy, retrospective [44]	71	175 (66)	4	VKA (INR not reported) and DOAC: 18 Riv, Api 24, Dab 24, Edo 5
Karwowski (2022) Poland, retrospective [30]	76.7 ± 8.2	5 (50)	4.58 ± 1.00	VKA (2 warfarin, 5 acenocoumarol; INR not reported) and DOAC 5 Api

Dab, dabigatran; Riv, rivaroxaban; Api, apixaban; Edo, edoxaban; INR, ≥ international normalized ratio.

4.3. LAAT Incidence under VKA and DOAC

A total of 13 cross-sectional analyses involving 8609 individuals were included in the first meta-analysis of the associations between VKA and DOAC use with LAAT. These studies were published between 2016 and 2022. Among the studies, four were conducted in Poland [27,29,30,33], two were conducted in Japan [25,32], two were conducted in the United States [24,28], and one was conducted each in the EU [26], Germany [31], Canada [21], Turkey [23], and Italy [22]. The selected studies were mostly conducted in Western countries, and only two were conducted in an Asian population (Japan). Six studies were prospective [21–23,26,29,33], and seven were retrospective [24,25,27,28,30–32].

Only the ENSURE-AF trial was a randomized, multicenter, global investigation [26]. The others were mainly single-center studies. Generally, patients were considered sufficiently anticoagulated after at least 3 or 4 weeks of administration of VKA or DOAC. TEE was performed for all study participants to detect the presence of LAAT. The allocation methods for VKA and DOAC were based on the clinician's decision.

The studies included different parameters for predicting the risk of LAAT. A higher CHA2DS2-VASc score [22,23,25,33], reduced left ventricular ejection fraction [22,23,27,33], reduced left atrial flow velocity [23,25,30,32], reduced B-type natriuretic peptide, and larger left atrium [25] were associated with the risk of LAAT. The demographic predictors of the risk of LAAT included aging, a lower body weight, lower creatinine clearance, heart failure, and diuretic treatment were also listed. [26].

4.4. LAAT Resolution Rate under VKA and DOAC Use

A total of 13 studies involving 922 patients were included in the second meta-analysis. These studies included three in Japan [25,38,40], five in European countries [30,37,39,43,44], three in China [34,36,42], and two in the United States [35,41]. One of the studies, the X-TRA study, was a multinational large-scale, prospective, single-arm, open-label, multicenter study. The X-TRA study evaluated a 6-week rivaroxaban treatment for left atrial and LAA thrombus resolution. Another study, the CLOT-AF study, retrospectively examined standard anticoagulation care provided to patients with left atrial and LAA thrombus for 3 to 12 weeks. These studies were published between 2015 and 2022. When the included study period was extended to 7 months, nine patients were identified as having LAAT [42]. When the included duration and dosage of anticoagulation were increased and considering the transition to DOAC, 12 patients (5%) were identified as experiencing LAAT resolution [41]. Of the 13 included studies, Kawabata, Karwowski, Durmaz, Feickert, and Shiraki studies contained data on the use of dabigatran, rivaroxaban, apixaban, and edoxaban, respectively. Edoxaban was used less frequently than the other three DOACs.

4.5. LAAT Incidence under VKA and DOAC Use

In the meta-analysis, 2963 patients were in the VKA arm, and 5646 were in the DOAC arm. The overall risk of LAAT under either VKA or DOAC treatment was 5.56% (479/8609). The risk ratio was derived from individual studies. The respective relative risks (RRs) with 95% CIs are listed on the right side of the forest plot (Figure 2A). The meta-analysis of 13 studies on LAAT incidence revealed significant superiority for DOACs (pooled RR = 0.65, 95% CI = 0.47–0.90, $p = 0.009$). Nearly moderate and significant heterogeneity ($Q_{12} = 22.97$, $p = 0.028$; $I^2 = 47.8\%$; $T^2 = 0.13$) was identified. The funnel plot (Figure 2B) revealed symmetric distribution. Egger's test (intercept = 0.684, t = 0.97, 2 tailed $p = 0.352$) and Begg's test (z = 0.18, $p = 0.855$) did not reveal any publication bias. The meta-regression analysis showed that none of the between-study variables significantly predicted the LAAT incidence under VKA and DOAC use (mean age of participants: $\beta = 0.071$, $p = 0.333$; male ratio: $\beta = 0.354$, $p = 0.808$).

4.6. LAAT Resolution Rate under VKA and DOAC Use

The VKA and DOAC arms included 484 and 438 patients, respectively. The summarized mean percentages of LAAT resolution for VKA and DOAC were 55.4% (268/484) and 67.6% (296/438), respectively. The odds ratio was derived from the individual studies (Figure 3A). This meta-analysis revealed that DOAC significantly increased the probability of LAAT resolution compared with VKA (pooled odds ratio = 1.52, 95% CI = 1.02–2.26, $p = 0.040$). In addition, no significant heterogeneity ($Q_{12} = 17.62$, $p = 0.128$; $I^2 = 31.9\%$; $T^2 = 0.16$) was identified. The funnel plot (Figure 3B), although slightly asymmetric, did not indicate a high risk of publication bias. Egger's test (intercept = 0.383, t = 0.44, 2-tailed $p = 0.671$) and Begg's test (z = 0.55, $p = 0.583$) did not reveal any publication bias. The meta-regression analysis showed that none of the between-study variables significantly predicted the LAAT resolution rate for VKA and DOAC (mean age of participants: $\beta = -0.042$, $p = 0.343$; male ratio: $\beta = 0.248$, $p = 0.860$).

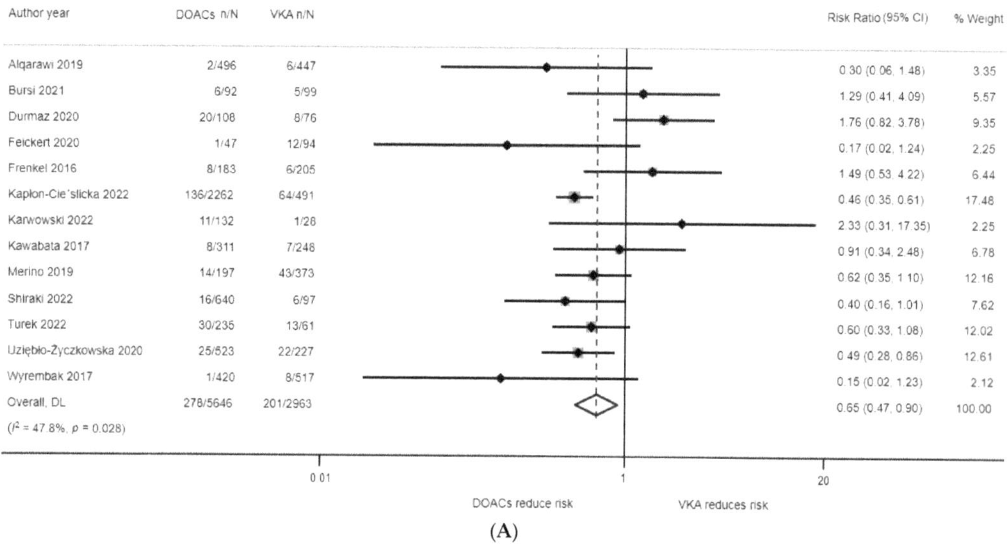

(A)

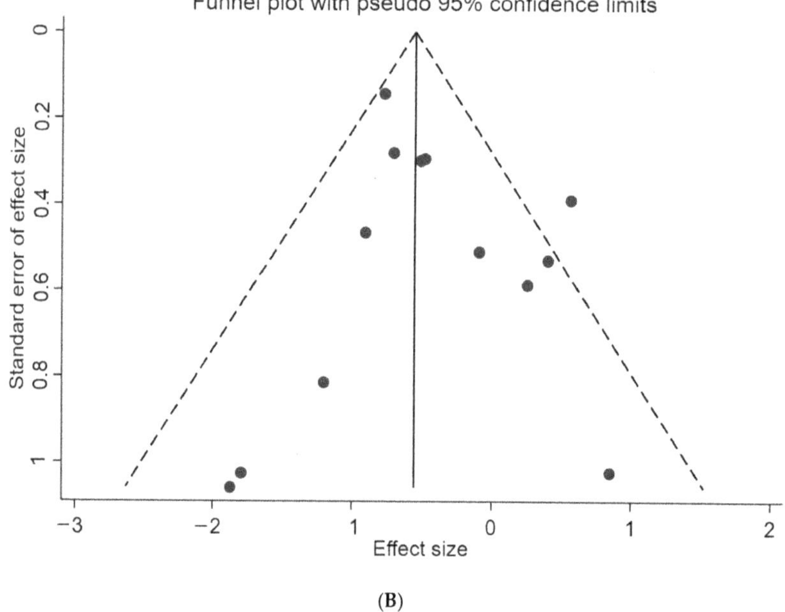

(B)

Figure 2. (**A**) The forest plot of random-effect meta-analysis of the incidence of left arterial appendage thrombus formation under the use of vitamin K antagonist (VKA) and direct oral anticoagulants (DOACs). (**B**) The funnel plot of the studies included in the meta-analysis of the incidence of left arterial appendage thrombus formation under the use of vitamin K antagonist (VKA) and direct oral anticoagulants (DOACs).

Author year	DOACs n/N	VKA n/N	Odds ratio (95% CI)	Weight %
Faggiano 2022	39/50	60/107	2.78 (1.29, 6.00)	13.33
Hao 2015	17/19	17/22	2.50 (0.42, 14.71)	4.25
Hussain 2019	17/22	17/23	1.20 (0.31, 4.69)	6.47
Karwowski 2022	2/5	5/7	0.27 (0.02, 3.02)	2.45
Kawabata 2017	1/1	6/9	1.62 (0.05, 51.11)	1.27
Ke 2019	32/40	28/40	1.71 (0.61, 4.79)	9.60
Lip 2015	22/60	60/127	0.65 (0.34, 1.21)	15.95
Manzur 2021	26/31	19/37	4.93 (1.55, 15.62)	8.24
Mitamura 2015	3/4	2/4	3.00 (0.15, 59.89)	1.66
Nelles 2021	28/49	11/29	2.18 (0.85, 5.58)	10.73
Niku 2019	21/32	16/29	1.55 (0.55, 4.36)	9.54
Wu 2018	11/20	14/24	0.87 (0.26, 2.89)	7.82
Yang 2019	34/55	10/17	1.13 (0.37, 3.43)	8.70
Overall, DL	296/438	268/484	1.52 (1.02, 2.26)	100.00

(I^2 = 31.9%, p = 0.128)

0.01 1 20

Favors VKA Favors DOACs

(A)

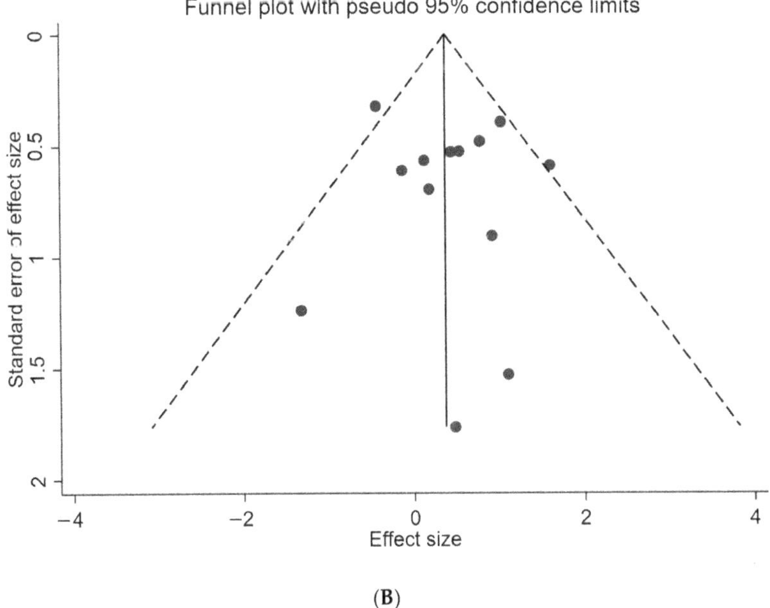

(B)

Figure 3. (**A**) The forest plot of random-effect meta-analysis of the likelihood of resolution of left arterial appendage thrombus under the use of VKA and DOACs. (**B**) The funnel plot of the studies included in the meta-analysis of the likelihood of resolution of left arterial appendage thrombus under the use of VKA and DOACs.

5. Discussion

Various guidelines recommend DOAC as a preferable anticoagulant option to VKA for stroke prevention in AF. However, evidence on optimal anticoagulant selection in patients with LAAT is lacking. Although DOACs are safer than VKAs, their efficacy remains unverified. After analyzing pooled data from studies conducted in the past decade, we discovered that DOACs showed superiority over VKAs in increasing the likelihood of LAAT resolution and reducing the chances of LAAT development in high-risk patients. By monitoring patients taking anticoagulants, the incidence of thrombus formation in patients taking DOACs was lower than in those taking VKAs. In addition, patients with LAAT showed a higher thrombus resolution rate with DOACs use in comparison to VKAs. The superiority of the safety of DOAC compared with VKA is well established; these findings provide evidence for developing future treatment recommendations.

AF is associated with a high incidence of LAAT [45]. Previous reports have demonstrated that the CHADS2 score is an independent predictor of LAAT, and the prevalence of LAAT increases with the CHADS2 score [46,47]. However, this parameter does not have a significant skew; therefore, its exclusion from this meta-analysis did not affect the risk of LAAT and resolution obtained from each included study.

Clinically, when the efficacy of anticoagulants in LAAT resolution is comparable, safety concerns, including the risk of bleeding and drug–drug interactions, become the priority of medicine choice. DOACs have not yet been approved for patients with mechanical mitral valves, thrombus in locations other than the LAA, and antiphospholipid syndrome. However, a growing body of evidence has demonstrated that DOACs lead to fewer bleeding complications than VKAs. Therefore, DOACs are more likely to be selected for the prevention of thromboembolism events in people with LAAT.

Many large-scale clinical phase III trials have demonstrated that the efficacy of DOACs in preventing stroke is superior to VKAs and that DOACs have lower rates of bleeding. However, conclusive data on the recommended type and duration of anticoagulant use in LAAT is limited. Two recent meta-analyses have demonstrated that DOACs are as efficacious as and safer than VKAs in the treatment of LAAT in patients with nonvalvular AF. In addition, two ongoing prospective randomized trial registry studies are seeking to compare DOAC and VKA in patients with LAAT. One of these is a randomized control trial in China (NCT03792152), in which the effectiveness of rivaroxaban and VKA are compared. The other randomized control trial (RE-LATED AF (NCT02256683)) is a comparison of dabigatran and VKA in patients with nonvalvular AF of LAAT.

The strength of this study is its analyses of both the risk of LAAT development in high-risk patients and the likelihood of LAAT resolution with the use of VKAs and DOACs. The development of LAAT is an indicator of subsequent systemic thromboembolism events, which require anticoagulation therapy. However, the presence of LAAT requires emergent anticoagulation therapy until the thrombus resolves. The comparable efficacy of DOAC and VKA indicates an opportunity to provide patients with safer prescriptions. This study has several limitations. First, no subgroup analyses were performed for the different classes of DOAC in the 18 studies due to the small number of studies for each DOAC. Further studies are warranted to investigate the effectiveness of different DOACs in patients with different comorbidities. Second, several studies reported adverse effects, bleeding risk, and thromboembolism events, which prevented these events from being included in the meta-analysis. Third, the relative weight of the included studies varied in analyses, which is why the random-effects model was employed. Fourth, although the anti-IIa/Xa activity effectiveness and possible drug–drug interactions are crucial in the issues investigated in the current study, it is not feasible to conduct relevant analyses considering the very limited information provided in the included studies. Another related concern is the ratio/period of the effective therapeutic range, such as the international normalized ratio (INR) values in patients taking VKA. The distinctive ways of reporting INRs in the included studies also hinder further analyses from examining the underlying origins of heterogeneity.

6. Conclusions

This meta-analysis of observation data revealed significant differences in LAAT development in high-risk patients and the likelihood of LAAT resolution in patients treated with DOAC or VKA. With respect to safety profiles, DOACs are preferable to VKAs in patients with LAAT and without absolute DOAC contraindication. As the role of DOACs expands, further studies should be conducted to provide clinicians with a practical reference for optimization of the selection of appropriate DOACs and the duration of treatment.

Supplementary Materials: The following supporting information can be downloaded at: https://www.mdpi.com/article/10.3390/jcdd9100351/s1, Table S1: Search history and results.

Author Contributions: Conceptualization, Y.-Y.C., S.T. and L.C.; methodology, S.T., C.-C.Y. and L.C.; software, C.-C.Y.; validation, C.-C.Y. and L.C.; formal analysis, S.T., C.-T.H., C.-C.Y. and L.C.; investigation, S.T. and C.-T.H.; resources, S.T. and C.-T.H.; data curation, Y.-Y.C., S.T. and C.-C.Y.; writing—original draft preparation, Y.-Y.C. and S.T.; writing—review and editing, C.-T.H., C.-C.Y. and L.C.; visualization, S.T. and C.-C.Y.; supervision, C.-T.H.; project administration, S.T., C.-T.H. and L.C. All authors have read and agreed to the published version of the manuscript.

Funding: This research received no external funding.

Institutional Review Board Statement: Ethical review and approval were waived for this study due to the nature of this meta-analytic study. All the included studies have been approved by their IRBs.

Informed Consent Statement: Patient consent was waived due to the nature of this meta-analytic study. No identifiable personal information was obtained in the included studies.

Data Availability Statement: The data presented in this study are available on request from the corresponding authors.

Conflicts of Interest: The authors report no conflict of interest.

References

1. Feigin, V.L.; Stark, B.A.; Johnson, C.O.; Roth, G.A.; Bisignano, C.; Abady, G.G.; Abbasifard, M.; Abbasi-Kangevari, M.; Abd-Allah, F.; Abedi, V.; et al. Global, regional, and national burden of stroke and its risk factors, 1990–2019: A systematic analysis for the Global Burden of Disease Study 2019. *Lancet Neurol.* **2021**, *20*, 795–820. [CrossRef]
2. Adams, H.P.; Bendixen, B.H.; Kappelle, L.J.; Biller, J.; Love, B.B.; Gordon, D.L.; Marsh, E.E. Classification of subtype of acute ischemic stroke. Definitions for use in a multicenter clinical trial. TOAST. Trial of Org 10172 in Acute Stroke Treatment. *Stroke* **1993**, *24*, 35–41. [CrossRef] [PubMed]
3. Wolf, P.A.; Abbott, R.D.; Kannel, W.B. Atrial fibrillation as an independent risk factor for stroke: The Framingham Study. *Stroke* **1991**, *22*, 983–988. [CrossRef] [PubMed]
4. Arboix, A.; Alio, J. Cardioembolic stroke: Clinical features, specific cardiac disorders and prognosis. *Curr. Cardiol. Rev.* **2010**, *6*, 150–161. [CrossRef]
5. Naksuk, N.; Padmanabhan, D.; Yogeswaran, V.; Asirvatham, S.J. Left Atrial Appendage: Embryology, Anatomy, Physiology, Arrhythmia and Therapeutic Intervention. *JACC Clin. Electrophysiol* **2016**, *2*, 403–412. [CrossRef]
6. Kamel, H.; Healey, J.S. Cardioembolic Stroke. *Circ. Res.* **2017**, *120*, 514–526. [CrossRef]
7. Campbell, B.C.V.; De Silva, D.A.; Macleod, M.R.; Coutts, S.B.; Schwamm, L.H.; Davis, S.M.; Donnan, G.A. Ischaemic stroke. *Nat. Rev. Dis. Primers* **2019**, *5*, 70. [CrossRef]
8. European Atrial Fibrillation Trial Study, G. Optimal oral anticoagulant therapy in patients with nonrheumatic atrial fibrillation and recent cerebral ischemia. *N. Engl. J. Med.* **1995**, *333*, 5–10. [CrossRef]
9. Kleindorfer, D.O.; Towfighi, A.; Chaturvedi, S.; Cockroft, K.M.; Gutierrez, J.; Lombardi-Hill, D.; Kamel, H.; Kernan, W.N.; Kittner, S.J.; Leira, E.C.; et al. 2021 Guideline for the Prevention of Stroke in Patients with Stroke and Transient Ischemic Attack: A Guideline from the American Heart Association/American Stroke Association. *Stroke* **2021**, *52*, e364–e467. [CrossRef]
10. Hart, R.G.; Pearce, L.A.; Aguilar, M.I. Meta-analysis: Antithrombotic therapy to prevent stroke in patients who have nonvalvular atrial fibrillation. *Ann. Intern. Med.* **2007**, *146*, 857–867. [CrossRef]
11. Ruff, C.T.; Giugliano, R.P.; Braunwald, E.; Hoffman, E.B.; Deenadayalu, N.; Ezekowitz, M.D.; Camm, A.J.; Weitz, J.I.; Lewis, B.S.; Parkhomenko, A.; et al. Comparison of the efficacy and safety of new oral anticoagulants with warfarin in patients with atrial fibrillation: A meta-analysis of randomised trials. *Lancet* **2014**, *383*, 955–962. [CrossRef]
12. Steffel, J.; Braunwald, E. Novel oral anticoagulants: Focus on stroke prevention and treatment of venous thrombo-embolism. *Eur. Heart J.* **2011**, *32*, 1968–1976. [CrossRef] [PubMed]

13. Hindricks, G.; Potpara, T.; Dagres, N.; Arbelo, E.; Bax, J.J.; Blomström-Lundqvist, C.; Boriani, G.; Castella, M.; Dan, G.A.; Dilaveris, P.E.; et al. 2020 ESC Guidelines for the diagnosis and management of atrial fibrillation developed in collaboration with the European Association for Cardio-Thoracic Surgery (EACTS): The Task Force for the diagnosis and management of atrial fibrillation of the European Society of Cardiology (ESC) Developed with the special contribution of the European Heart Rhythm Association (EHRA) of the ESC. *Eur. Heart J.* **2021**, *42*, 373–498. [CrossRef]
14. Higgins, J.P.; Thompson, S.G.; Deeks, J.J.; Altman, D.G. Measuring inconsistency in meta-analyses. *BMJ* **2003**, *327*, 557–560. [CrossRef]
15. Borenstein, M.; Hedges, L.V.; Higgins, J.P.; Rothstein, H.R. A basic introduction to fixed-effect and random-effects models for meta-analysis. *Res. Synth. Methods* **2010**, *1*, 97–111. [CrossRef]
16. Higgins, J.P.T.; Altman, D.G.; Gøtzsche, P.C.; Jüni, P.; Moher, D.; Oxman, A.D.; Savović, J.; Schulz, K.F.; Weeks, L.; Sterne, J.A.C. The Cochrane Collaboration's tool for assessing risk of bias in randomised trials. *BMJ* **2011**, *343*, d5928. [CrossRef]
17. Egger, M.; Smith, G.D. Misleading meta-analysis. *BMJ* **1995**, *310*, 752–754. [CrossRef]
18. Egger, M.; Davey Smith, G.; Schneider, M.; Minder, C. Bias in meta-analysis detected by a simple, graphical test. *BMJ* **1997**, *315*, 629–634. [CrossRef]
19. Begg, C.B.; Mazumdar, M. Operating Characteristics of a Rank Correlation Test for Publication Bias. *Biometrics* **1994**, *50*, 1088–1101. [CrossRef]
20. Page, M.J.; McKenzie, J.E.; Bossuyt, P.M.; Boutron, I.; Hoffmann, T.C.; Mulrow, C.D.; Shamseer, L.; Tetzlaff, J.M.; Akl, E.A.; Brennan, S.E.; et al. The PRISMA 2020 statement: An updated guideline for reporting systematic reviews. *BMJ* **2021**, *372*, n71. [CrossRef]
21. Alqarawi, W.; Birnie, D.H.; Spence, S.; Ramirez, F.D.; Redpath, C.J.; Lemery, R.; Nair, G.M.; Nery, P.B.; Davis, D.R.; Green, M.S.; et al. Prevalence of left atrial appendage thrombus detected by transoesophageal echocardiography before catheter ablation of atrial fibrillation in patients anticoagulated with non-vitamin K antagonist oral anticoagulants. *EP Eur.* **2019**, *21*, 48–53. [CrossRef]
22. Bursi, F.A.-O.; Santangelo, G.; Ferrante, G.; Massironi, L.; Carugo, S. Prevalence of left atrial thrombus by real time three-dimensional echocardiography in patients undergoing electrical cardioversion of atrial fibrillation: A contemporary cohort study. *Echocardiography* **2021**, *38*, 518–524. [CrossRef] [PubMed]
23. Durmaz, E.A.-O.; Karpuz, M.H.; Bilgehan, K.; Ikitimur, B.; Ozmen, E.; Ebren, C.; Polat, F.; Koca, D.; Tokdil, K.O.; Kandemirli, S.G.; et al. Left atrial thrombus in patients with atrial fibrillation and under oral anticoagulant therapy; 3-D transesophageal echocardiographic study. *Int. J. Cardiovasc. Imaging* **2020**, *36*, 1097–1103. [CrossRef] [PubMed]
24. Frenkel, D.; D'Amato, S.A.; Al-Kazaz, M.; Markowitz, S.M.; Liu, C.F.; Thomas, G.; Ip, J.E.; Sharma, S.K.; Yang, H.; Singh, P.; et al. Prevalence of Left Atrial Thrombus Detection by Transesophageal Echocardiography: A Comparison of Continuous Non-Vitamin K Antagonist Oral Anticoagulant Versus Warfarin Therapy in Patients Undergoing Catheter Ablation for Atrial Fibrillation. *JACC Clin. Electrophysiol.* **2016**, *2*, 295–303. [CrossRef]
25. Kawabata, M.; Goya, M.; Sasaki, T.; Maeda, S.; Shirai, Y.; Nishimura, T.; Yoshitake, T.; Shiohira, S.; Isobe, M.; Hirao, K. Left Atrial Appendage Thrombi Formation in Japanese Non-Valvular Atrial Fibrillation Patients During Anticoagulation Therapy—Warfarin vs. Direct Oral Anticoagulants. *Circ. J.* **2017**, *81*, 645–651. [CrossRef] [PubMed]
26. Merino, J.L.; Lip, G.Y.H.; Heidbuchel, H.; Cohen, A.A.; De Caterina, R.; de Groot, J.R.; Ezekowitz, M.D.; Le Heuzey, J.Y.; Themistoclakis, S.; Jin, J.; et al. Determinants of left atrium thrombi in scheduled cardioversion: An ENSURE-AF study analysis. *Europace* **2019**, *21*, 1633–1638. [CrossRef] [PubMed]
27. Uziębło-Życzkowska, B.; Krzesiński, P.; Jurek, A.; Kapłon-Cieślicka, A.; Gorczyca, I.; Budnik, M.; Gielerak, G.; Kiliszek, M.; Gawałko, M.; Scisło, P.; et al. Left Ventricular Ejection Fraction Is Associated with the Risk of Thrombus in the Left Atrial Appendage in Patients with Atrial Fibrillation. *Cardiovasc. Ther.* **2020**, *2020*, 3501749. [CrossRef]
28. Wyrembak, J.; Campbell, K.B.; Steinberg, B.A.; Bahnson, T.D.; Daubert, J.P.; Velazquez, E.J.; Samad, Z.; Atwater, B.D. Incidence and Predictors of Left Atrial Appendage Thrombus in Patients Treated with Nonvitamin K Oral Anticoagulants Versus Warfarin Before Catheter Ablation for Atrial Fibrillation. *Am. J. Cardiol.* **2017**, *119*, 1017–1022. [CrossRef]
29. Kapłon-Cieślicka, A.; Gawałko, M.; Budnik, M.; Uziębło-życzkowska, B.; Krzesiński, P.; Starzyk, K.; Gorczyca-Głowacka, I.; Daniłowicz-Szymanowicz, L.; Kaufmann, D.; Wójcik, M.; et al. Left Atrial Thrombus in Atrial Fibrillation/Flutter Patients in Relation to Anticoagulation Strategy: LATTEE Registry. *J. Clin. Med.* **2022**, *11*, 2705. [CrossRef]
30. Karwowski, J.; Rekosz, J.; Mączyńska-Mazuruk, R.; Wiktorska, A.; Wrzosek, K.; Loska, W.; Szmarowska, K.; Solecki, M.; Sumińska-Syska, J.; Dłużniewski, M. Left atrial appendage thrombus in patients with atrial fibrillation who underwent oral anticoagulation. *Cardiol. J.* **2022**. [CrossRef]
31. Feickert, S.; D'Ancona, G.; Ince, H.; Graf, K.; Kugel, E.; Murero, M.; Safak, E. Routine transesophageal Echocardiography in atrial fibrillation before electrical cardioversion to detect left atrial thrombosis and echocontrast. *J. Atr. Fibrillation* **2020**, *13*, 2364. [CrossRef]
32. Shiraki, H.; Tanaka, H.; Yamauchi, Y.; Yoshigai, Y.; Yamashita, K.; Tanaka, Y.; Sumimoto, K.; Shono, A.; Suzuki, M.; Yokota, S.; et al. Characteristics of non-valvular atrial fibrillation with left atrial appendage thrombus who are undergoing appropriate oral anticoagulation therapy. *Int. J. Cardiovasc. Imaging* **2022**, *38*, 941–951. [CrossRef]

33. Turek, Ł.; Sadowski, M.; Janion-Sadowska, A.; Kurzawski, J.; Jaroszyński, A. Left atrial appendage thrombus in patients referred for electrical cardioversion for atrial fibrillation: A prospective single-center study. *Pol. Arch. Intern. Med.* **2022**, *132*, 16214. [CrossRef] [PubMed]
34. Hao, L.; Zhong, J.Q.; Zhang, W.; Rong, B.; Xie, F.; Wang, J.T.; Yue, X.; Zheng, Z.T.; Zhu, Q.; Zhang, Y. Uninterrupted dabigatran versus warfarin in the treatment of intracardiac thrombus in patients with non-valvular atrial fibrillation. *Int. J. Cardiol.* **2015**, *190*, 63–66. [CrossRef] [PubMed]
35. Hussain, A.; Katz, W.E.; Genuardi, M.V.; Bhonsale, A.; Jain, S.K.; Kancharla, K.; Saba, S.; Shalaby, A.A.; Voigt, A.H.; Wang, N.C. Non-vitamin K oral anticoagulants versus warfarin for left atrial appendage thrombus resolution in nonvalvular atrial fibrillation or flutter. *Pacing. Clin. Electrophysiol.* **2019**, *42*, 1183–1190. [CrossRef]
36. Ke, H.-H.; He, Y.; Lv, X.-W.; Zhang, E.-H.; Wei, Z.; Li, J.-Y. Efficacy and safety of rivaroxaban on the resolution of left atrial/left atrial appendage thrombus in nonvalvular atrial fibrillation patients. *J. Thromb. Thrombolysis* **2019**, *48*, 270–276. [CrossRef]
37. Lip, G.Y.; Hammerstingl, C.; Marin, F.; Cappato, R.; Meng, I.L.; Kirsch, B.; Morandi, E.; van Eickels, M.; Cohen, A. Rationale and design of a study exploring the efficacy of once-daily oral rivaroxaban (X-TRA) on the outcome of left atrial/left atrial appendage thrombus in nonvalvular atrial fibrillation or atrial flutter and a retrospective observational registry providing baseline data (CLOT-AF). *Am. Heart J.* **2015**, *169*, 464–471. [CrossRef]
38. Mitamura, H.; Nagai, T.; Watanabe, A.; Takatsuki, S.; Okumura, K. Left atrial thrombus formation and resolution during dabigatran therapy: A Japanese Heart Rhythm Society report. *J. Arrhythm.* **2015**, *31*, 226–231. [CrossRef]
39. Nelles, D.; Lambers, M.; Schafigh, M.; Morais, P.; Schueler, R.; Vij, V.; Tiyerili, V.; Weber, M.; Schrickel, J.W.; Nickenig, G.; et al. Clinical outcomes and thrombus resolution in patients with solid left atrial appendage thrombi: Results of a single-center real-world registry. *Clin. Res. Cardiol.* **2021**, *110*, 72–83. [CrossRef]
40. Niku, A.D.; Shiota, T.; Siegel, R.J.; Rader, F. Prevalence and Resolution of Left Atrial Thrombus in Patients with Nonvalvular Atrial Fibrillation and Flutter with Oral Anticoagulation. *Am. J. Cardiol.* **2019**, *123*, 63–68. [CrossRef]
41. Wu, M.S.; Gabriels, J.; Khan, M.; Shaban, N.; D'Amato, S.A.; Liu, C.F.; Markowitz, S.M.; Ip, J.E.; Thomas, G.; Singh, P.; et al. Left atrial thrombus despite continuous direct oral anticoagulant or warfarin therapy in patients with atrial fibrillation: Insights into rates and timing of thrombus resolution. *J. Interv. Card. Electrophysiol.* **2018**, *53*, 159–167. [CrossRef] [PubMed]
42. Yang, Y.; Du, X.; Dong, J.; Ma, C. Outcome of Anticoagulation Therapy of Left Atrial Thrombus or Sludge in Patients with Nonvalvular Atrial Fibrillation or Flutter. *Am. J. Med. Sci.* **2019**, *358*, 273–278. [CrossRef]
43. Mazur, E.S.; Mazur, V.V.; Bazhenov, N.D.; Orlov, Y.A. Efficiency of the left atrial appendage thrombus dissolution in patients with persistent nonvalvular atrial fibrillation with warfarin or direct oral anticoagulants therapy. *Ration. Pharmacother. Cardiol.* **2021**, *17*, 724–728. [CrossRef]
44. Faggiano, P.; Dinatolo, E.; Morco, A.; De Chiara, B.; Sbolli, M.; Musca, F.; Curnis, A.; Belli, O.; Giannattasio, C.; Tomasi, C.; et al. Prevalence and Rate of Resolution of Left Atrial Thrombus in Patients with Non-Valvular Atrial Fibrillation: A Two-Center Retrospective Real-World Study. *J. Clin. Med.* **2022**, *11*, 1520. [CrossRef] [PubMed]
45. McCready, J.W.; Nunn, L.; Lambiase, P.D.; Ahsan, S.Y.; Segal, O.R.; Rowland, E.; Lowe, M.D.; Chow, A.W. Incidence of left atrial thrombus prior to atrial fibrillation ablation: Is pre-procedural transoesophageal echocardiography mandatory? *Europace* **2010**, *12*, 927–932. [CrossRef] [PubMed]
46. Scherr, D.; Dalal, D.; Chilukuri, K.; Dong, J.U.N.; Spragg, D.; Henrikson, C.A.; Nazarian, S.; Cheng, A.; Berger, R.D.; Abraham, T.P.; et al. Incidence and Predictors of Left Atrial Thrombus Prior to Catheter Ablation of Atrial Fibrillation. *J. Cardiovasc. Electrophysiol.* **2009**, *20*, 379–384. [CrossRef]
47. Puwanant, S.; Varr, B.C.; Shrestha, K., Hussain, S.K.; Tang, W.H.; Gabriel, R.S.; Wazni, O.M.; Bhargava, M.; Saliba, W.I.; Thomas, J.D.; et al. Role of the CHADS2 score in the evaluation of thromboembolic risk in patients with atrial fibrillation undergoing transesophageal echocardiography before pulmonary vein isolation. *J. Am. Coll. Cardiol.* **2009**, *54*, 2032–2039. [CrossRef]

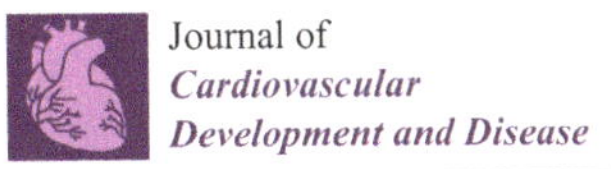
Journal of
Cardiovascular Development and Disease

MDPI

Review

Antithrombotic Therapy in Peripheral Artery Disease: Current Evidence and Future Directions

Mario Enrico Canonico [1,2,*], Raffaele Piccolo [2], Marisa Avvedimento [2,3], Attilio Leone [2], Salvatore Esposito [2], Anna Franzone [2], Giuseppe Giugliano [2], Giuseppe Gargiulo [2], Connie N. Hess [1], Scott D. Berkowitz [1], Judith Hsia [1], Plinio Cirillo [2], Giovanni Esposito [2] and Marc P. Bonaca [1]

1 CPC Clinical Research, Department of Medicine, University of Colorado, Aurora, CO 80045, USA
2 Department of Advanced Biomedical Sciences, University of Naples Federico II, 80131 Naples, Italy
3 Quebec Heart and Lung Institute, Laval University, Quebec City, QC G1V0A6, Canada
* Correspondence: marioenrico.canonico@cpcmed.org; Tel.: +1-(303)-860-9900

Abstract: Patients with peripheral artery disease (PAD) are at an increased risk of major adverse cardiovascular events, and those with disease in the lower extremities are at risk of major adverse limb events primarily driven by atherothrombosis. Traditionally, PAD refers to diseases of the arteries outside of the coronary circulation, including carotid, visceral and lower extremity peripheral artery disease, and the heterogeneity of PAD patients is represented by different atherothrombotic pathophysiology, clinical features and related antithrombotic strategies. The risk in this diverse population includes systemic risk of cardiovascular events as well as risk related to the diseased territory (e.g., artery to artery embolic stroke for patients with carotid disease, lower extremity artery to artery embolism and atherothrombosis in patients with lower extremity disease). Moreover, until the last decade, clinical data on antithrombotic management of PAD patients have been drawn from subanalyses of randomized clinical trials addressing patients affected by coronary artery disease. The high prevalence and related poor prognosis in PAD patients highlight the pivotal role of tailored antithrombotic therapy in patients affected by cerebrovascular, aortic and lower extremity peripheral artery disease. Thus, the proper assessment of thrombotic and hemorrhagic risk in patients with PAD represents a key clinical challenge that must be met to permit the optimal antithrombotic prescription for the various clinical settings in daily practice. The aim of this updated review is to analyze different features of atherothrombotic disease as well as current evidence of antithrombotic management in asymptomatic and secondary prevention in PAD patients according to each arterial bed.

Keywords: peripheral artery disease (PAD); antithrombotic therapy; dual pathway inhibition (DPI); major adverse cardiovascular events (MACE); major adverse limb events (MALE)

Citation: Canonico, M.E.; Piccolo, R.; Avvedimento, M.; Leone, A.; Esposito, S.; Franzone, A.; Giugliano, G.; Gargiulo, G.; Hess, C.N.; Berkowitz, S.D.; et al. Antithrombotic Therapy in Peripheral Artery Disease: Current Evidence and Future Directions. *J. Cardiovasc. Dev. Dis.* **2023**, *10*, 164. https://doi.org/10.3390/jcdd10040164

Academic Editor: Aleš Blinc

Received: 10 March 2023
Revised: 7 April 2023
Accepted: 8 April 2023
Published: 10 April 2023

1. Introduction

Peripheral artery disease (PAD) encompasses a variety of non-coronary artery diseases, and its prevalence varies based on screening approaches and clinical features. Recent data reveal a global prevalence of 80 million strokes, the majority (87%) of which are ischemic [1]. Estimated global recent prevalence of abdominal aortic aneurysm (AAA) recognizes a cohort of 35 million patients with AAA, whereas more than 230 million people are affected by lower extremity peripheral artery disease (LEPAD) with increasing prevalence over time due to lack of awareness and consequently underdiagnosis and undertreatment [2,3]. Moreover, epidemiologic data indicate a prevalence of stroke of 3% with 800,000 new/recurrent strokes annually in the United States, with a higher risk in women (20–21%) compared to men (14–17%) for patients aged 55 years or older, and a prevalence of AAA of 0.92% in people aged 30–79 years with a 4:1 ratio for men vs. women [1,3]. The prevalence of LEPAD in men ranges from 6.5% in patients aged 60–69 years to 29.4% in those aged >80 years; in the same age groups, the prevalence of LEPAD in women increases from 5.3%

to 24.7% [4,5]. In addition, probably based on genetic and risk factor exposure, LEPAD is more prevalent in black patients than in white patients, whereas prevalence is lowest in Asian and Hispanic patients [4,5]. According to the American College of Cardiology (ACC)/American Heart Association (AHA) and European Society of Cardiology (ESC) guidelines, the PAD definition includes, for each arterial district, a ≥50% stenosis of the extracranial internal carotid artery assessed with the North American Symptomatic Carotid Endarterectomy Trial (NASCET) method, AAA with aortic diameter ≥ 3 cm, and for LEPAD, ankle-brachial index ≤ 0.90, history of claudication, acute limb ischemia (ALI), chronic limb-threatening ischemia (CLTI), amputation for vascular causes or previous lower extremity revascularization (LER) [4–6]. Compared with myocardial infarction (MI), PAD shows a more variable clinical presentation, from vague to fatal signs that often lead to delayed diagnosis and treatment [2]. Carotid artery disease manifests a spectrum of different clinical features ranging from asymptomatic cases to hemispheric symptoms such as weakness, numbness, aphasia or face, arm and leg contralateral paresthesia resulting from transient ischemic attack (TIA) or ischemic stroke (IS) [5]. AAA often represents an incidental finding during other imaging tests (e.g., abdominal ultrasound or CT/MRA scan) with usually no specific clinical features, even in patients with more than 5 cm diameter [7]. Life-threatening complications of AAA include aortic rupture with or without previous chronic dissection [7]. LEPAD represents the majority of PAD observed with mild to severe clinical presentations. Claudication represents a mild manifestation of LEPAD, including muscle fatigue, discomfort, cramping or pain triggered by exercise with recovery upon rest [4,5]. ALI represents one of the life-threatening conditions of LEPAD characterized by acute (within 2 weeks) severe limb hypoperfusion with pain, pallor, pulselessness, paresthesia and often paralysis with impaired prognosis in terms of all-cause death and amputation for vascular causes [4,5]. In contrast, CLTI is characterized by chronic (more than 2 week duration) ischemic rest pain, leg nonhealing wound/ulcers or gangrene caused by arterial occlusive disease with poor prognosis [4,5]. Overall, several studies showed an increased risk of major adverse cardiovascular events (MACE), including MI, IS and cardiovascular (CV) death among PAD patients, along with a heightened risk of major adverse limb events (MALE), which is usually defined as severe limb ischemia leading to an intervention or major vascular amputation [4,5]. PAD treatment includes medical therapy, supervised exercise and revascularization (e.g., endovascular, surgical or hybrid) based on anatomical features, patient characteristics and local expertise [4,5,8]. Antithrombotic therapy represents a milestone in PAD management, given that atherosclerosis represents the common pathophysiologic feature in arterial beds [4–6,8]. The purpose of this review is to highlight current evidence and future directions on antithrombotic therapy in PAD patients with an overview of the principal trials in this field.

2. Role of Atherothrombosis in the Progression and Complications of Non-Coronary Artery Disease

Atherothrombosis represents the common pathophysiologic process in coronary and non-coronary artery disease in patients affected by atherosclerotic damage. PAD involves the same CV risk factors as coronary artery disease (CAD), including arterial hypertension, diabetes mellitus, dyslipidemia, smoking, history of CV disease, chronic kidney disease, life habits, history of radiation therapy, psycho-social and genetic factors [4,5]. These shared factors explain the common finding of polyvascular artery disease, defined by the concomitant presence of relevant atherosclerotic disease in at least two vascular beds [4,5]. However, at variance with CAD and IS, smoking is the risk factor most strongly associated with LEPAD [9]. Finally, differences in atherothrombosis pathophysiology have to be considered among carotid artery disease, abdominal aortic disease and LEPAD.

2.1. Carotid Artery Disease

Extracranial carotid atherosclerosis can be readily detected with non-invasive assessment such as high-resolution B-mode carotid ultrasonography (Duplex US), CT and MRI scan also able to detect subclinical atherosclerosis [4,5]. Carotid atherosclerosis leads to 25% of IS associated with disability and impaired prognosis [10]. Stenosis degree is one of the most important risk factors of ipsilateral IS, along with hemodynamic factors. Although hypoperfusion plays a role in the pathogenesis of IS, the majority of stroke events are attributed to embolization from unstable atherosclerotic plaque or carotid artery acute occlusion with thrombus distally detected [10]. As in coronary arteries, vulnerable plaque characteristics include a lipid-rich necrotic core with a thin/ruptured fibrous cap, ulceration and intraplaque hemorrhage (IPH) associated with the presence of inflammatory cells [10]. Carotid stenosis progression is also recently considered a marker of vulnerability, contributing to distal embolization and subsequent TIA [10]. IPH represents one of the plaque progression factors with increased rupture risk and future risk events [10]. Further pathophysiologic findings on vulnerable plaque highlight the role of inflammation in atherosclerosis with intraplaque angiogenesis and hypoxia in cerebral adverse events. Hybrid imaging, such as PET/CT or PET/MRI, can detect plaque rupture features [10]. Some morphologic characteristics, such as ulceration and IPH, are also associated with the occurrence of ischemic events, independently of the degree of stenosis. Recent data from the American Society of Neuroradiology showed that the annualized event rates of ipsilateral stroke in those with IPH are higher than in patients without IPH irrespective of stenosis degree: 9.0% versus 0.7% (<50% stenosis), 18.1% versus 2.1% (50–69% stenosis) and 29.3% versus 1.5% (70–99% stenosis), confirming IPH as an independent predictor of ipsilateral stroke (Hazard Ratio—HR 3.3; 95% confidence interval—CI, 1.4–7.8) [10]. Plaque calcification represents a stabilizing factor in carotid artery disease with less inflammation, neovascularization and IPH and lower likelihood of rupture [10]. Furthermore, several atherosclerosis-related factors such as aging, inflammation and ischemia increase circulating levels and deposition of amyloid-beta (Aβ) in intracranial arteries contributing to different types of dementia with impaired cognitive performance. Moreover, among Aβ peptides, Aβ1-40 was independently associated with impaired vasodilating properties, higher IMT, low ABI, as well as coronary and aorta arterial damage, with a worse prognosis in elderly patients [11]. These findings highlight the interplay between dementia and CVD, particularly driven by diffuse atherosclerosis, although current Aβ pathophysiology and therapeutic options are still uncertain areas.

2.2. *Abdominal Aortic Disease*

Atherosclerosis is frequently associated with abdominal aortic disease, especially in polyvascular disease [7,12]. While the pathophysiologic role of atherosclerosis in medium and small arteries is well-known, the relationship with abdominal aortic disease is incompletely understood. Acute abdominal aortic thrombosis is a fatal and rare condition, and abdominal aortic disease is mostly represented by AAA, which arises as a pathological response to aortic atherosclerosis [12]. In animal models, inflammatory pathways, along with aortic matrix degradation and hemodynamic forces, lead to AAA development [12]. During intraluminal stenosis development, the atherosclerotic process includes compensatory chronic inflammatory changes in the media with extracellular matrix remodeling promoting artery diameter growth leading to the development of an aortic aneurysm [12]. Moreover, aortic media chronic inflammation driven by myo-fibroblast favors aortic false lumen development with chronic aortic dissection origin [7]. To date, the interplay of chronic dissection and aneurysm is not completely understood. Chronic aortic dissection leads to more rapid aortic aneurysm growth than non-dissected aorta [7]. Arterial pressure and relative wall tension drive false lumen propagation in the aortic axis with a high rupture risk, which overcomes the remodeling capability of aneurysmatic artery wall [7]. In addition, partial chronic abdominal aortic thrombosis is a common finding in patients with chronic aortic dissection and/or aneurysm [6]. Often, aortic thrombus shows a multi-layered morphology with dense fibrin and inflammatory cells such as leukocytes and platelet-derived proteins with proteolytic proprieties and increased risk of peripheral embolism [6]. Around 40% of chronic aortic dissection patients require urgent revascularization for aortic rupture and/or branch vessel hypoperfusion [7]. New understandings are evolving from combining 3- and 4-dimensional CT morphology data, MRI flow data, computer simulation of fluid dynamics and the fields of biomechanics and mechanobiology, which may help to better comprehend the physiopathologic key elements leading to false lumen degeneration and aneurysm development and facilitate the development of novel treatments and appropriate timing for them in patients affected by chronic aortic dissection and aneurysm [13,14].

2.3. *Lower Extremity Peripheral Artery Disease*

Atherosclerosis is a common LEPAD feature that can explain symptoms and signs related to different clinical presentations, from claudication to ALI/CLTI [15]. Lower extremity peripheral arteries represent a very diverse arterial bed with several differences and related clinical scenarios between itself. One difference is driven by anatomical factors (e.g., arterial diameter) considering large vessels (e.g., iliac–femoral axis, popliteal artery) and smaller vessels below the knee (BTK) [16]. Consequently, flow characteristics and atherosclerotic complications will be different. Overall, compared to cerebrovascular disease and CAD, the role of atherothrombosis in the progression and complications of LEPAD is less clear and studied. Atherosclerosis causes claudication, which represents the clinical manifestation of significant atherosclerotic stenosis during exercise and relief within 10 min rest. Particularly, symptoms stem from the muscles perfused by the stenosed artery [5,7]. Similar to chronic CAD, claudication represents the chronic manifestation of LEPAD, with management depending mostly on CV risk factor and physical exercise management [5,7]. Considering atherothrombotic complications of LEPAD, approximately 10% of patients with claudication develop CLTI within 5 years, contributing to poor prognosis, including 1-year rates of mortality of 25% and 1-year rates of amputation of 30% [15]. The main difference from CAD atherothrombosis is the occurrence of thrombotic events even in the absence of significant atherosclerotic disease. Histopathological analysis on LEPAD presenting with CLTI shows that thrombotic occlusion in the BTK district is the main cause of disease even in patients without significant atherosclerosis, while significant atherosclerotic lesions were more often detected in the femoral-popliteal artery [16]. On the contrary, ALI and related thrombus often occur in patients with both significant and non-significant atherosclerosis, while small vessel obliteration is driven by media calcification, intimal fibrosis and superimposed cholesterol emboli [16]. Similar to acute MI, ALI is characterized by a sudden decrease in limb perfusion that often results in tissue loss and requires early intervention. However, in contrast to atherothrombotic acute coronary events, ALI in patients with PAD is driven not only by atherothrombosis but also by emboli from the heart and proximal vessels and graft occlusion in patients with previous lower extremity revascularization (LER) [17]. Several pieces of evidence support the thromboembolic origin of CLTI/ALI-affected popliteal and BTK artery rather than stenotic atherosclerotic disease. The embolic source is often an aorto-iliac-femoral atherosclerotic plaque with subsequent lumen obliteration of a distal smaller artery [16]. The main differences between ALI and CTLI are represented by the duration of symptoms (less vs. more than two weeks), clinical presentation (acute vs. chronic), presence of collateral arteries in CTLI and timing of revascularization (urgent in order to address the high risk of amputation vs. non-urgent in order to minimize tissue loss) [18].

3. Approaches to Antithrombotic Therapy in Peripheral Artery Disease

Guideline-directed medical therapy (GDMT) includes antithrombotic therapy as one of the cornerstones of multidimensional management, which includes structured exercise and lifestyle modification in order to reduce MACE and MALE [4–6]. The well-recognized role of atherosclerosis and its related complications in PAD patients explains the need for antithrombotic therapy in each arterial bed, even in asymptomatic patients [4,5]. Antithrombotic management in PAD represents a challenge due to different evidence for each arterial bed, symptoms assessment and personal expertise. While the importance of antithrombotic therapy in CAD has been well-recognized over the past decades and includes dedicated randomized clinical trials (RCTs), evidence supporting antithrombotic therapy in PAD has been based, until recently, on subgroup analyses of coronary artery disease trials, often with slim and conflicting data. In the last years, a new antithrombotic strategy emerged in PAD research that combines an anticoagulant with standard antiplatelet therapy. The recognized role of embolic source of many cases of LEPAD promoted a new target therapy called dual pathway inhibition (DPI) to inhibit thrombus formation via dual pathways: platelet activation and thrombin generation [19]. In addition, thrombo-hemorrhagic risk has to be assessed in order to choose the right antithrombotic regimen. Due to concomitant diseases, PAD patients have a high bleeding risk compared to the CAD population, but there is less evidence on which to develop a bleeding score risk assessment. To date, Thrombolysis in Myocardial Infarction (TIMI) and the International Society on Thrombosis and Hemostasis (ISTH) bleeding are the most common safety outcomes assessed in PAD trials [20]. Pharmacodynamic targets of antithrombotic drugs in peripheral artery disease by thrombotic pathway are summarized in Figure 1.

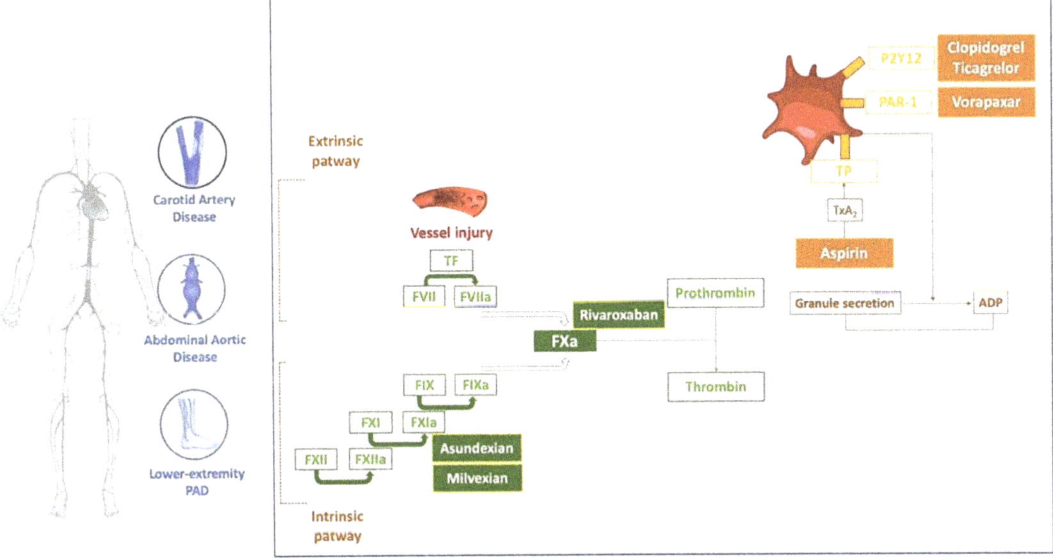

Figure 1. Pharmacodynamic targets of antithrombotic drugs in peripheral artery disease by thrombotic pathway. TF = tissue factor, TxA_2 = Thromboxane A2, TP = thromboxane prostanoid, PAR-1 = protease-activated receptor-1, ADP = adenosine diphosphate.

3.1. *Carotid Artery Disease*

3.1.1. Asymptomatic Patients

GDMT includes antithrombotic therapy in carotid artery disease in the presence of a $\geq$50% stenosis since no RCT has assessed antithrombotic therapy in non-significant carotid stenosis [5,19]. In asymptomatic patients with significant carotid stenosis, GDMT recommends single antiplatelet therapy (SAPT) either with aspirin (75–100 mg) or clopidogrel (75 mg) (Class IIa) for primary prevention of MACE if bleeding risk is low [5,21]. The use of aspirin in overall PAD patients was assessed in AntiThrombotic Trialists (ATT) meta-analysis, including six primary prevention RCTs assessing different doses of aspirin as well as other antiplatelet agents such as picotamide. In primary prevention RCTs, aspirin reduced serious vascular events (including MI, stroke and vascular death) by 12% (HR 0.88; 95% CI 0.82–0.94 without benefit on CAD death, HR 0.95; 95% CI 0.78–1.15 or stroke death, HR 1.21; 95% CI 0.84–1.74), associated with an increase of hemorrhagic stroke HR, 1.32; 95%

CI 1.00–1.75 and major extracranial bleeding (HR 1.54; 95% CI 1.30–1.82) [22]. These results confirm the uncertain net benefit of aspirin in PAD primary prevention in the absence of concomitant diseases. Recently, an analysis on stroke risk in Cardiovascular Outcomes for People Using Anticoagulation Strategies (COMPASS) trial, comparing rivaroxaban 2.5 mg twice daily plus aspirin vs. rivaroxaban 5 mg twice daily vs. aspirin in stable CAD or PAD patients, proved the efficacy of rivaroxaban 2.5 mg plus aspirin with a 53% relative reduction in the risk of ischemic/unknown stroke in high-risk patients without a history of stroke (HR, 0.57; 95% CI 0.39–0.84), with no significant increase in the risk of hemorrhagic stroke for rivaroxaban plus aspirin vs. aspirin alone (HR, 1.76; 95% CI 0.59–5.24) [23]. Therefore, a low dose of rivaroxaban plus aspirin could represent a new antithrombotic regimen for polyvascular disease patients with CAD and/or LEPAD without prior history of stroke, especially among those without high bleeding risk features. However, appropriately designed trials are needed to share light in this field.

3.1.2. Secondary Prevention

Antithrombotic therapy is recommended in patients with symptomatic carotid artery disease to prevent recurrent cerebrovascular events [8,19]. GDMT recommends lifelong SAPT with aspirin or clopidogrel in patients with prior IS or TIA due to large artery disease [5,8]. SAPT showed a better safety profile in major bleeding outcomes compared to oral anticoagulation. Data from the ATT meta-analysis among 16 secondary prevention RCTs, including 10 with previous stroke/TIA, highlighted the benefit of aspirin on major coronary events (HR 0.80; 95% CI 0.73–0.88) as well as IS (HR 0.78; 95% CI 0.61–0.99) and serious vascular events (HR 0.81; 95% CI 0.75–0.87), with an increase of major extracranial bleeding (HR 2.69; 95% CI 1.25–5.76) but not for rates of hemorrhagic stroke (HR 1.67; 95% CI 0.97–2.90) [22]. Different RCTs compared oral anticoagulation vs. SAPT in PAD patients (Table 1). In the European/Australasian Stroke Prevention in Reversible Ischemia Trial (ESPRIT), oral anticoagulation (either phenprocoumon, warfarin or acenocoumarol) was compared with aspirin in patients with recent (within 6 months) TIA or minor stroke. Overall, there was no significant difference between the two groups for recurrent CV events (HR 1.02; 95% CI 0.77–1.35). However, an excess of major bleeding among patients randomized to oral anticoagulation was observed (HR 2.56; 95% CI 1.48–4.43) [24]. The Warfarin Antiplatelet Vascular Evaluation (WAVE) compared a combination therapy with an antiplatelet agent (either aspirin, ticlopidine or clopidogrel) plus oral anticoagulant agent (either warfarin or acenocoumarol) vs. antiplatelet therapy alone in PAD patients. No differences were detected in the primary efficacy outcome of MACE (RR 0.92; 95% CI 0.73–1.16), although the risk of life-threatening bleeding was three-fold higher in the combination therapy arm (RR 3.41; 95% CI 1.84–6.35) [25]. Hence, unless indicated for other clinical circumstances, anticoagulation therapy is not recommended for secondary prevention after TIA/stroke [5,8]. Other antiplatelet drugs, such as P2Y12 receptor inhibitors, have been assessed in secondary IS prevention. In the Clopidogrel and Aspirin for Reduction of Emboli in Symptomatic Carotid Stenosis (CARESS) trial, dual antiplatelet therapy (DAPT) with aspirin and clopidogrel was compared to aspirin among patients with recently symptomatic $\geq$ 50% carotid stenosis assessing microembolic signals (MES) by transcranial Doppler (TCD). DAPT was associated with a relative reduction of 40% in the risk of asymptomatic embolization and stroke (relative risk reduction—RRR 39.8%; 95% CI 13.8–58.0) on day 7 with no significant difference in bleeding adverse events among DAPT (3.9%) vs. aspirin alone (1.8%) group [26]. Similarly, in the clopidogrel plus aspirin versus aspirin alone for reducing embolization in patients with acute symptomatic cerebral or carotid artery stenosis (CLAIR) trial, DAPT with aspirin and clopidogrel compared with aspirin alone was associated with similar risk reduction (about 42%) of MES in patients with acute IS or TIA (RRR 42.4%; 95% CI 4.6–65.2) with only two minor bleeding events in the DAPT group [27]. The Clopidogrel in High-Risk Patients with Acute Nondisabling Cerebrovascular Events (CHANCE) trial assessed a 3-month strategy of DAPT with aspirin plus clopidogrel vs. aspirin alone among patients with acute high-risk TIA or minor ischemic stroke. During 90 days of follow-up, DAPT reduced the occurrence of stroke (ischemic or hemorrhagic) by 32% compared to aspirin alone (HR 0.68; 95% CI 0.57–0.81) with a non-significant increased rate of any bleeding events in DAPT (2.3%) vs. aspirin group (1.6%) (HR 1.41; 95% CI 0.95–2.10) [28]. Like the CHANCE trial, the Platelet-Oriented Inhibition in New TIA and Minor Ischemic Stroke (POINT) trial assessed DAPT with aspirin and clopidogrel vs. aspirin alone in patients affected by minor IS or high-risk TIA for 90 days. DAPT reduced the composite primary efficacy outcome of major ischemic events defined as IS, MI and death due to ischemic vascular events (HR 0.75, 95% CI 0.59–0.95), which was counterbalanced by an increased risk of major bleeding in DAPT arm of HR (2.32, 95% CI 1.10–4.87) [29]. A pooled analysis of the CHANCE and POINT trials showed a reduced risk of new stroke with DAPT (HR 0.69; 95% CI 0.60–80), with a non-significantly higher risk of major bleeding (HR 1.67, 95% CI 0.93–2.99) [30]. A risk benefit-analysis of the two trials showed that the recurrent ischemic events with DAPT are mainly prevented within the first 2 weeks after randomization, whereas the risk of major bleeding was small and constant throughout the follow-up. In view of these findings, the AHA/ASA guidelines recommend DAPT with clopidogrel and aspirin up to 3 months after a minor stroke [8]. In analogy with the issues related to clopidogrel use in the field of CAD (delayed onset of action, large interindividual variability and irreversibility of inhibitory action), data from the CHANCE trial showed that the benefit of a clopidogrel-based DAPT was essentially confined to extensive clopidogrel metabolizer phenotype, whereas there was no benefit with DAPT in poor or intermediate clopidogrel metabolizers [31]. Cilostazol is another drug assessed in the secondary prevention of IS. Cilostazol is a selective inhibitor of phosphodiesterase type 3, which increases the cyclic adenosine monophosphate (cAMP) levels that lead to inhibition of platelet aggregation. "Dual Antiplatelet therapy using Cilostazol for Secondary Prevention in Patients with high-risk ischemic stroke in Japan", a multicenter RCT, assessed the safety and efficacy of cilostazol to prevent stroke recurrence in a DAPT strategy with either aspirin or clopidogrel vs. SAPT with aspirin or clopidogrel. DAPT with cilostazol reduced IS recurrence (3%) vs. SAPT (7%) (HR 0.49; 95% CI 0.31–0.76) with no differences in severe or life-threatening bleeding among study groups (HR 0.66; 95% CI 0.27–1.60) [32]. However, the RCT enrolled 47% of the planned sample size (n = 4000) and included only Japanese people. Moreover, there were a very limited number of patients (n = 93) with primary efficacy outcomes that could probably reduce the statistical accuracy [32]. AHA/ASA guidelines recommend cilostazol with a class of recommendation 2b in IS secondary prevention with aspirin or clopidogrel due to limited evidence and known side effects such as headache, palpitations and tachycardia. In addition, Cilostazol is contraindicated in patients affected by heart failure (HF) treated by phosphodiesterase 3 inhibitors [8]. The efficacy and safety of a different antiplatelet strategy after IS/TIA were also assessed with newer $P2Y_{12}$ receptor inhibitors. The benefit of ticagrelor compared to aspirin in patients with acute cerebral ischemia (non-severe IS or high-risk TIA) was assessed among 13,199 patients enrolled in the Acute Stroke or Transient Ischemic Attack Treated with Aspirin or Ticagrelor and Patient Outcomes (SOCRATES) trial. Compared with aspirin, ticagrelor monotherapy (90 mg twice daily) was not superior to aspirin with respect to the primary outcome of stroke, MI or death at 90 days (HR 0.89; 95% CI 0.87–1.01). No differences were detected in major bleeding (according to PLATO classification) comparing the two study groups (HR 0.83; 95% CI 0.52–1.34) [33]. However, in a sub-analysis of the SOCRATES trial in which the primary outcome data were stratified by randomization arm and prior use of aspirin within 7 days before randomization, the benefit of ticagrelor was present among patients with prior aspirin use (HR 0.76; 95% CI 0.61–0.95) [34]. Recently, DAPT, including the more potent P2Y12 receptor inhibitor (ticagrelor 90 mg twice daily) on top of background therapy, was investigated in The Acute Stroke or Transient Ischemic Attack Treated with Ticagrelor and ASA (acetylsalicylic acid) for Prevention of Stroke and Death (THALES) trial, which enrolled patients with no more than moderate and non cardioembolic IS. DAPT with ticagrelor reduced the risk of IS or death within 30 days (HR 0.83; 95% CI 0.71–0.96) with no differences in disability. However, severe bleeding occurred more frequently in patients randomized to the experimental arm according to the Global Use of Strategies to Open Occluded Coronary Arteries (GUSTO) classification (HR 3.99, 95% CI 1.74–9.14) [35]. Considering the thromboembolic component of IS, recent data from a post-hoc analysis on DPI comes from RCT. In particular, the analysis of stroke outcomes in the COMPASS trial among patients with prior stroke showed a benefit of low-dose rivaroxaban plus aspirin compared with aspirin alone in MACE prevention (HR 0.57; 95% CI 0.34–0.96) without differences in ISTH major bleeding (HR 1.06; 95% CI 0.72–1.56) [23]. Nevertheless, these results cannot be extrapolated to the early phase after IS. In the early phase of a cerebrovascular event, DAPT represents the antithrombotic strategy in order to minimize the risk of asymptomatic cerebral embolization and stroke [19]. Antithrombotic therapy in secondary prevention of carotid artery disease was also assessed after CAS/CEA revascularization. Data derived from two small RCTs showed benefit from the DAPT regimen vs. SAPT after CAS in reducing cerebrovascular events. "The Benefits of Combined Anti-platelet Treatment in Carotid Artery Stenting" RCT compared DAPT with aspirin plus clopidogrel vs. aspirin plus 24 h of heparin in a cohort of patients undergoing CAS. At 30 days of follow-up, neurological complications, including all amaurosis fugax, TIA and all stroke, occurred in 0% of the DAPT group and in 25% of heparin group ($p = 0.02$) without any difference in major bleeding or groin complication (9% in DAPT group vs. 17% in heparin group, p = NS) [36]. Similar findings were noted in "Dual Antiplatelet Regime Versus Acetyl-acetic Acid for Carotid Artery Stenting" RCT in patients after CAS assessing a DAPT regimen (325 mg of aspirin plus 250 mg of ticlopidine) vs. 325 mg of aspirin plus 24 h of heparin as control group. After 30 days, DAPT significantly reduced minor IS/TIA (2% vs. 16%, $p < 0.05$) without any major bleeding in either group, and no difference in groin complications was observed (2% in DAPT vs. 4% in control group, p = NS) [37]. The optimal SAPT strategy after CEA was tested in low-dose and high-dose acetylsalicylic acid for patients undergoing carotid endarterectomy: a randomized controlled trial (ACE RCT) where different doses of aspirin (i.e., 81–325 mg vs. 650–1300 mg) were compared at 3 months after CEA. The occurrence of IS was lower in low-dose aspirin (3.2% vs. 6.9%), while hemorrhagic stroke was numerically more frequent in the high-dose aspirin group (RR 1.68, 95% CI 0.77–3.68) [38]. Moreover, after CEA, SAPT management is recommended over a DAPT regimen in view of a lack of benefit in fatal stroke prevention and heightened risk of major bleeding with DAPT, as shown in a systematic metanalysis involving three RCTs and seven observational studies with DAPT vs. SAPT risk difference (RD) in major bleeding 0.00; 95% CI 0.00–0.01 and neck hematoma (RD 0.04, 95% CI 0.01–0.06) [39]. The need to reduce thrombotic adverse events without increasing bleeding risk drives the development of newer antithrombotic drugs with a different target in atherosclerotic cardiovascular disease (ASCVD), including carotid artery disease for non-cardioembolic IS. Considering the key role of factor XI (FXI) in pathological thrombosis as it amplifies thrombin generation, new data provide the first clinical benefits of factor XI inhibitors [40]. Moreover, patients with FXI deficiency are known to have a lower risk of IS, while those with high FXI levels have an increased risk of recurrent IS [40]. The Factor XIa inhibition with asundexian after acute non-cardioembolic ischemic stroke (PACIFIC-Stroke) is a phase 2b RCT, the first study to report on the efficacy and safety of factor Xia inhibition in secondary prevention of non-cardioembolic IS. Patients with minor-moderate IS were randomized in a 1:1:1:1 study to receive either asundexian 10 mg, 20 mg, or 50 mg or placebo, in addition to SAPT, according to a local investigator. After 26 weeks from randomization, no differences were observed in primary efficacy outcome (i.e., symptomatic recurrent IS and incident covert brain infarct detected by MRI) among each asundexian arm vs. placebo: HR 0.99; 95% CI 0.79–1.24 for 10 mg, HR 1.15; 95% CI 0.93–1.43 for 20 mg and HR 1.06; 95% CI 0.85–1.32 for 50 mg. Considering the primary safety composite outcome of ISTH major and clinically relevant non-major bleeding, no significant differences were found among each asundexian arm and even considering all doses vs. placebo (HR 1.57, 95% CI 0.91–2.71) [41]. Despite the PACIFIC-stroke trial not finding a significant difference in the primary efficacy endpoint, in a post-hoc analysis, asundexian 50 mg reduced recurrent IS and TIA in patients with known atherosclerosis. This reinforced the rationale for assessing this hypothesis in a larger phase 3 RCT. Another XIa inhibitor, milvexian, was more recently tested in the Antithrombotic treatment with factor XIa inhibition to Optimize Management of Acute Thromboembolic events for Secondary Stroke Prevention (AXIOMATIC-SSP) phase 2 RCT in patients with acute IS or high-risk TIA compared to matched placebo on top of DAPT with aspirin and clopidogrel until day 21 from randomization followed by aspirin alone. The primary efficacy endpoint included incident IS during the treatment period or new covert brain infarction detected by the comparison of 90-day and baseline MRIs. Although the full results are not yet available at the time of this writing, preliminary data from the ESC 2022 Congress showed no difference between milvexian and placebo in the primary efficacy endpoint at 90 days. As it relates to the safety endpoint (Bleeding Academic Research Consortium—BARC classification), milvexian did not increase adverse events: milvexian 25 mg daily vs. 25 mg BID vs. 50 mg BID vs. 100 mg BID vs. 200 mg BID vs. placebo, was: 10.8% vs. 8.6% vs. 12.3% vs. 13.1% vs. 10.2% vs. 7.9% ($p > 0.05$) [42]. To date, GDMT and the recent ESC consensus document recommend at least DAPT for 1-month with aspirin and clopidogrel after CAS and SAPT after CEA (class I) [4,5,19]. An operative proposal algorithm for antithrombotic management of carotid artery disease according to asymptomatic or secondary prevention patients is depicted in Figure 2.

Table 1. Key antithrombotic trials in a primary population with peripheral artery disease.

Study	Enrolled Patients	Population	Treatment	Follow-Up	Primary Endpoint	
					Efficacy	Safety
			Carotid Artery Disease			
ESPRIT Trial [24]	1068	Recent TIA or minor stroke (within 6 months)	Oral anticoagulation (phenprocoumon, warfarin or acenocoumarol) vs. Aspirin	4.6 years	No difference in the composite outcome of all-cause death, non-fatal stroke, non-fatal MI (HR 1.02; 95% CI 0.77–1.35)	Increased major bleeding (HR 2.56; 95% CI 1.48–4.43)
CARESS Trial [26]	107	Symptomatic ≥50% carotid stenosis	DAPT (Aspirin + Clopidogrel) vs. Aspirin	7 days	Reduction in the risk of asymptomatic embolization (RR 39.8%; 95% CI 13.8–58.0)	Bleeding adverse events 3.9% vs. 1.8%, p = NS
CLAIR Trial [27]	100	Acute IS or TIA	DAPT (Aspirin + Clopidogrel) vs. Aspirin	7 days	Reduction of microembolic signals (RR 42.4%; 95% CI 4.6–65.2)	No difference in any bleeding complications
CHANCE Trial [28]	5170	Minor IS or high-risk TIA	DAPT (Aspirin + Clopidogrel) vs. Aspirin + Placebo	90 days	Reduction of stroke rate in the DAPT group (HR 0.68; 95% CI 0.57–0.81)	No difference in bleeding complications (HR, 1.41; 95% CI 0.95–2.10)
POINT Trial [29]	4881	Minor IS or high-risk TIA	DAPT (Aspirin + Clopidogrel) vs. Aspirin	90 days	Reduction of major ischemic events (IS, MI and death due to ischemic vascular events) (HR 0.75; 95% CI 0.59–0.95)	Increased hemorrhagic complications (HR 2.32; 95% CI 1.10–4.87)
Dual Antiplatelet therapy using Cilostazol for Secondary Prevention in Patients with high-risk ischemic stroke in Japan [32]	1884	High-risk non-cardioembolic IS	Aspirin/Clopidogrel + Cilostazol vs. Aspirin or Clopidogrel	1.4 years	Reduction of IS in the DAPT group (HR 0.49; 95% CI 0.31–0.76)	No difference in severe or life-threatening bleeding (HR 0.66; 95% CI 0.27–1.60)

Table 1. *Cont.*

Study	Enrolled Patients	Population	Treatment	Follow-Up	Primary Endpoint	
					Efficacy	Safety
SOCRATES Trial [33]	13,199	Non-severe IS or high-risk TIA	Ticagrelor vs. Aspirin	90 days	No difference in the composite outcome of stroke, MI or death (HR 0.89; 95% CI 0.87–1.01)	No difference in major bleeding complications (HR 0.83; 95% CI 0.52–1.34)
THALES Trial [35]	11,016	Mild-to-moderate acute noncardioembolic IS or TIA	DAPT (Aspirin + Ticagrelor) vs. Aspirin	30 days	Reduction of composite of stroke or death in the DAPT group (HR 0.83; 95% CI 0.71–0.96)	Increased severe bleeding (HR 3.99; 95% CI 1.74–9.14)
The Benefits of Combined Antiplatelet Treatment in Carotid Artery Stenting [36]	47	Patients undergoing carotid artery stenting	DAPT (Aspirin + Clopidogrel) vs. Aspirin + 24-h heparin	30 days	Neurological events (amaurosis fugax, TIA and all stroke) 0% vs. 25%	No difference in major bleeding 9% vs. 17%, p = NS
Dual Antiplatelet Regime Versus Acetyl-acetic Acid for Carotid Artery Stenting [37]	100	Patients undergoing carotid artery stenting	DAPT (Aspirin 325 mg + Ticlopidine) vs. Aspirin 325 mg + 24-h heparin	30 days	Reduction of minor IS/TIA in the DAPT group, 2% vs. 16%, $p < 0.05$	No difference in major bleeding
ACE RCT [38]	2849	Patients undergoing carotid endarterectomy	Aspirin 81–325 mg vs. Aspirin 650–1300 mg	90 days	Lower rate of IS, MI and death in low-dose group 6.2% vs. 8.4%, $p = 0.03$	Increased hemorrhagic stroke in high dose group (RR, 1.68; 95% CI 0.77–3.68)
PACIFIC Stroke Trial [41]	1808	Acute non-cardioembolic IS	Asundexian 10 mg vs. Asundexian 20 mg vs. Asundexian 50 mg vs. Placebo	26 weeks	No differences in IS and incident covert brain infarct detected by MRI (HR 0.99; 95% CI, 0.79–1.24) 10 mg (HR 1.15; 95% CI 0.93–1.43) 20 mg (HR 1.06; 95% CI 0.85–1.32) 50 mg	No difference in major or clinically relevant non-major bleeding (HR 1.57; 95% CI 0.91–2.71)
AXIOMATIC-SPP [42]	2366	Acute non-lacunar IS	Milvexian 25 mg vs. Milvexian 50 mg vs. Milvexian 100 mg vs. Milvexian 200 mg vs. Placebo	90 days	No differences in IS and incident covert brain infarct detected by MRI 25 mg, 16.2% and 18.5% 50 mg, 14.1% 100 mg, 14.7% 200 mg, 16.4% Placebo, 16.6%	No differences in rates of BARC 3 or 5 25 mg, 0.6% 50 mg, 1.5% 100 mg, 1.6% 200 mg, 1.5% Placebo, 0.6%
			Abdominal Aortic Disease			
TicAAA Trial [43]	144	Patients with AAA and with a maximum aortic diameter 35–49 mm	Ticagrelor vs. Placebo	12 months	No differences were found in AAA volume increase assessed by MRI (HR 1.013; 95% CI 0.993–1.034)	Increased bleeding events rate in ticagrelor group (33% vs. 11%, $p = 0.002$)

Table 1. *Cont.*

Study	Enrolled Patients	Population	Treatment	Follow-Up	Primary Endpoint	
					Efficacy	Safety
			Lower-extremity PAD			
POPADAD Trial [44]	1276	Patients with diabetes with ABI < 0.99	Aspirin vs. Placebo	6.7 years	No difference in the composite of death due to CAD or stroke, non-fatal MI or stroke, or amputation (HR 0.98; 95% CI 0.76–1.26)	No difference in gastrointestinal bleeding (HR 0.90; 95% CI 0.53–1.52)
AAA Trial [45]	3350	General population with ABI ≤ 0.95	Aspirin vs. Placebo	8.2 years	No difference in the composite of fatal or non-fatal coronary event, stroke, or revascularization (HR 1.03; 95% CI 0.84–1.27)	Increased major bleeding (HR 1.71;95% CI 0.99–2.97)
WAVE Trial [25]	2161	PAD	Antiplatelet therapy (aspirin, ticlopidine or clopidogrel) + Oral anticoagulation (warfarin or acenocoumarol) vs. Antiplatelet therapy alone	35 months	No difference in the composite outcome of all-cause death, stroke and MI (RR 0.92; 95% CI 0.73–1.16)	Increased life-threatening or moderate bleeding (RR 3.21; 95% CI 2.02–5.08)
EUCLID Trial [46]	13,885	Symptomatic LEPAD	Ticagrelor vs. Clopidogrel	30 months	Ticagrelor not superior to clopidogrel for MACE reduction (HR 1.02; 95% CI 0.92–1.13)	No increase in bleeding (HR 1.1; 95% CI 0.84–1.43)
Dutch BOA [47]	2690	Patients with LEPAD after infrainguinal arterial grafting	Oral anticoagulant (phenprocoumon or acenocoumarol; coumarin derivatives) vs. aspirin equivalent	21 months	No difference in graft occlusion (HR 0.95; 95% CI, 0.82–1.11) No difference in the composite of vascular mortality, MI, stroke, or amputation (HR 0.89; 95% CI 0.75–1.06)	Increase in severe bleeding (HR 1.96; 95% CI 1.42–2.71)
CASPAR Trial [48]	851	Patients with LEPAD undergoing below-knee bypass grafting	Aspirin + Clopidogrel vs. Aspirin + Placebo	24 months	No reduction in the composite of graft occlusion, revascularization, major amputation, or death (HR 0.98; 95% CI 0.78–1.23)	No difference in severe bleeding (2.1% vs. 1.2%)
Cilostazol reduces restenosis after endovascular therapy in patients with femoropopliteal lesions [49]	127	Patients with LEPAD after endovascular LER	Aspirin + Cilostazol vs. Aspirin + Ticlopidine	36 months	Higher patency lesion rates at 12, 24, 36 months in Cilostazol group (87%, 82%, 73%) vs. Ticlopidine group (65%, 60%, 51%), $p = 0.013$	No difference in bleeding $p = 0.72$

Table 1. *Cont.*

Study	Enrolled Patients	Population	Treatment	Follow-Up	Primary Endpoint	
					Efficacy	Safety
MIRROR Study [50]	80	Patients with LEPAD undergoing endovascular LER	Clopidogrel + Aspirin vs. Aspirin + Placebo	6 months	Decreased risk of target lesion revascularization (5% vs. 8%, $p = 0.04$) at 6 months but no difference at 1 year (25% vs. 32%, $p = 0.35$)	No increase in bleeding (2.5% vs. 5%, $p = 0.56$)
ePAD Trial [51]	203	Patients with LEPAD after endovascular LER	Edoxaban + Aspirin vs. Aspirin + Clopidogrel	3 months	No difference in restenosis or reocclusion of femoropopliteal targets (HR 0.89; 95% CI 0.59–1.34)	No difference in bleeding (RR 0.56; 95% CI 0.19–1.62)
VOYAGER-PAD Trial [52]	6564	Patients with LEPAD after LER	Aspirin + Rivaroxaban vs. Aspirin + Placebo	3 years	Composite of reduction of MACE and MALE (HR 0.85; 95% CI 0.76–0.96)	No difference in TIMI major bleeding (HR 1.43; 95% CI 0.97–2.10) increase in ISTH major bleeding (HR 1.42; 95% CI 1.10–1.84)

HR = hazard ratio; CI = confidence interval; NS = not significant; RR = risk ratio; TIA = transient ischemic attack; DAPT = dual antiplatelet therapy; IS = ischemic stroke; MI = myocardial infarction; AAA = abdominal aortic aneurysm; MRI = magnetic resonance imaging; ABI = ankle brachial index; PAD = peripheral artery disease; CAD = coronary artery disease; MACE = major adverse cardiovascular events; MALE = male adverse limb events; TIMI = thrombolysis in myocardial infarction; ISTH = International Society on Thrombosis and Hemostasis.

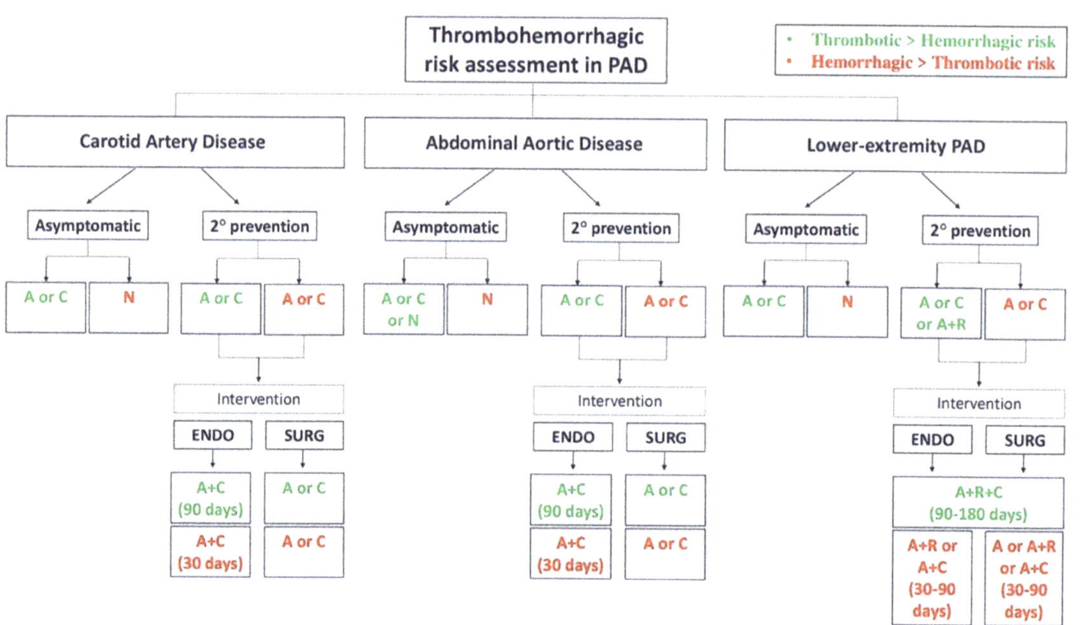

Figure 2. Proposed algorithm for antithrombotic management in patients with PAD asymptomatic or in secondary prevention. A = low-dose aspirin, C = clopidogrel 75 mg/day, N = no medical therapy, R = rivaroxaban 2.5 mg/twice daily, ENDO = endovascular revascularization, SURG = surgical revascularization.

3.2. Abdominal Aortic Disease

3.2.1. Asymptomatic Patients

GDMT does not recommend SAPT in asymptomatic abdominal aortic atherosclerosis overall [6,19]. Despite the common finding of abdominal aortic atherosclerosis in middle-aged patients, SAPT therapy should be avoided unless complex and high-risk plaque such as atheroma dimension >3 mm or >2 mm with mobile/ulcerated features, which confer an increased risk of MACE is observed [19]. Considering the relationship between intra-luminal thrombus and AAA, a hypothesis of antithrombotic treatment efficacy on the growth rate of AAA was assessed in The Efficacy of Ticagrelor on Abdominal Aortic Aneurysm (AAA) Expansion (TicAAA), where ticagrelor was compared with placebo for reducing AAA over 12 months. No differences were found in AAA volume increase with ticagrelor vs. placebo as assessed by MRI (HR 1.013; 95% CI 0.993–1.034) and in diameter change assessed by US for ticagrelor (2.3 mm) vs. placebo (2.2 mm), $p = 0.778$ [43]. Moreover, patients in the ticagrelor group showed an increased rate of bleeding events (33% vs. 11%, $p = 0.002$) (Table 1) [43]. A recent review and metanalysis of seven studies on antiplatelet therapy (including aspirin) in AAA confirms the absence of benefit on aneurysm growth by antiplatelet therapy with an overall standardized mean difference (SMD) of −0.36 mm/year, 95% CI −0.75–0.02 [53]. In addition, compared to placebo, aspirin use in AAA was not associated with reduced all-cause mortality (HR 0.91; 95% CI 0.75–1.11) or abdominal aortic rupture events (HR 0.98; 95% CI 0.37–2.59) [53]. Given the higher bleeding risk, DAPT or oral anticoagulation are not indicated for primary prevention of abdominal aortic aneurysm [19].

3.2.2. Secondary Prevention

After a peripheral embolic event from abdominal aortic plaque, GDMT recommends SAPT (aspirin or clopidogrel), whereas little evidence is available for DAPT in this clinical scenario [6,19]. Due to the higher risk of MACE in patients affected by AAA, unless contraindicated, SAPT is a reasonable option [19]. As mentioned before, intraluminal thrombus is a common finding in patients with AAA. Parenteral administration of factor Xa/II inhibitors in experimental aortic aneurysms and atherosclerosis were assessed in an animal model. Reduction in the severity of aortic aneurism and atherosclerosis was detected in mice treated with enoxaparin or fondaparinux [54]. Even in the setting of AAA complications (e.g., aortic dissection), SAPT should not be withdrawn in order to prevent thrombosis origin and propagation [19]. Data on antithrombotic therapy after AAA repair are still limited and mostly of an observational nature. In patients undergoing intervention (either EVAR or surgical) for AAA, pooled data from observational studies showed no effect on all-cause mortality with antithrombotics compared to placebo/no treatment (HR 1.00; 95% CI 0.81–1.22) [53]. However, studies comparing aspirin vs. placebo/no treatment show a reduction in all-cause mortality in the aspirin arm (HR 0.78; 95% CI 0.68–0.89) with apparent early endoleak risk (<30 days) increase in patients on antithrombotics treatment (HR 1.63; 95% CI 1.17–2.27) [53]. In a small prospective cohort of AAA patients undergoing EVAR, a DAPT strategy showed low complication rates (i.e., 30 day-mortality and endoleak) [55]. During perioperative management of isolated AAA by endovascular treatment, a short-term DAPT (1–3 months) may be indicated [56]. Moreover, DAPT strategy after EVAR showed a good efficacy and safety profile in an observational cohort of patients who underwent recent percutaneous coronary intervention (PCI) compared to SAPT considering BARC major bleeding (0% vs. 1.1%, p = NS), endoleak (0% vs. 3.4%, p = NS) and MI (2.4% vs. 0%, p = NS) [57]. Very limited experience with more potent DAPT (e.g., with ticagrelor or prasugrel) exhibited an increased bleeding risk after AAA repair, irrespective of intervention type [56]. Moreover, recent data on the Safety of Chronic Anticoagulation Therapy After Endovascular Abdominal Aneurysm Repair registry underlined the use of anticoagulation drugs (i.e., vitamin K antagonists/heparin) is independently associated with an increased risk of endoleak (HR 1.6; 95% CI 1.23–2.07) and reintervention (HR 1.8; 95% CI: 1.31–2.48) compared to patients with SAPT [58].

An operative proposal algorithm for the antithrombotic management of abdominal aortic disease according to asymptomatic or secondary prevention patients is represented in Figure 2.

3.3. *Lower Extremity Peripheral Artery Disease*

3.3.1. Asymptomatic Patients

Antithrombotic management recommendations in patients with isolated asymptomatic LEPAD are conflicting. ESC guidelines and a recent ESC position paper do not recommend SAPT in asymptomatic LEPAD, whereas ACA/AHC guidelines acknowledge SAPT as a reasonable option in asymptomatic patients with abnormal ABI ($\leq$0.90) to reduce MI, IS and vascular death risk [4,5,19]. Two RCTs assessed antiplatelet therapy as primary prevention in LEPAD patients. In the prevention of progression of arterial disease and diabetes (POPADAD) trial, aspirin did not show benefit vs. placebo for MACE or major amputation (HR 0.98; 95% CI 0.76–1.26 in primary prevention in diabetes patients with ABI $\leq$ 0.99). No differences in gastrointestinal bleeding were detected (HR 0.90; 95% CI 0.53–1.52) [44]. The Aspirin for Prevention of Cardiovascular Events in a General Population Screened for a Low Ankle Brachial Index (AAA) trial assessed the role of aspirin in a general population with impaired ABI $\leq$ 0.95. Considering fatal or non-fatal coronary events or stroke or revascularization, no statistically significant differences were found between aspirin vs. placebo (HR 1.03; 95% CI 0.84–1.27) but there was increased major bleeding in the aspirin arm (HR, 1.71; 95% CI 0.99–2.97) [45]. SAPT seems a reasonable option in primary prevention in patients with impaired ABI with low bleeding risk [4,5,19].

3.3.2. Secondary Prevention

GDMT supports antithrombotic therapy for the secondary prevention of LEPAD with SAPT (aspirin or clopidogrel) with a class I recommendation [4,5,59]. ACC/AHA guidelines endorse DAPT (aspirin plus clopidogrel) and vorapaxar added on top of DAPT in class IIb recommendation in symptomatic PAD [4]. Moreover, a recent ESC consensus paper endorses aspirin plus rivaroxaban 2.5 mg with or without clopidogrel after LER in patients without high bleeding risk [19]. Several new lines of evidence are available considering different clinical scenarios (e.g., medical therapy management alone vs. LER) balancing thrombo-hemorrhagic risk (Table 1) [19]. Historically, data on LEPAD antithrombotic therapy were extracted by subgroup analysis of non-dedicated RCTs on CAD patients [15,60]. More than 25 years ago, subgroup analysis on clopidogrel versus aspirin in patients at risk of ischemic events (CAPRIE), including 6452 symptomatic PAD patients, highlighted MACE prevention in clopidogrel-allocated subjects over aspirin (HR 0.78; 95% CI 0.65–0.93). Moreover, in the overall cohort of the study, clopidogrel showed a favorable gastrointestinal bleeding profile compared to aspirin (1.99% vs. 2.66%, $p < 0.05$) with no differences in intracranial hemorrhage (0.35% vs. 0.49%, p = NS) [61]. In the LEPAD patients subgroup analysis of The Clopidogrel for High Atherothrombotic Risk and Ischemic Stabilization, Management and Avoidance (CHARISMA) trial on the efficacy and safety of clopidogrel plus aspirin as compared with aspirin alone in patients at high risk for a cardiovascular event, no benefit was observed for DAPT over SAPT considering MACE (HR 0.85; 95% CI 0.66–1.08) [19]. However, subsequent analyses from the CHARISMA trial showed a favorable effect of DAPT vs. SAPT with aspirin in high-risk patients with prior MI (HR 0.78; 95% CI 0.61–0.98), prior IS (HR 0.78; 95% CI 0.62–0.97) and also with a history of LEPAD (HR 0.83; 95% CI 0.72–0.95). The overall rates of moderate-severe/fatal bleeding did not differ between the groups, whereas minor bleeding increased with DAPT vs. SAPT (HR 1.99; 95% CI 1.69–2.34) [62]. As in carotid artery disease, the feasibility of oral anticoagulation on top of SAPT vs. SAPT alone was assessed in the WAVE study, where LEPAD patients made up around 80% of the overall cohort. No differences were found in the primary efficacy outcome of MACE (RR 0.92; 95% CI 0.73–1.16), but there was a significant increase in life-threatening bleeding with combination therapy (RR 3.41; 95% CI 1.84–6.35) [25]. Unless prescribed for another indication, oral anticoagulation on top of SAPT in symptomatic LEPAD patients is not recommended due to increased bleeding risk [4,5,19]. A DAPT regimen of aspirin and ticagrelor for secondary prevention among PAD patients was evaluated in the subanalysis of the Prevention of Cardiovascular Events in Patients With Prior Heart Attack Using Ticagrelor Compared to Placebo on a Background of Aspirin—Thrombolysis In Myocardial Infarction 54 (PEGASUS-TIMI 54) trial. This RCT assessed the efficacy of ticagrelor 90 mg twice daily, ticagrelor 60 mg twice daily or placebo on top of aspirin in secondary prevention 1 to 3 years after MI. In the LEPAD subgroup ticagrelor 60 mg reduced CV mortality compared to placebo (HR 0.47; 95% CI 0.25–0.86). Considering pooled doses, ticagrelor reduced ALI (HR 0.56; 95% CI 0.23–1.37) and peripheral revascularization for limb ischemia (HR 0.63; 95% CI 0.43–0.93). However, the pooled ticagrelor dose group had a numerical increase of TIMI major bleeding (HR 1.32; 95% CI 0.41–4.29) [63]. DAPT with ticagrelor was also assessed in The Effect of Ticagrelor on Health Outcomes in Diabetes Mellitus Patients Intervention Study (THEMIS) RCT, which enrolled diabetic patients without prior MI or IS to ticagrelor 90 mg twice daily plus aspirin or placebo plus aspirin. Ticagrelor reduced MACE (HR 0.90; 95% CI 0.81–0.99) and MALE (HR 0.45; 95% CI 0.23–0.86) counterbalanced by increased TIMI major bleeding (HR 2.32; 95% CI 1.82–2.94) [64]. In the THEMIS-PAD substudy, DAPT significantly reduced limb events (defined as peripheral revascularization, ALI, major amputation) by 1.3% vs. 1.6% with SAPT (p = 0.022) counterbalanced by increased TIMI major bleeding of 2.0% vs. 1.1%, $p < 0.0001$. The overall benefit was greater among PAD patients compared to those without [64]. Ticagrelor was also compared to clopidogrel in the Examining Use of Ticagrelor in Peripheral Artery Disease (EUCLID) trial that investigated the efficacy of ticagrelor, compared to clopidogrel, for reduction of MACE in patients with symptomatic LEPAD. No difference was found for MACE (HR 1.02; 95% CI 0.92–1.13) or major bleeding (HR 1.10, 95% CI 0.84–1.43) between groups, highlighting ticagrelor as not superior to clopidogrel in symptomatic LEPAD patients [46]. Unless for another indication, DAPT therapy is not indicated vs. SAPT in symptomatic LEPAD patients managed with medical therapy alone. Another therapeutic target assessed in the last years is represented by thrombin inhibition on top of low-dose aspirin therapy in DPI. Vorapaxar, a competitive and selective antagonist of thrombin receptor PAR-1, was assessed in Thrombin Receptor Antagonist in Secondary Prevention of Atherothrombotic Ischemic Events–Thrombolysis in Myocardial Infarction 50 (TRA 2°P–TIMI 50) trial, where vorapaxar 2.5 mg daily was compared against placebo on top of standard therapy (e.g., aspirin or clopidogrel) in the secondary prevention of patients with a history of MI, IS or PAD [65]. Overall, vorapaxar reduced the risk of CV death, MI, stroke or recurrent ischemia leading to revascularization (HR 0.88; 95% CI 0.82–0.95) with an increased rate of moderate–severe bleeding than placebo (HR 1.66; 95% CI 1.43–1.93), including intracranial hemorrhage [65]. Subsequent analysis of the PAD population showed greater benefits of vorapaxar in limb events with a 42% reduction in hospitalization for ALI (HR 0.58; 95% CI 0.39–0.86) and a significant reduction in peripheral artery revascularization (HR 0.84; 95% CI 0.73–0.97) although evidence of increased GUSTO moderate–severe bleeding in the vorapaxar arm (HR 1.62; 95% CI 1.21–2.18) [66]. Further data among patients with PAD confirmed the clinical benefits of vorapaxar on MACE in patients with concomitant CAD (HR 0.85; 95% CI 0.73–0.99) and MALE in those with prior LER (HR 0.67; 95% CI 0.49–0.91) accompanied by an increase of ISTH major bleeding by 39% (HR 1.39; 95% CI 1.12–1.71) [67]. Vorapaxar is currently approved by the United States Food and Drug Administration as secondary prevention in patients with CAD or PAD. However, the marketing authorization was withdrawn in the European Union by the European Medicine Agency. In keeping with the targeting of thrombin, dual pathway inhibition with low-dose rivaroxaban was tested in the Cardiovascular Outcomes for People Using Anticoagulation Strategies (COMPASS) RCT in combination with aspirin in patients with stable atherosclerotic vascular disease. Rivaroxaban 2.5 mg plus aspirin vs. aspirin reduced MACE (HR 0.76; 95% CI 0.66–0.86), while there was a higher rate of major bleeding in the rivaroxaban plus aspirin arm (HR 1.70; 95% CI 1.40–2.05), there was no significant difference in intracranial or fatal bleeding between groups [68]. However, the benefits outweighed the risks, especially in patients with diabetes, renal dysfunction, HF or polyvascular disease [68]. In the LEPAD

COMPASS substudy, rivaroxaban 2.5 mg plus aspirin prevented MACE prevention than aspirin alone (HR 0.74; 95% CI 0.58–0.92) and also prevented MALE (HR 0.55; 95% CI 0.35–0.85). ISTH major bleeding was increased in the combination therapy vs. aspirin alone (HR 1.69; 95% CI 1.18–2.40) with a numerical increase of fatal/critical organ bleeding (HR 1.56; 95% CI 0.78–3.39) [69]. Antithrombotic therapy in secondary prevention in LEPAD patients was assessed in RCTs after endovascular or surgical revascularization. The first trials on surgical revascularization assessed the efficacy of oral anticoagulation and aspirin or DAPT. Efficacy of oral anticoagulants compared with aspirin after infrainguinal bypass surgery (Dutch BOA) compared oral anticoagulation (i.e., phenprocoumon or acenocoumarol) vs. aspirin in patients who underwent infrainguinal grafting. No difference was detected between groups for the composite outcome of vascular death, MI, stroke or amputation (HR 0.89; 95% CI 0.75–1.06) or graft occlusion (HR 0.95; 95% CI 0.82–1.11) with an increased risk of major bleeding in patients treated with oral anticoagulation (HR 1.96; 95% CI 1.42–2.71) [47]. There was a similar finding from placebo-controlled clopidogrel and acetylsalicylic acid in bypass surgery for peripheral arterial disease (CASPAR) trial where aspirin plus clopidogrel vs. aspirin and placebo after BTK bypass graft did not show better outcomes considering index-graft occlusion or revascularization, above-ankle amputation of the affected limb, or death (HR 0.98; 95% CI 0.78–1.23), with increased total bleeding rate in the DAPT arm (16.7% vs. 7.1%, $p < 0.001$) without differences in severe bleeding (2.1% vs. 1.2%, p = NS) according to the GUSTO classification [48]. Cilostazol was investigated in LEPAD patients who underwent LER due to femoropopliteal lesions. In the "Cilostazol reduces restenosis after endovascular therapy in patients with femoropopliteal lesions" RCT, cilostazol was compared to ticlopidine in addition to aspirin in a DAPT strategy. Patency rates were higher in the cilostazol (73%) vs. ticlopidine arm (51%) at 36 months of follow-up ($p = 0.013$) with no differences in bleeding adverse events [49]. However, this was a small RCT with 127 patients enrolled and so not powered for clinical endpoints. Moreover, cilostazol is not currently used as an antithrombotic agent [19]. Cilostazol was assessed in LEPAD patients with moderate to severe claudication in a three-arm RCT compared to pentoxifylline and placebo. After 24 weeks, the improvement in the maximal walking distance by treadmill test was significantly greater in the cilostazol arm (107 $\pm$ 158 m) vs. pentoxifylline (64 $\pm$ 127 m with $p < 0.001$) and placebo (65 $\pm$ 135 m with $p < 0.001$) with more side effects as headache, diarrhea and palpitation in cilostazol group ($p < 0.001$) [70]. In subsequent metanalyses, cilostazol confirmed the improvement of maximal and pain-free walking distance of 15% in LEPAD patients affected by claudication and the main effect of cilostazol is represented by vasodilatation due to the increased level of cAMP [71]. However, the use of cilostazol is limited in patients with HF when associated with other phosphodiesterase 3 inhibitors and has known side effects such as headache, palpitations and tachycardia [4]. ACC/AHA GDMT suggests cilostazol with a class I recommendation as an effective therapy to improve symptoms and walking distance in LEPAD patients with claudication [4]. In the management of peripheral arterial interventions with mono or dual antiplatelet therapy (the MIRROR study), DAPT with aspirin plus clopidogrel vs. aspirin was assessed in patients who underwent endovascular LER. DAPT showed a lower rate of target lesion revascularization at 6 months (5% versus 8%, $p = 0.04$) but this benefit disappeared by 1 year (25% versus 32%, $p = 0.35$). Moreover, the bleeding complication rate at 6 months after LER was comparable among groups (2.5% vs. 5.0%, p = NS) [50]. The Edoxaban Plus Aspirin vs. Dual Antiplatelet Therapy in Endovascular Treatment of Patients With Peripheral Artery Disease (ePAD Trial) assessed edoxaban plus aspirin vs. standard DAPT with low-dose aspirin and clopidogrel. The trial was not powered for efficacy, and the risk of restenosis/reocclusion of femoropopliteal target lesion was not different between the two groups (HR 0.89; 95% CI 0.59–1.34). No significant excess in TIMI bleeding was observed with edoxaban, although the confidence interval was wide (HR 0.56; 95% CI 0.19–1.62) [51]. The efficacy of DPI after LER was assessed in the more recent Efficacy and Safety of Rivaroxaban in Reducing the Risk of Major Thrombotic Vascular Events in Subjects With Symptomatic Peripheral Artery Disease Undergoing Peripheral

Revascularization Procedures of the Lower Extremities (VOYAGER-PAD) trial where PAD patients with LER were randomized to rivaroxaban 2.5 mg twice daily plus aspirin vs. placebo plus aspirin. Rivaroxaban plus aspirin showed a benefit in the primary efficacy outcome, including ALI, major amputation for vascular causes, MI, IS or death from CV causes (HR 0.85; 95% CI 0.76–0.96) [52]. No differences in TIMI major bleeding occurred between groups (HR 1.43; 95% CI 0.97–2.10), whereas ISTH major bleeding was higher in the rivaroxaban arm (HR 1.42; 95% CI 1.10–1.84) [52]. Subsequent analysis from the VOYAGER-PAD trial highlighted the benefits of rivaroxaban in reducing the first and subsequent events of the primary endpoint (HR 0.86; 95% CI 0.75–0.98) and vascular events (HR 0.86; 95% CI 0.79–0.95) [72]. In the VOYAGER-PAD, DPI benefit was observed in the primary efficacy outcome regardless of clopidogrel use (HR 0.85; 95% CI 0.71–1.01) or without clopidogrel (HR 0.86; 95% CI 0.73–1.01). Clopidogrel did not influence the main safety outcome of ISTH major bleeding with (HR 1.36; 95% CI 0.96–1.92) and without clopidogrel (HR 1.50; 95% CI 1.02–2.20) [73]. It was also noted in VOYAGER PAD that treatment with rivaroxaban reduced the risk for ALI (HR 0.67; 95% CI 0.55–0.82), particularly within the first month from randomization attesting an early benefit [74]. It is important to highlight that despite this tradeoff between efficacy and safety outcomes in VOYAGER PAD, when data are interpreted on an absolute rather than relative risk scale, the number of efficacy events prevented by rivaroxaban was exceedingly larger than the number of safety outcomes associated with a DPI strategy (181 events of efficacy outcomes prevented by rivaroxaban plus aspirin vs. 29 more safety events with the same strategy). Of interest, the median time from revascularization to randomization in the VOYAGER PAD trial was 5 days, suggesting that rivaroxaban should be started early after LER. It is noteworthy that LER in VOYAGER PAD was endovascular in approximately 65% of cases and surgical in the remaining 35%, therefore supporting the use of rivaroxaban irrespective of the type of treatment. Finally, femoro-popliteal revascularization in VOYAGER PAD represented about 90% of revascularization procedures, and therefore, this vascular district is probably key for the combination therapy of rivaroxaban and low-dose aspirin. A recent meta-analysis, including LEPAD patients from the COMPASS and VOYAGER trials, showed a favorable effect of rivaroxaban on the efficacy outcome of CV death, MI, IS, ALI or major amputation (HR 0.79; 95% CI 0.65–0.95) with an increased risk of ISTH major bleeding (HR 1.51; 95% CI 1.22–1.87). However, significant fatal or critical organ bleeding was similar between rivaroxaban and placebo (0.5% vs. 0.4% by year, p = NS) [75]. While the efficacy and safety of rivaroxaban were consistent regardless of DAPT use in VOYAGER PAD, few data are available on a direct comparison between DPI with aspirin and rivaroxaban vs. DAPT with aspirin and clopidogrel. A post-hoc analysis from VOYAGER PAD assessed the impact of aspirin plus rivaroxaban for a "CASPAR-like outcome", including ALI, unplanned limb revascularization, amputation or CV death at 1 year in patients who underwent surgical LER according to CASPAR trial [48]. Considering 2185 patients of surgical LER group of VOYAGER PAD, rivaroxaban reduced the composite CASPAR-like endpoint at 1 year (HR 0.76; 95% CI 0.62–0.95) and increased ISTH major bleeding compared to placebo (HR 1.37; 95% CI 0.83–2.25) [76]. Although the aim of this analysis was not to compare the two trials, rivaroxaban showed benefit for a CASPAR-like outcome suggesting a role for thrombin inhibition in limb adverse events prevention even in surgical patients where the results in the CASPAR trial were neutral with DAPT [48]. An operative algorithm for antithrombotic management in LEPAD according to asymptomatic or secondary prevention patients is illustrated in Figure 2.

3.4. Antithrombotic Therapy in Peripheral Artery Disease Patients with Polyvascular Disease and Cardiovascular Comorbidities

Antithrombotic therapy in PAD patients affected by polyvascular disease and CV comorbidities may represent a challenge in daily practice. Considering patients with previous MI, the benefit of a DAPT strategy, including aspirin and ticagrelor, provided an absolute risk reduction of 4.1% for MACE and a 35% of MALE reduction counterbalanced by an increase of TIMI major bleeding (HR 1.32; 95% CI 0.41–4.29) in the PAD subgroup of PEGASUS-TIMI 54 RCT [63]. A secondary prevention strategy, including aspirin and low-dose rivaroxaban (DPI), confirmed a 26% MACE and 45% MALE reduction in an LEPAD COMPASS substudy with an increase of ISTH major bleeding vs. aspirin alone (HR 1.69; 95% CI 1.18–2.40). In this LEPAD COMPASS subanalysis, more than 50% of patients had a known history of CAD [69]. A different scenario is represented by PAD patients with recent PCI/acute coronary syndrome. Subanalyses of Dual Antiplatelet Therapy (DAPT) and Prolonging Dual Antiplatelet Treatment After Grading Stent-Induced Intimal Hyperplasia Study—PRODIGY RCTs demonstrated the efficacy of DAPT, even in PAD patients. In the DAPT trial, extended DAPT (with aspirin and either clopidogrel or prasugrel) reduced MI/stent thrombosis (HR 0,63; 95% CI 0.32–1.22) with a numerical increase of bleeding according to GUSTO classification (HR 1.82; 95% CI 0.87–3.83) among PAD patients [77]. The PRODIGY trial showed a significant MACE reduction comparing prolonged (24 months) vs. shorter DAPT (6 months) with aspirin and clopidogrel (HR 0.54; 95% CI 0.31–0.95) in PAD patients, particularly for those presenting with acute coronary syndrome (ACS) without significant differences in BARC $\geq$ 2 bleeding (HR 0.77; 95% CI 0.27–2.21) [60]. Therefore, in PAD patients with concomitant CAD, and particularly for those with recent ACS/PCI, GDMT supports 6–12 months of DAPT strategy with aspirin and either ticagrelor or prasugrel in patients without high bleeding risk, while a shorter DAPT (up to 6 months) with aspirin and clopidogrel in those with high bleeding risk [4,5]. Newer evidence is now available on the efficacy and safety of different duration of $P2Y_{12}$ inhibitors in daily practice. Until last years, after DAPT discontinuation, the SAPT strategy included aspirin but new data suggest $P2Y_{12}$ monotherapy as an effective antithrombotic strategy after short DAPT (up to 6 months) in ACS patients with high bleeding risk features [78]. DAPT with aspirin and clopidogrel up to 3 months is recommended by AHA/ASA guidelines after acute minor stroke due to carotid stenosis [8]. Atrial fibrillation (AF) is another common comorbidity among PAD patients due to several known common risk factor such as chronic kidney disease (CKD), type 2 diabetes mellitus, hypertension and advanced age [19]. In the subanalysis on PAD patients of Rivaroxaban versus Warfarin in Nonvalvular Atrial Fibrillation (ROCKET AF) RCT, rivaroxaban (15/20 mg daily) was compared to warfarin. Among patients with PAD, there were similar efficacy outcome rates considering stroke/systemic embolism for rivaroxaban vs. warfarin (HR 1.19; 95% CI 0.63–2.22) with significant interaction for major or clinically non-major relevant bleeding in PAD patients (HR 1.40; 95% CI 1.06–1.86) compared to those without PAD (HR 1.03; 95% CI 0.95–1.11), p-interaction 0.037 [79]. However, the difference in major bleeding rates between rivaroxaban and vitamin K antagonists (VKAs) was not confirmed in subsequent studies involving PAD patients (HR 1.13; 95% CI 0.84–1.52) [80]. The efficacy in AF prevention by direct oral anticoagulants compared to warfarin in PAD patients was confirmed by subanalysis of The Apixaban for Reduction in Stroke and Other Thromboembolic Events in Atrial Fibrillation (ARISTOTLE) trial. The risk of stroke/systemic embolism was similar for apixaban vs. warfarin in patients with PAD (HR 0.63; 95% CI 0.32–1.25) with no differences in major or clinically non-major relevant bleeding (HR 1.05; 95% CI 0.69–1.58) [81]. More recently, the safety and effectiveness of edoxaban vs. standard practice of dosing with warfarin in patients with atrial fibrillation (ENGAGE AF-TIMI 48) trial, evaluated edoxaban vs. warfarin on systemic embolism and major bleeding in AF patients with concomitant PAD. Even in PAD patients, edoxaban was not inferior compared to warfarin considering systemic embolism (HR 1.16; 95% CI 0.42–3.20) and for ISTH major bleeding (HR 0.96; 95% CI 0.54–1.70) [82]. Hence, in PAD patients with concomitant AF, GDMT recommends oral anticoagulation [4,5,19]. In case of endovascular LER with stenting or CAS, adding SAPT to oral anticoagulation for 1 month may represent an option in AF patients without high bleeding risk [19]. Patients with end-stage CKD/dialysis are often affected by ASCVD, including PAD and therefore are considered a very high CV risk subgroup patients for MACE, MALE and bleeding adverse events [83]. Moreover, the presence of end-stage CKD/dialysis dependence in patients is not an isolated disease and often coexists with

PAD and major tissue loss (i.e., Rutherford category 6) [83]. These patients are usually excluded from RCTs on antithrombotic therapy, even in a PAD setting. However, new data on the safety and efficacy of DOACs in CKD are available from an RCT (Valkyrie study), which compared rivaroxaban vs. VKA in dialysis patients with AF. There, 132 patients were randomized among three groups, VKA, rivaroxaban (10 mg daily) or rivaroxaban plus vitamin K2 for 18 months. Compared to VKA, rivaroxaban reduced the primary efficacy endpoint of fatal and non-fatal CV events (HR 0.41; 95% CI 0.25–0.68) in the rivaroxaban group and HR 0.34 (95% CI 0.19–0.61) in the rivaroxaban and vitamin K2 group. Symptomatic limb ischemia occurred more frequently in VKA (45%) than in rivaroxaban groups (22%), $p = 0.02$. There were fewer life-threatening and major bleeding adverse events in the rivaroxaban arm (HR 0.39; 95% CI 0.17–0.90) and in the rivaroxaban plus vitamin K2 arm (HR 0.48; 95% CI 0.22–1.08) compared to the VKA arm [84]. Further data from larger RCTs are needed to determine the optimal anticoagulation strategy in dialysis patients with AF. Moreover, two recent phase 2 RCTs on factor XI inhibitors (IONIS-FXI$_{Rx}$ and Xisomab 3G3) were conducted in dialysis patients with promising results on FXI inhibition with no evidence of impaired major bleeding rates compared to placebo [85]. Other ongoing RCTs will assess the efficacy and safety of factor XI inhibitors in a larger cohort of dialysis patients [85]. Hence, established safety and efficacy data of newer antithrombotic therapy in PAD patients, such as DPI, are not available in patients with end-stage CKD/dialysis and related complications. Despite the fact that having end-stage CKD/dialysis per se is not an exclusion criterion for aspirin and P2Y$_{12}$, such as clopidogrel and ticagrelor, there are few available data, particularly on safety profiles in PAD patients. Balancing thrombo/hemorrhagic risk, an SAPT strategy with aspirin or clopidogrel is recommended according to Kidney Disease Improving Global Outcomes guidelines in patients with end-stage renal disease/dialysis with concomitant ASCVD [86].

4. Conclusions

PAD patients are at increased risk of MACE and MALE events that can be reduced by antithrombotic therapy. We have provided an overview of current evidence in different clinical settings in PAD and propose an algorithm for antithrombotic therapy management in daily practice. In patients undergoing revascularization, evidence supports more aggressive antithrombotic therapy, specifically, dual pathway inhibition after LER, irrespective of the type of intervention. The optimal management of patients undergoing revascularization for carotid and abdominal aortic disease is less clear. The development of newer antithrombotic strategies (i.e., factor XIa inhibition) may play an important role in the future.

Author Contributions: Conceptualization, M.E.C., R.P. and M.A.; methodology, A.L. and S.E.; investigation, A.F., G.G. (Giuseppe Giugliano) and G.G. (Giuseppe Gargiulo); writing—original draft preparation, M.E.C., C.N.H. and S.D.B.; writing—review and editing, J.H. and P.C.; supervision, G.E. and M.P.B. All authors have read and agreed to the published version of the manuscript.

Funding: This research received no external funding.

Institutional Review Board Statement: Not applicable.

Informed Consent Statement: Not applicable.

Data Availability Statement: All data underlying this article will be shared on reasonable request to the corresponding author.

Conflicts of Interest: The authors declare no conflict of interest.

References

1. Saini, V.; Guada, L.; Yavagal, D.R. Global Epidemiology of Stroke and Access to Acute Ischemic Stroke Interventions. *Neurology* **2021**, *97* (Suppl. S2), S6–S16. [CrossRef] [PubMed]

2. Criqui, M.H.; Matsushita, K.; Aboyans, V.; Hess, C.N.; Hicks, C.W.; Kwan, T.W.; Ujueta, F. Lower Extremity Peripheral Artery Disease: Contemporary Epidemiology, Management Gaps, and Future Directions: A Scientific Statement From the American Heart Association. *Circulation* **2021**, *144*, e171–e191. [CrossRef] [PubMed]
3. Song, P.; He, Y.; Adeloye, D.; Zhu, Y.; Ye, X.; Yi, Q.; Global Health Epidemiology Research Group. The Global and Regional Prevalence of Abdominal Aortic Aneurysms: A Systematic Review and Modelling Analysis. *Ann. Surg.* **2022**, *10*, 1097. [CrossRef] [PubMed]
4. Gerhard-Herman, M.D.; Gornik, H.L.; Barrett, C.; Barshes, N.R.; Corriere, M.A.; Drachman, D.E.; Fleisher, L.A.; Fowkes, F.G.R.; Hamburg, N.; Kinlay, S.; et al. 2016 AHA/ACC Guideline on the Management of Patients with Lower Extremity Peripheral Artery Disease: Executive Summary: A Report of the American College of Cardiology/American Heart Association Task Force on Clinical Practice Guidelines. *Circulation* **2017**, *135*, e686–e725. [CrossRef]
5. Aboyans, V.; Björck, M.; Brodmann, M.; Collet, J.P.; Czerny, M.; De Carlo, M.; Naylor, A.R.; Roffi, M.; Tendera, M.; Vlachopoulos, C.; et al. 2017 ESC Guidelines on the Diagnosis and Treatment of Peripherasl Arterial Diseases, in collaboration with the European Society for Vascular Surgery (ESVS). *Eur. Heart J.* **2018**, *39*, 763–816. [CrossRef]
6. Isselbacher, E.M.; Preventza, O.; Black, I.J.H.; Augoustides, J.G.; Beck, A.W.; Bolen, M.A.; Braverman, A.C.; Bray, B.E.; Brown-Zimmerman, M.M.; Chen, E.P.; et al. 2022 ACC/AHA Guideline for the Diagnosis and Management of Aortic Disease: A Report of the American Heart Association/American College of Cardiology Joint Committee on Clinical Practice Guidelines. *Circulation* **2022**, *80*, e223–e393. [CrossRef]
7. Fleischmann, D.; Afifi, R.O.; Casanegra, A.I.; Elefteriades, J.A.; Gleason, T.G.; Hanneman, K.; Roselli, E.E.; Willemink, M.J.; Fischbein, M.P. Imaging and Surveillance of Chronic Aortic Dissection: A Scientific Statement From the American Heart Association. *Circ. Cardiovasc. Imaging* **2022**, *15*, e000075. [CrossRef]
8. Kleindorfer, D.O.; Towfighi, A.; Chaturvedi, S.; Cockroft, K.M.; Gutierrez, J.; Lombardi-Hill, D.; Kamel, H.; Kernan, W.N.; Kittner, S.J.; Leira, E.C.; et al. 2021 Guideline for the Prevention of Stroke in Patients with Stroke and Transient Ischemic Attack: A Guideline from the American Heart Association/American Stroke Association. *Stroke* **2021**, *52*, e364–e467. [CrossRef]
9. Ding, N.; Sang, Y.; Chen, J.; Ballew, S.H.; Kalbaugh, C.A.; Salameh, M.J.; Blaha, M.J.; Allison, M.; Heiss, G.; Selvin, E.; et al. Cigarette Smoking, Smoking Cessation, and Long-Term Risk of 3 Major Atherosclerotic Diseases. *J. Am. Coll. Cardiol.* **2019**, *74*, 498–507. [CrossRef]
10. Saba, L.; Brinjikji, W.; Spence, J.; Wintermark, M.; Castillo, M.; de Borst, G.; Yang, Q.; Yuan, C.; Buckler, A.; Edjlali, M.; et al. Roadmap Consensus on Carotid Artery Plaque Imaging and Impact on Therapy Strategies and Guidelines: An International, Multispecialty, Expert Review and Position Statement. *Am. J. Neuroradiol.* **2021**, *42*, 1566–1575. [CrossRef]
11. Stakos, D.A.; Stamatelopoulos, K.; Bampatsias, D.; Sachse, M.; Zormpas, E.; Vlachogiannis, N.I.; Tual-Chalot, S.; Stellos, K. The Alzheimer's Disease Amyloid-Beta Hypothesis in Cardiovascular Aging and Disease. *J. Am. Coll. Cardiol.* **2020**, *75*, 952–967. [CrossRef]
12. Golledge, J.; Norman, P.E. Atherosclerosis and Abdominal Aortic Aneurysm. *Arter. Thromb. Vasc. Biol.* **2010**, *30*, 1075–1077. [CrossRef]
13. Zhao, S.; Gu, H.; Chen, B.; Cheng, Z.; Yang, S.; Duan, Y.; Ghavamian, A.; Wang, X. Dynamic Imaging Features of Retrospective Cardiac Gating CT Angiography Influence Delayed Adverse Events in Acute Uncomplicated Type B Aortic Dissections. *Cardiovasc. Interv. Radiol.* **2020**, *43*, 620–629. [CrossRef]
14. Burris, N.S.; Patel, H.J.; Hope, M.D. Retrograde flow in the false lumen: Marker of a false lumen under stress? *J. Thorac. Cardiovasc. Surg.* **2019**, *157*, 488–491. [CrossRef]
15. Hess, C.N.; Bonaca, M.P. Contemporary Review of Antithrombotic Therapy in Peripheral Artery Disease. *Circ. Cardiovasc. Interv.* **2020**, *13*, e009584. [CrossRef]
16. Narula, N.; Dannenberg, A.J.; Olin, J.W.; Bhatt, D.L.; Johnson, K.W.; Nadkarni, G.; Min, J.; Torii, S.; Poojary, P.; Anand, S.S.; et al. Pathology of Peripheral Artery Disease in Patients with Critical Limb Ischemia. *J. Am. Coll. Cardiol.* **2018**, *72*, 2152–2163. [CrossRef]
17. Narula, N.; Olin, J.W.; Narula, N. Pathologic Disparities Between Peripheral Artery Disease and Coronary Artery Disease. *Arterioscler. Thromb. Vasc. Biol.* **2020**, *40*, 1982–1989. [CrossRef]
18. King, R.W.; Canonico, M.E.; Bonaca, M.P.; Hess, C.N. Management of Peripheral Arterial Disease: Lifestyle Modifications and Medical Therapies. *J. Soc. Cardiovasc. Angiogr. Interv.* **2022**, *1*, 100513. [CrossRef]
19. Aboyans, V.; Bauersachs, R.; Mazzolai, L.; Brodmann, M.; Palomares, J.F.R.; Debus, S.; Collet, J.-P.; Drexel, H.; Espinola-Klein, C.; Lewis, B.S.; et al. Antithrombotic therapies in aortic and peripheral arterial diseases in 2021: A consensus document from the ESC working group on aorta and peripheral vascular diseases, the ESC working group on thrombosis, and the ESC working group on cardiovascular pharmacotherapy. *Eur. Heart J.* **2021**, *42*, 4013–4024.
20. Kansal, A.; Huang, Z.; Rockhold, F.W.; Baumgartner, I.; Berger, J.; Blomster, J.I.; Fowkes, F.G.; Katona, B.; Mahaffey, K.W.; Norgren, L.; et al. Impact of Procedural Bleeding in Peripheral Artery Disease. *Circ. Cardiovasc. Interv.* **2019**, *12*, e008069. [CrossRef]
21. Arnett, D.K.; Blumenthal, R.S.; Albert, M.A.; Buroker, A.B.; Goldberger, Z.D.; Hahn, E.J.; Himmelfarb, C.D.; Khera, A.; Lloyd-Jones, D.; McEvoy, J.W.; et al. 2019 ACC/AHA Guideline on the Primary Prevention of Cardiovascular Disease: A Report of the American College of Cardiology/American Heart Association Task Force on Clinical Practice Guidelines. *Circulation* **2019**, *140*, e596–e646. [CrossRef] [PubMed]

22. Baigent, C.; Blackwell, L.; Collins, R.; Emberson, J.; Godwin, J.; Peto, R.; Buring, J.; Hennekens, C.; Kearney, P.; Meade, T. Aspirin in the primary and secondary prevention of vascular disease: Collaborative meta-analysis of individual participant data from randomised trials. *Lancet* **2009**, *373*, 1849–1860. [PubMed]
23. Sharma, M.; Hart, R.G.; Connolly, S.J.; Bosch, J.; Shestakovska, O.; Ng, K.K.H.; Catanese, L.; Keltai, K.; Aboyans, V.; Alings, M.; et al. Stroke Outcomes in the COMPASS Trial. *Circulation* **2019**, *139*, 1134–1145. [CrossRef] [PubMed]
24. ESPRIT Study Group; Halkes, P.H.; van Gijn, J.; Kappelle, L.J.; Koudstaal, P.J.; Algra, A. Medium intensity oral anticoagulants versus aspirin after cerebral ischaemia of arterial origin (ESPRIT): A randomised controlled trial. *Lancet Neurol.* **2007**, *6*, 115–124. [PubMed]
25. Warfarin Antiplatelet Vascular Evaluation Trial Investigators; Anand, S.S.; Yusuf, S.; Xie, C.; Pogue, J.; Eikelboom, J.; Budaj, A.; Sussex, B.; Liu, L.; Guzman, R.; et al. Oral Anticoagulant and Antiplatelet Therapy and Peripheral Arterial Disease. *N. Engl. J. Med.* **2007**, *357*, 217–227.
26. Markus, H.S.; Droste, D.W.; Kaps, M.; Larrue, V.; Lees, K.R.; Siebler, M.; Ringelstein, E.B. Dual Antiplatelet Therapy with Clopidogrel and Aspirin in Symptomatic Carotid Stenosis Evaluated Using Doppler Embolic Signal Detection. *Circulation* **2005**, *111*, 2233–2240. [CrossRef]
27. Wong, K.S.L.; Chen, C.; Fu, J.; Chang, H.M.; Suwanwela, N.C.; Huang, Y.N.; Han, Z.; Tan, K.S.; Ratanakorn, D.; Chollate, P.; et al. Clopidogrel plus aspirin versus aspirin alone for reducing embolisation in patients with acute symptomatic cerebral or carotid artery stenosis (CLAIR study): A randomised, open-label, blinded-endpoint trial. *Lancet Neurol.* **2010**, *9*, 489–497. [CrossRef]
28. Wang, Y.; Pan, Y.; Zhao, X.; Li, H.; Wang, D.; Johnston, S.C.; Liu, L.; Meng, X.; Wang, A.; Wang, C.; et al. Clopidogrel with Aspirin in Acute Minor Stroke or Transient Ischemic Attack. *N. Engl. J. Med.* **2013**, *369*, 11–19. [CrossRef]
29. Johnston, S.C.; Easton, J.D.; Farrant, M.; Barsan, W.; Conwit, R.A.; Elm, J.J.; Kim, A.S.; Lindblad, A.S.; Palesch, Y.Y.; Neurological Emergencies Treatment Trials Network; et al. Clopidogrel and Aspirin in Acute Ischemic Stroke and High-Risk TIA. *N. Engl. J. Med.* **2018**, *379*, 215–225. [CrossRef]
30. Pan, Y.; Elm, J.J.; Li, H.; Easton, J.D.; Wang, Y.; Farrant, M.; Meng, X.; Kim, A.S.; Zhao, X.; Meurer, W.J.; et al. Outcomes Associated with Clopidogrel-Aspirin Use in Minor Stroke or Transient Ischemic Attack. *JAMA Neurol.* **2019**, *76*, 1466. [CrossRef]
31. Wang, Y.; Zhao, X.; Lin, J.; Li, H.; Johnston, S.C.; Lin, Y.; Pan, Y.; Liu, L.; Wang, D.; Wang, C.; et al. Association Between *CYP2C19* Loss-of-Function Allele Status and Efficacy of Clopidogrel for Risk Reduction Among Patients with Minor Stroke or Transient Ischemic Attack. *JAMA* **2016**, *316*, 70–78. [CrossRef]
32. Toyoda, K.; Uchiyama, S.; Yamaguchi, T.; Easton, J.D.; Kimura, K.; Hoshino, H.; Sakai, N.; Okada, Y.; Tanaka, K.; Origasa, H.; et al. Dual antiplatelet therapy using cilostazol for secondary prevention in patients with high-risk ischaemic stroke in Japan: A multicentre, open-label, randomised controlled trial. *Lancet Neurol.* **2019**, *18*, 539–548. [CrossRef]
33. Johnston, S.C.; Amarenco, P.; Albers, G.W.; Denison, H.; Easton, J.D.; Evans, S.R.; Held, P.; Jonasson, J.; Minematsu, K.; Molina, C.A.; et al. Ticagrelor versus Aspirin in Acute Stroke or Transient Ischemic Attack. *N. Engl. J. Med.* **2016**, *375*, 35–43. [CrossRef]
34. Wong, K.L.; Amarenco, P.; Albers, G.W.; Denison, H.; Easton, J.D.; Evans, S.R.; Held, P.; Himmelmann, A.; Kasner, S.E.; Knutsson, M.; et al. Efficacy and Safety of Ticagrelor in Relation to Aspirin Use within the Week before Randomization in the SOCRATES Trial. *Stroke* **2018**, *49*, 1678–1685. [CrossRef]
35. Johnston, S.C.; Amarenco, P.; Denison, H.; Evans, S.R.; Himmelmann, A.; James, S.; Knutsson, M.; Ladenvall, P.; Molina, C.A.; Wang, Y. Ticagrelor and Aspirin or Aspirin Alone in Acute Ischemic Stroke or TIA. *N. Engl. J. Med.* **2020**, *383*, 207–217. [CrossRef]
36. McKevitt, F.M.; Randall, M.S.; Cleveland, T.J.; Gaines, P.A.; Tan, K.T.; Venables, G.S. The Benefits of Combined Anti-platelet Treatment in Carotid Artery Stenting. *Eur. J. Vasc. Endovasc. Surg.* **2005**, *29*, 522–527. [CrossRef]
37. Dalainas, I.; Nano, G.; Bianchi, P.; Stegher, S.; Malacrida, G.; Tealdi, D.G. Dual Antiplatelet Regime Versus Acetyl-acetic Acid for Carotid Artery Stenting. *Cardiovasc. Interv. Radiol.* **2006**, *29*, 519–521. [CrossRef]
38. Taylor, D.W.; Barnett, H.J.; Haynes, R.B.; Ferguson, G.G.; Sackett, D.L.; E Thorpe, K.; Simard, D.; Silver, F.L.; Hachinski, V.; Clagett, G.P.; et al. Low-dose and high-dose acetylsalicylic acid for patients undergoing carotid endarterectomy: A randomised controlled trial. *Lancet* **1999**, *353*, 2179–2184. [CrossRef]
39. Barkat, M.; Hajibandeh, S.; Hajibandeh, S.; Torella, F.; Antoniou, G.A. Systematic Review and Meta-analysis of Dual Versus Single Antiplatelet Therapy in Carotid Interventions. *Eur. J. Vasc. Endovasc. Surg.* **2017**, *53*, 53–67. [CrossRef]
40. De Caterina, R.; Prisco, D.; Eikelboom, J.W. Factor XI inhibitors: Cardiovascular perspectives. *Eur. Heart J.* **2023**, *44*, 280–292. [CrossRef]
41. Shoamanesh, A.; Mundl, H.; Smith, E.E.; Masjuan, J.; Milanov, I.; Hirano, T.; Agafina, A.; Campbell, B.; Caso, V.; Mas, J.-L.; et al. Factor XIa inhibition with asundexian after acute non-cardioembolic ischaemic stroke (PACIFIC-Stroke): An international, randomised, double-blind, placebo-controlled, phase 2b trial. *Lancet* **2022**, *400*, 997–1007. [CrossRef] [PubMed]
42. Sharma, M.; Molina, C.A.; Toyoda, K.; Bereczki, D.; Kasner, S.E.; Lutsep, H.L.; Tsivgoulis, G.; Ntaios, G.; Czlonkowska, A.; Shuaib, A.; et al. Rationale and design of the AXIOMATIC-SSP phase II trial: Antithrombotic treatment with factor XIa inhibition to Optimize Management of Acute Thromboembolic events for Secondary Stroke Prevention. *J. Stroke Cerebrovasc. Dis.* **2022**, *31*, 106742. [CrossRef] [PubMed]
43. Wanhainen, A.; Mani, K.; Kullberg, J.; Svensjö, S.; Bersztel, A.; Karlsson, L.; Holst, J.; Gottsäter, A.; Linné, A.; Gillgren, P.; et al. The effect of ticagrelor on growth of small abdominal aortic aneurysms—A randomized controlled trial. *Cardiovasc. Res.* **2019**, *116*, 450–456. [CrossRef] [PubMed]

44. Belch, J.; MacCuish, A.; Campbell, I.; Cobbe, S.; Taylor, R.; Prescott, R.; Lee, R.; Bancroft, J.; MacEwan, S.; Shepherd, J.; et al. The prevention of progression of arterial disease and diabetes (POPADAD) trial: Factorial randomised placebo controlled trial of aspirin and antioxidants in patients with diabetes and asymptomatic peripheral arterial disease. *BMJ* **2008**, *337*, a1840. [CrossRef] [PubMed]
45. Fowkes, F.G.R.; Price, J.F.; Stewart, M.C.W.; Butcher, I.; Leng, G.C.; Pell, A.C.H.; Sandercock, P.; Fox, K.; Lowe, G.D.O.; Murray, G.; et al. Aspirin for prevention of cardiovascular events in a general population screened for a low ankle brachial index: A randomized controlled trial. *JAMA* **2010**, *303*, 841–848. [CrossRef]
46. Hiatt, W.R.; Fowkes, F.G.R.; Heizer, G.; Berger, J.S.; Baumgartner, I.; Held, P.; Katona, B.G.; Mahaffey, K.W.; Norgren, L.; Jones, W.S. Ticagrelor versus Clopidogrel in Symptomatic Peripheral Artery Disease. *N. Engl. J. Med.* **2017**, *376*, 32–40. [CrossRef]
47. Dutch Bypass Oral Anticoagulants or Aspirin (BOA) Study Group. Efficacy of oral anticoagulants compared with aspirin after infrainguinal bypass surgery (The Dutch Bypass Oral Anticoagulants or Aspirin study): A randomised trial. *Lancet* **2000**, *355*, 346–351. [CrossRef]
48. Belch, J.J.; Dormandy, J. Results of the randomized, placebo-controlled clopidogrel and acetylsalicylic acid in bypass surgery for peripheral arterial disease (CASPAR) trial. *J. Vasc. Surg.* **2010**, *52*, 825–833.e2. [CrossRef]
49. Iida, O.; Nanto, S.; Uematsu, M.; Morozumi, T.; Kitakaze, M.; Nagata, S. Cilostazol reduces restenosis after endovascular therapy in patients with femoropopliteal lesions. *J. Vasc. Surg.* **2008**, *48*, 144–149. [CrossRef]
50. Tepe, G.; Bantleon, R.; Brechtel, K.F.M.; Schmehl, J.-M.; Zeller, T.; Claussen, C.D.; Strobl, F.F.X. Management of peripheral arterial interventions with mono or dual antiplatelet therapy—The MIRROR study: A randomised and double-blinded clinical trial. *Eur. Radiol.* **2012**, *22*, 1998–2006. [CrossRef]
51. Moll, F.; Baumgartner, I.; Jaff, M.; Nwachuku, C.; Tangelder, M.; Ansel, G.; Adams, G.; Zeller, T.; Rundback, J.; Grosso, M.; et al. Edoxaban Plus Aspirin vs. Dual Antiplatelet Therapy in Endovascular Treatment of Patients With Peripheral Artery Disease: Results of the ePAD Trial. *J. Endovasc. Ther.* **2018**, *25*, 158–168. [CrossRef]
52. Bonaca, M.P.; Bauersachs, R.M.; Anand, S.S.; Debus, E.S.; Nehler, M.R.; Patel, M.R.; Hiatt, W.R. Rivaroxaban in Peripheral Artery Disease after Revascularization. *N. Engl. J. Med.* **2020**, *382*, 1994–2004. [CrossRef]
53. Wong, K.H.F.; Zlatanovic, P.; Bosanquet, D.C.; Saratzis, A.; Kakkos, S.K.; Aboyans, V.; Twine, C.P. Antithrombotic Therapy for Aortic Aneurysms: A Systematic Review and Meta-Analysis. *Eur. J. Vasc. Endovasc. Surg.* **2022**, *64*, 544–556. [CrossRef]
54. Moran, C.S.; Seto, S.-W.; Krishna, S.M.; Sharma, S.; Jose, R.J.; Biros, E.; Wang, Y.; Morton, S.K.; Golledge, J. Parenteral administration of factor Xa/IIa inhibitors limits experimental aortic aneurysm and atherosclerosis. *Sci. Rep.* **2017**, *7*, 43079. [CrossRef]
55. De Bruin, J.L.; Brownrigg, J.R.; Patterson, B.O.; Karthikesalingam, A.; Holt, P.J.; Hinchliffe, R.J.; Loftus, I.M.; Thompson, M.M. The Endovascular Sealing Device in Combination with Parallel Grafts for Treatment of Juxta/Suprarenal Abdominal Aortic Aneurysms: Short-term Results of a Novel Alternative. *Eur. J. Vasc. Endovasc. Surg.* **2016**, *52*, 458–465. [CrossRef]
56. Wanhainen, A.; Verzini, F.; Van Herzeele, I.; Allaire, E.; Bown, M.; Cohnert, T.; Dick, F.; van Herwaarden, J.; Karkos, C.; Koelemay, M.; et al. Editor's Choice—European Society for Vascular Surgery (ESVS) 2019 Clinical Practice Guidelines on the Management of Abdominal Aorto-iliac Artery Aneurysms. *Eur. J. Vasc. Endovasc. Surg.* **2019**, *57*, 8–93. [CrossRef]
57. He, R.-X.; Zhang, L.; Zhou, T.-N.; Yuan, W.-J.; Liu, Y.-J.; Fu, W.-X.; Jing, Q.-M.; Liu, H.-W.; Wang, X.-Z. Safety and Necessity of Antiplatelet Therapy on Patients Underwent Endovascular Aortic Repair with Both Stanford Type B Aortic Dissection and Coronary Heart Disease. *Chin. Med. J.* **2017**, *130*, 2321–2325.
58. De Rango, P.; Verzini, F.; Parlani, G.; Cieri, E.; Simonte, G.; Farchioni, L.; Isernia, G.; Cao, P. Safety of Chronic Anticoagulation Therapy after Endovascular Abdominal Aneurysm Repair (EVAR). *Eur. J. Vasc. Endovasc. Surg.* **2014**, *47*, 296–303. [CrossRef]
59. Canonico, M.E.; Hsia, J.; Hess, C.N.; Bonaca, M.P. Sex differences in guideline-directed medical therapy in 2021–22 among patients with peripheral artery disease. *Vasc. Med.* **2023**. [CrossRef]
60. Franzone, A.; Piccolo, R.; Gargiulo, G.; Ariotti, S.; Marino, M.; Santucci, A.; Baldo, A.; Magnani, G.; Moschovitis, A.; Windecker, S.; et al. Prolonged vs. Short Duration of Dual Antiplatelet Therapy after Percutaneous Coronary Intervention in Patients with or without Peripheral Arterial Disease. *JAMA Cardiol.* **2016**, *1*, 795. [CrossRef]
61. CAPRIE Steering Committee. A randomised, blinded, trial of clopidogrel versus aspirin in patients at risk of ischaemic events (CAPRIE). *Lancet* **1996**, *348*, 1329–1339. [CrossRef] [PubMed]
62. Cacoub, P.P.; Bhatt, D.L.; Steg, P.G.; Topol, E.J.; Creager, M.A. Patients with peripheral arterial disease in the CHARISMA trial. *Eur. Heart J.* **2008**, *30*, 192–201. [CrossRef] [PubMed]
63. Bonaca, M.P.; Bhatt, D.L.; Storey, R.F.; Steg, P.h.G.; Cohen, M.; Kuder, J.; Goodrich, E.; Nicolau, J.C.; Parkhomenko, A.; López-Sendón, J.; et al. Ticagrelor for Prevention of Ischemic Events after Myocardial Infarction in Patients with Peripheral Artery Disease. *J. Am. Coll. Cardiol.* **2016**, *67*, 2719–2728. [CrossRef] [PubMed]
64. Steg, P.G.; Bhatt, D.L.; Simon, T.; Fox, K.; Mehta, S.R.; Harrington, R.A.; Held, C.; Andersson, M.; Himmelmann, A.; Ridderstråle, W.; et al. Ticagrelor in Patients with Stable Coronary Disease and Diabetes. *N. Engl. J. Med.* **2019**, *381*, 1309–1320. [CrossRef]
65. Morrow, D.A.; Braunwald, E.; Bonaca, M.P.; Ameriso, S.F.; Dalby, A.J.; Fish, M.P.; Fox, K.A.A.; Lipka, L.J.; Liu, X.; Nicolau, J.C.; et al. Vorapaxar in the Secondary Prevention of Atherothrombotic Events. *N. Engl. J. Med.* **2012**, *366*, 1404–1413. [CrossRef]
66. Bonaca, M.P.; Scirica, B.M.; Creager, M.A.; Olin, J.; Bounameaux, H.; Dellborg, M.; Lamp, J.M.; Murphy, S.A.; Braunwald, E.; Morrow, D.A. Vorapaxar in Patients with Peripheral Artery Disease. *Circulation* **2013**, *127*, 1522–1529. [CrossRef]

67. Qamar, A.; Morrow, D.A.; Creager, M.A.; Scirica, B.M.; Olin, J.W.; Beckman, J.A.; Murphy, S.A.; Bonaca, M.P. Effect of vorapaxar on cardiovascular and limb outcomes in patients with peripheral artery disease with and without coronary artery disease: Analysis from the TRA 2°P-TIMI 50 trial. *Vasc. Med.* **2020**, *25*, 124–132. [CrossRef]
68. Eikelboom, J.W.; Connolly, S.J.; Bosch, J.; Dagenais, G.R.; Hart, R.G.; Shestakovska, O.; Diaz, R.; Alings, M.; Lonn, E.M.; Anand, S.S.; et al. Rivaroxaban with or without Aspirin in Stable Cardiovascular Disease. *N. Engl. J. Med.* **2017**, *377*, 1319–1330. [CrossRef]
69. Kaplovitch, E.; Eikelboom, J.W.; Dyal, L.; Aboyans, V.; Abola, M.T.; Verhamme, P.; Avezum, A.; Fox, K.A.A.; Berkowitz, S.D.; Bangdiwala, S.I.; et al. Rivaroxaban and Aspirin in Patients with Symptomatic Lower Extremity Peripheral Artery Disease. *JAMA Cardiol.* **2020**, *6*, 21–29. [CrossRef]
70. Dawson, D.L.; Cutler, B.S.; Hiatt, W.R.; Hobson, R.W.; Martin, J.D.; Bortey, E.B.; Forbes, W.P.; Strandness, D. A comparison of cilostazol and pentoxifylline for treating intermittent claudication. *Am. J. Med.* **2000**, *109*, 523–530. [CrossRef]
71. Bonaca, M.P.; Hamburg, N.M.; Creager, M.A. Contemporary Medical Management of Peripheral Artery Disease. *Circ. Res.* **2021**, *128*, 1868–1884. [CrossRef]
72. Bauersachs, R.M.; Szarek, M.; Brodmann, M.; Gudz, I.; Debus, E.S.; Nehler, M.R.; VOYAGER PAD Committees and Investigators. Total Ischemic Event Reduction with Rivaroxaban after Peripheral Arterial Revascularization in the VOYAGER PAD Trial. *J. Am. Coll. Cardiol.* **2021**, *78*, 317–326. [CrossRef]
73. Hiatt, W.R.; Bonaca, M.P.; Patel, M.R.; Nehler, M.R.; Debus, E.S.; Anand, S.S.; Capell, W.H.; Brackin, T.; Jaeger, N.; Hess, C.N.; et al. Rivaroxaban and Aspirin in Peripheral Artery Disease Lower Extremity Revascularization. *Circulation* **2020**, *142*, 2219–2230. [CrossRef]
74. Hess, C.N.; Debus, E.S.; Nehler, M.R.; Anand, S.S.; Patel, M.R.; Szarek, M.; Capell, W.H.; Hsia, J.; Beckman, J.A.; Brodmann, M.; et al. Reduction in Acute Limb Ischemia with Rivaroxaban Versus Placebo in Peripheral Artery Disease after Lower Extremity Revascularization: Insights from VOYAGER PAD. *Circulation* **2021**, *144*, 1831–1841. [CrossRef]
75. Anand, S.S.; Hiatt, W.; Dyal, L.; Bauersachs, R.; Berkowitz, S.D.; Branch, K.R.H.; Debus, S.; Fox, K.A.A.; Liang, Y.; Muehlhofer, E.; et al. Low-dose rivaroxaban and aspirin among patients with peripheral artery disease: A meta-analysis of the COMPASS and VOYAGER trials. *Eur. J. Prev. Cardiol.* **2022**, *29*, e181–e189. [CrossRef]
76. Bonaca, M.P.; Szarek, M.; Debus, E.S.; Nehler, M.R.; Patel, M.R.; Anand, S.S.; Muehlhofer, E.; Berkowitz, S.D.; Haskell, L.P.; Bauersachs, R.M. Efficacy and safety of rivaroxaban versus placebo after lower extremity bypass surgery: A post hoc analysis of a "CASPAR like" outcome from VOYAGER PAD. *Clin. Cardiol.* **2022**, *45*, 1143–1146. [CrossRef]
77. Secemsky, E.A.; Yeh, R.W.; Kereiakes, D.J.; Cutlip, D.E.; Steg, P.G.; Massaro, J.M.; Apruzzese, P.K.; Mauri, L. Extended Duration Dual Antiplatelet Therapy after Coronary Stenting Among Patients with Peripheral Arterial Disease. *JACC Cardiovasc. Interv.* **2017**, *10*, 942–954. [CrossRef]
78. Capodanno, D.; Angiolillo, D.J. Timing, Selection, Modulation, and Duration of P2Y12 Inhibitors for Patients with Acute Coronary Syndromes Undergoing PCI. *JACC Cardiovasc. Interv.* **2023**, *16*, 1–18. [CrossRef]
79. Jones, W.S.; Hellkamp, A.S.; Halperin, J.; Piccini, J.P.; Breithardt, G.; Singer, D.E.; Patel, M.R. Efficacy and safety of rivaroxaban compared with warfarin in patients with peripheral artery disease and non-valvular atrial fibrillation: Insights from ROCKET AF. *Eur. Heart J.* **2014**, *35*, 242–249. [CrossRef]
80. Coleman, C.I.; Baker, W.L.; Meinecke, A.K.; Eriksson, D.; Martinez, B.K.; Bunz, T.J.; Alberts, M.J. Effectiveness and safety of rivaroxaban vs. warfarin in patients with non-valvular atrial fibrillation and coronary or peripheral artery disease. *Eur. Heart J. Cardiovasc. Pharmacother.* **2020**, *6*, 159–166. [CrossRef]
81. Hu, P.T.; Lopes, R.D.; Stevens, S.R.; Wallentin, L.; Thomas, L.; Alexander, J.H.; Jones, W.S. Efficacy and Safety of Apixaban Compared with Warfarin in Patients with Atrial Fibrillation and Peripheral Artery Disease: Insights From the ARISTOTLE Trial. *J. Am. Heart Assoc.* **2017**, *6*, e004699. [CrossRef] [PubMed]
82. Bonaca, M.P.; Antman, E.M.; Cunningham, J.W.; Wiviott, S.D.; Murphy, S.A.; Halperin, J.L.; Weitz, J.I.; Grosso, M.A.; Lanz, H.J.; Braunwald, E.; et al. Ischaemic and bleeding risk in atrial fibrillation with and without peripheral artery disease and efficacy and safety of full- and half-dose edoxaban vs. warfarin: Insights from ENGAGE AF-TIMI 48. *Eur. Heart J. Cardiovasc. Pharmacother.* **2022**, *8*, 695–706. [CrossRef] [PubMed]
83. Anantha-Narayanan, M.; Sheikh, A.B.; Nagpal, S.; Smolderen, K.G.; Regan, C.; Ionescu, C.; Mena-Hurtado, C. Systematic review and meta-analysis of outcomes of lower extremity peripheral arterial interventions in patients with and without chronic kidney disease or end-stage renal disease. *J. Vasc. Surg.* **2021**, *73*, 331–340.e4. [CrossRef] [PubMed]
84. De Vriese, A.S.; Caluwé, R.; Van Der Meersch, H.; De Boeck, K.; De Bacquer, D. Safety and Efficacy of Vitamin K Antagonists versus Rivaroxaban in Hemodialysis Patients with Atrial Fibrillation: A Multicenter Randomized Controlled Trial. *J. Am. Soc. Nephrol.* **2021**, *32*, 1474–1483. [CrossRef]
85. Greco, A.; Laudani, C.; Spagnolo, M.; Agnello, F.; Faro, D.C.; Finocchiaro, S.; Legnazzi, M.; Mauro, M.S.; Mazzone, P.M.; Occhipinti, G.; et al. Pharmacology and Clinical Development of Factor XI Inhibitors. *Circulation* **2023**, *147*, 897–913. [CrossRef]
86. Rossing, P.; Caramori, M.L.; Chan, J.C.N.; Heerspink, H.J.L.; Hurst, C.; Khunti, K.; de Boer, I.H. KDIGO 2022 Clinical Practice Guideline for Diabetes Management in Chronic Kidney Disease. *Kidney Int.* **2022**, *102*, S1–S127. [CrossRef]

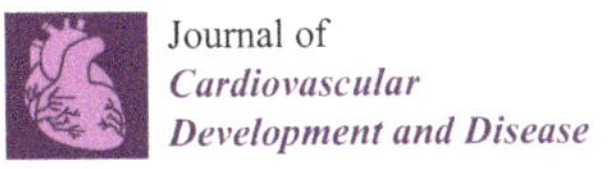
Journal of
Cardiovascular
Development and Disease

MDPI

Article

Metformin Directly Binds to MMP-9 to Improve Plaque Stability

Xianda Chen [1,2,3,4,5,†], Shuaixing Wang [1,2,3,4,5,†], Wenli Xu [1,2,3,4,5], Mingming Zhao [1,2,3,4,5], Youyi Zhang [1,2,3,4,5] and Han Xiao [1,2,3,4,5,*]

1 Department of Cardiology, Institute of Vascular Medicine, Peking University Third Hospital, Beijing 100191, China
2 NHC Key Laboratory of Cardiovascular Molecular Biology and Regulatory Peptides, Beijing 100191, China
3 Key Laboratory of Molecular Cardiovascular Science, Ministry of Education, Beijing 100191, China
4 Beijing Key Laboratory of Cardiovascular Receptors Research, Beijing 100191, China
5 Research Unit of Medical Science Research Management/Basic and Clinical Research of Metabolic Cardiovascular Diseases, Chinese Academy of Medical Sciences, Beijing 100191, China
* Correspondence: xiaohan@bjmu.edu.cn
† These authors contributed equally to this work.

Abstract: Vulnerable atherosclerotic plaque rupture is the principal mechanism that accounts for myocardial infarction and stroke. High matrix metalloproteinase-9 (MMP-9) expression and activity have been proven to lead to plaque instability. Metformin, a first-line treatment for type 2 diabetes, is beneficial to plaque vulnerability. However, the mechanism underlying its anti-atherogenic effect remains unclear. Molecular docking and surface plasmon resonance experiments showed that metformin directly interacts with MMP-9, and incubated MMP-9 overexpressing HEK293A cells with metformin (1 μmol$\cdot$L^{-1}) significantly attenuates MMP-9's activity using zymography and MMP activity assays. Moreover, metformin treatment drives MMP-9 degradation. Next, we constructed a carotid artery atherosclerotic plaque model and administered consecutive 14-day metformin (200 mg$\cdot$kg$^{-1}\cdot$d^{-1}) treatment by intragastric gavage. Immunofluorescence staining of the right carotid common artery and serum MMP activity assay results showed that metformin treatment decreased local plaque MMP-9 protein level and circulating MMP-9 activity, respectively. Histochemical staining revealed that after metformin treatment, the collagen content in plaque was significantly preserved, and the plaque vulnerability index decreased. These findings suggested that metformin improved atherosclerotic plaque stability by directly binding to MMP-9 and driving its degradation.

Keywords: surface plasmon resonance; metformin; matrix metalloproteinase-9; plaque instability; atherosclerosis

Citation: Chen, X.; Wang, S.; Xu, W.; Zhao, M.; Zhang, Y.; Xiao, H. Metformin Directly Binds to MMP-9 to Improve Plaque Stability. *J. Cardiovasc. Dev. Dis.* **2023**, *10*, 54. https://doi.org/10.3390/jcdd10020054

Academic Editors: Giovanni Cimmino and Plinio Cirillo

Received: 22 December 2022
Revised: 19 January 2023
Accepted: 27 January 2023
Published: 30 January 2023

1. Introduction

Nearly 17.5 million people die each year from atherosclerosis-related diseases (31% of the global mortality). Of these, approximately 7.4 million died from coronary heart disease and 6.7 million from stroke [1]. Vulnerable atherosclerotic plaque rupture is the principal mechanism that accounts for myocardial infarction and stroke [2]. Therefore, there is a clinical need for plaque stabilization drugs.

Metformin, a biguanide, is the top choice of oral agent for the treatment of type 2 diabetes owing to its glucose-lowering effectiveness, safety, favorable effect on body weight, and low cost [3]. Moreover, metformin has been associated with decreased all-cause mortality and a reduced incidence of cardiovascular disease among patients with diabetes. A clinical trial investigated the effect of long-term metformin use and lifestyle at a diabetes prevention program and found that metformin was protective against atherosclerotic vascular disease early in diabetes development and potentially extended the range of this action to include high-risk male prediabetic subjects [4]. A recent meta-analysis showed the association between metformin and decreased cardiovascular mortality (95% CI, OR 0.44 [0.34−0.57]) or incidence of cardiovascular diseases (95% CI, OR 0.73 [0.59−0.90]) among patients with diabetes [5]. However, the molecular mechanism by which metformin improves atherosclerosis plaque stability remains unclear.

Vulnerable plaques are characterized by fragile, thin fibrous caps, massive lipid cores, intraplaque hemorrhage, immune activation, and increased levels of pro-inflammatory mediators (cytokines, chemokines, and matrix metalloproteinases) [6]. IL-6 stimulates the expression of adhesive molecules and results in an increase in the production and reactivity of acute phase indicators, such as C-reactive protein and TNF-α [7,8]. IL-18 and TNF-α are crucial for atherosclerotic plaque development and stability [9,10]. All of the cytokines mentioned above play a significant role in the formation and destabilization of atherosclerotic plaques. Mature plaques mainly comprise endothelial cells, vascular smooth muscle cells, macrophages, and fibrous caps containing extracellular matrix (ECM) components [11]. Among these components, the ECM is especially important for plaque stability [12]. Proteases have been implicated in the development and progression of atherosclerosis due to their ability to cause focal destruction of the ECM of blood vessels. Matrix metalloproteinase (MMP)-9, also known as gelatinase B, is a widely studied member of the MMP family. Histopathological studies have shown that MMP-9 is mainly distributed in the shoulder area, necrotic core, and fibrous cap area of atherosclerotic plaques, and the level and activity of MMP-9 in unstable plaques are higher than those in stable plaques [13–15]. Moreover, many studies have shown that high MMP-9 expression can be used as a predictor of atherosclerotic plaque instability, whereas its overexpression may lead to plaque instability [16–18]. Therefore, MMP-9 is a potential target for improving atherosclerotic plaque stability. However, whether metformin can target MMP-9 and inhibit its activity to stabilize plaque remains unclear.

Here, we report a novel mechanism by which metformin directly binds to MMP-9 and inhibits its activity to improve atherosclerotic plaque stability.

2. Materials and Methods

2.1. Mice

The investigations conformed to the US National Institutes of Health Guide for the Care and Use of Laboratory Animals (NIH Publication No. 85-23, revised 1996). Animal experiments were approved by the Committee of Peking University on Ethics of Animal Experiments (LA 2018-112) and conducted in accordance with the Guidelines for Animal Experiments, Peking University Health Science Center. Male ApoE knockout mice (ApoE-/-, C57BL/6J background) were purchased from Cyagen Biosciences Inc. (Suzhou, China) and used for the experiments. From 8 weeks of age, ApoE-/- mice were fed a high-fat, high-cholesterol diet containing 40 kcal% fat and 1.25% cholesterol (D12108C; Research Diets, New Brunswick, NJ, USA) for 14 weeks. All mice were housed in a specific pathogen-free environment under a 12 h/12 h light-dark cycle.

2.2. Carotid Collar Placement and Drug Treatment

Male ApoE knockout mice (8 weeks of age, C57BL/6J background) were fed a high-fat diet containing 40 kcal% fat and 1.25% cholesterol (D12108C; Research Diets, New Brunswick, NJ, USA) for 2 weeks. Carotid collar placement was performed 2 weeks later, and the operation process is briefly described as follows [19]: the mice were weighed, anesthetized by an intraperitoneal injection of 2% pentobarbital sodium (50 mg·kg^{-1}), and their limbs were fixed on a thermostatic operating table. Erythromycin eye ointment was applied to the eyes of the mice to prevent dry eyes. Hair removal ointment was applied to remove neck and chest fur and fully expose the neck and chest surgical field. The epidermis was cut off at the median line of the neck using scissors, the right common carotid artery (RCCA) was bluntly separated with forceps, and the accompanying nerves and vessels were not damaged. A silicone collar with an inner diameter of about 0.3 mm ($\approx$30% stenosis) was placed on the lateral side of the RCCA. The collar was fixed with a 6-0 silk thread and sutured for disinfection. Meloxicam (1.5 mg·kg^{-1}) was injected intraperitoneally for analgesia after surgery and resuscitated on heat mats. High-fat feeding was continued for more than 3 months until plaque formation. Subsequently, metformin (Sigma-Aldrich, St. Louis, MO, USA; 200 mg·kg^{-1} body weight) or saline was administered by intragastric gavage for 14 consecutive days.

2.3. Histopathology and Immunofluorescence

The RCCAs from mice were harvested and embedded in an OCT compound (Lot# 4583; Tissue-Tek, USA). The OCT-embedded vascular tissue was sequentially sliced into slices approximately 6–8 μm thick using a microtome (Leica, Wetzlar, Germany), and placed on polylysine-coated glass slides. For all subsequent pathological staining (including immunofluorescence, oil red O, and Sirius red), 2–4 frozen sections of each vascular tissue with an interval of more than 50 μm were stained, and the average of the statistical values from the same sample was used as the final result [20].

To analyze plaque stability, serial sections (8 μm thick) were stained with picrosirius red to detect collagen deposition and oil red O to detect lipid deposition; both stains were analyzed by quantifying the positive area per total plaque area. Slices were incubated with primary antibodies against the macrophage marker CD-68 (1:50 dilution; Abcam, ab53444, Cambridge, UK) and smooth muscle cell marker α-SMA (1:50 dilution; Abcam, ab124964, Cambridge, UK), followed by incubation with fluorescence-conjugated secondary antibodies. The sections were mounted with 4′, 6-diamidino-2-phenylindole (DAPI; Abcam, ab104139, Cambridge, UK) for nuclei visualization.

To further characterize the carotid arteries, slices were incubated with the following primary antibodies: anti-MMP-9 (1:50 dilution; Invitrogen, MA5-15886, Carlsbad, CA, USA), anti-active MMP-9 (1:50 dilution; NOVUS, NBP2-13173, Carlsbad, CA, USA), anti-MMP-2 (1:50 dilution; Abcam, ab92536, Cambridge, UK), anti-MMP-12 (1:50 dilution; Proteintech, 22989-1-AP, Rosemont, IL, USA), anti-IL-1β (1:50 dilution; Bioss, bs0812R, Peking, China), anti-IL-6 (1:50 dilution; Proteintech, 66146-1-Ig, Rosemont, IL, USA), and anti-TNF-α (1:50 dilution; Abcam, ab1793, Cambridge, UK), followed by incubation with fluorescence-conjugated secondary antibodies. The sections were mounted with 4′, 6-diamidino-2-phenylindole (DAPI; Abcam, ab104139, Cambridge, UK) for nuclei visualization.

2.4. Western Blotting

Liver tissues and cell lines were lysed in a RIPA lysis buffer containing 1 mmol·L^{-1} phenylmethanesulfonyl fluoride (Beyotime Institute of Biotechnology, Beijing, China) at 4 °C for 30 min. The lysates were then centrifuged at 15,000× g for 10 min at 4 °C and their protein concentrations were determined using the BCA Protein Assay (Beyotime Institute of Biotechnology, Beijing, China). Samples were mixed with 5× SDS loading buffer, boiled for 5 min, and 50 μg of total protein was subjected to SDS-PAGE in 10% gels and transferred to nitrocellulose membranes. After blocking, the membranes were incubated overnight at 4 °C with the following primary antibodies: anti-MMP-9 (1:1000 dilution; Invitrogen, MA5-15886, Carlsbad, CA, USA), anti-p-AMPK (1:1000 dilution; CST, #2535, Danvers, MA, USA), anti-AMPK (1:1000 dilution; CST, #2532, Danvers, MA, USA), and anti-GAPDH (1:5000; CST, #2118, Danvers, MA, USA). The membranes were washed with Tris-buffered saline/0.1% Tween 20 (TBST) and incubated with secondary antibodies for 1 h at 25 °C. Signals were detected using Pierce™ ECL Western Blotting Substrate (Thermo Fisher Scientific, Waltham, MA, USA). Protein levels were quantified by calculating the grayscale value of each band using ImageJ (version 1.43, National Institutes of Health, Bethesda, MD, USA) software.

2.5. Matrix Metalloproteinases (MMPs) Activity Assay

Matrix metalloproteinases (MMPs) activity in mouse serum and cell culture supernatant was measured using Invitrogen DQ™ luciferase-conjugated gelatin substrate (D12054; Invitrogen, Carlsbad, CA, USA), a fluorescent substrate that can detect protease activity with high sensitivity. The substrate consists of highly quenched fluorescein-labelled gelatin. After proteolytic digestion, the exhibited bright green fluorescence can be used to measure enzyme activity. Increased fluorescence intensity was monitored using a fluorescent microplate reader or fluorimeter. After receiving the cell supernatant, the cells were incubated with DQ gelatin, and a zinc-ion-containing buffer was added. After standing at room temperature and away from light for 24 h, the fluorescence intensity of each well was measured using a fluorescence microplate reader (TECAN, Männedorf, Switzerland).

2.6. Molecular Docking and Dynamics Simulation

The ligand metformin was processed using the Schrödinger 10.2 software (Schrödinger, LLC, NY, USA) LigPrep module. An OPLS3 force field was adopted for energy minimization. The crystal structure of MMP-9 was obtained from the RCSB Protein Data Bank. The crystallographic structure of 4WZV was prepared using the Protein Preparation Wizard module. A glide was applied to predict the potential binding mode of metformin with the MMP-9 protein. Following the docking results, an independent 50 ns molecular dynamics simulation was performed using Desmond. Na^+ and Cl^- ions were each added at the physiological concentration of 0.15 mol·L^{-1} to ensure the overall neutrality of the systems. Simulations were conducted using an OPLS3 force field and a TIP3P explicit solvent model. The final size of the solvated system was approximately 20,000 atoms. A 5 ps recording interval was selected, and the NPT ensemble was employed with a fixed temperature of 300 K and pressure of 1.01 bar. The analysis tool of the simulation interactions diagram was used to monitor ligand–protein interactions.

2.7. Cell Culture, Plasmids, and Transfection

HEK 293A cells were obtained from the Cell Resource Center, Peking Union Medical College (which is the headquarter of National Science & Technology Infrastructure–National BioMedical Cell-Line Resource, NSTI-BMCR). Cells were maintained at 37 °C, with 5% CO_2 in DMEM supplemented with 10% FBS and 10^4 U·mL^{-1} Pen/Strep. MMP-9 was overexpressed using an MMP-9-pcDNA3.1(+)-3Xflag plasmid synthesized by Ruibiotech (Beijing, China). Control plasmid did not contain sequences homologous to those of humans, mice, or rats. HEK 293A cells were seeded into 6-well plates (1.0×10^6 cells/well) for 24 h and transfected with MMP-9 or control plasmid using lipofectamine 3000 (Invitrogen, Waltham, MA, USA) for 24 h, according to the manufacturer's instructions. Furthermore, the transfected cells were incubated with metformin (Sigma-Aldrich, St. Louis, MO, USA; 1 μmol·L^{-1}) for an additional 24 h. For the degradation experiment, the transfected HEK293A cells were pretreated with metformin for half an hour and incubated with cycloheximide (CHX; MedChemExpress, HY-12320; 10 μmol·L^{-1}) to block protein synthesis for the indicated periods (0, 1, 2, 3 h). Lysates are harvested from the cells and analyzed by Western blotting.

2.8. Quantitative Real-Time PCR

Total RNA was extracted from the cell line using TRIzol reagent (Invitrogen, Carlsbad, CA, USA), according to the manufacturer's protocol. Relative quantitation by real-time PCR was performed using SYBR Green to detect PCR products in real-time using the QuantStudioTM3 system (Applied Biosystems). A melting curve analysis was performed at the end of each PCR reaction. MMP-9 gene expression was expressed as a ratio to that of GAPDH, a housekeeping gene. Oligonucleotide primer sequences were as follows: Mmp-9, forward 5′-GGACCCGAAGCGGACATTG-3′ and reverse 5′-CGTCGTCGAAATGGGCATCT-3′; Gapdh, forward 5′-TGGATTTGGACGCATTGGTC-3′ and reverse 5′-TTTGCACTGGTACGTGTTGAT-3′.

2.9. Surface Plasmon Resonance (SPR) Spectroscopy

Experiments were performed at 25 °C using a Biacore T200, and the data were analyzed using Biacore T200 evaluation software 2.0 (GE Healthcare, Stockholm, Sweden). Human

MMP-9 recombinant protein (911-MP; R&D Systems Incorporated, Minneapolis, MN, USA) was covalently coupled to a CM5 chip (GE Healthcare). All measurements were performed at 25 °C, using a TCNB buffer: 50 mmol·L^{-1} Tris, 10 mmol·L^{-1} $CaCl_2$, 150 mmol·L^{-1} NaCl, 0.05% Brij-35 (w/v), and pH 7.5, and metformin was injected in a two-fold dilution concentration series (range, 0.0156–15.6 μmol·L^{-1}). Steady-state values were calculated from the sensorgrams and plotted against concentrations. Data were fitted into a single-site binding model to calculate the KD value.

2.10. Zymography

Gelatinase activity was detected in HEK293A supernatants and recombinant human MMP-9 protein (911-MP; R&D Systems Incorporated, Minneapolis, MN, USA) after metformin incubation for 24 h. Zymography was performed according to the manufacturer's instructions (Applygen, P1700, Beijing, China). Following electrophoresis, the gels were washed twice with 2.5% Triton X-100 to remove sodium dodecyl sulfate and further washed with 50 mmol·L^{-1} Tris–HCl pH 8.0. Gels were incubated for the following 20 h in an activation buffer (50 mmol·L^{-1} Tris–HCl supplemented with 5 mmol·L^{-1} $CaCl_2$). The gels were stained with Coomassie brilliant blue R-250 and de-stained with 20% methanol and 10% acetic acid in distilled water until clear bands were visualized.

2.11. Statistics

Data are expressed as mean ± SD. All samples were independent, including those measured over time in the experiments. For parametric data, Student's t-test or an analysis of variance (ANOVA) was used to analyze intergroup differences for normally distributed data. For parametric data with unequal variances, ANOVA with Tukey's post hoc test was used. For non-parametric data, the Mann–Whitney U test with the exact method was used to analyze intergroup differences. A Kruskal–Wallis ANOVA combined with post hoc Tukey's multiple comparison tests was performed when more than two groups were evaluated. Data were analyzed using GraphPad Prism software (version 8.0; GraphPad Software Inc., San Diego, CA, USA), and $p < 0.05$ was considered statistically significant.

3. Results

3.1. Matrix Metalloproteinase-9 (MMP-9) Is Predicted to Bind Directly to Metformin

We hypothesized that metformin inhibits MMP-9 activity through its direct interaction with MMP-9. Molecular modeling was performed to rationalize the activities of metformin against MMP-9. Metformin was situated in the active cavity, engaging in several interactions with MMP-9 (Figure 1a). Two hydrogen bonds were between the urea moiety and Pro-246 and Glu 227. Additionally, the protonated imine group formed an ionic bond with Glu-227. Notably, metal coordination was observed between metformin and the zinc ions, which might have strengthened the binding affinity. As shown in the protein–ligand contact histogram, the results were consistent with those of the docking study. The two hydrogen bonds formed by Pro-246 and Glu-227 were maintained at 76% and 30% of the simulation time, respectively (Figure 1b,c). A powerful coordination bond was formed between the nitrogen atom of metformin and the zinc metal ions. In addition, the amino group formed a hydrogen bond network through a water bridge with Ala-189. Further molecular dynamic (MD) simulation analysis revealed that the complex was stable during a 50 ns simulation (Figure 1d). Overall, these findings provided a better understanding of the metformin mechanisms and may facilitate a future search for optimized MMP-9 inhibitors.

3.2. Metformin Directly Interacts with MMP-9 and Attenuates Its Activity

To verify whether metformin directly binds to MMP-9, we conducted surface plasmon resonance (SPR) experiments. The findings of the SPR-based assay suggested that the binding of metformin to MMP-9 occurred with a KD of 0.6950 μmol·L^{-1} (Figure 2a,b). To examine the ability of metformin to inhibit MMP-9 activity, we constructed an overexpression plasmid for human MMP-9 and transfected it, or a control plasmid into HEK293A cells using lipofectamine (Figure S1). Next, we incubated the transfected cells with metformin (1 μmol·L^{-1}) for 24 h, and the MMP-9 activity in the cultured supernatant was detected using zymography and an MMP activity assay. Both results indicated that metformin incubation significantly attenuated the activity of MMP-9 (Figure 2c–e).

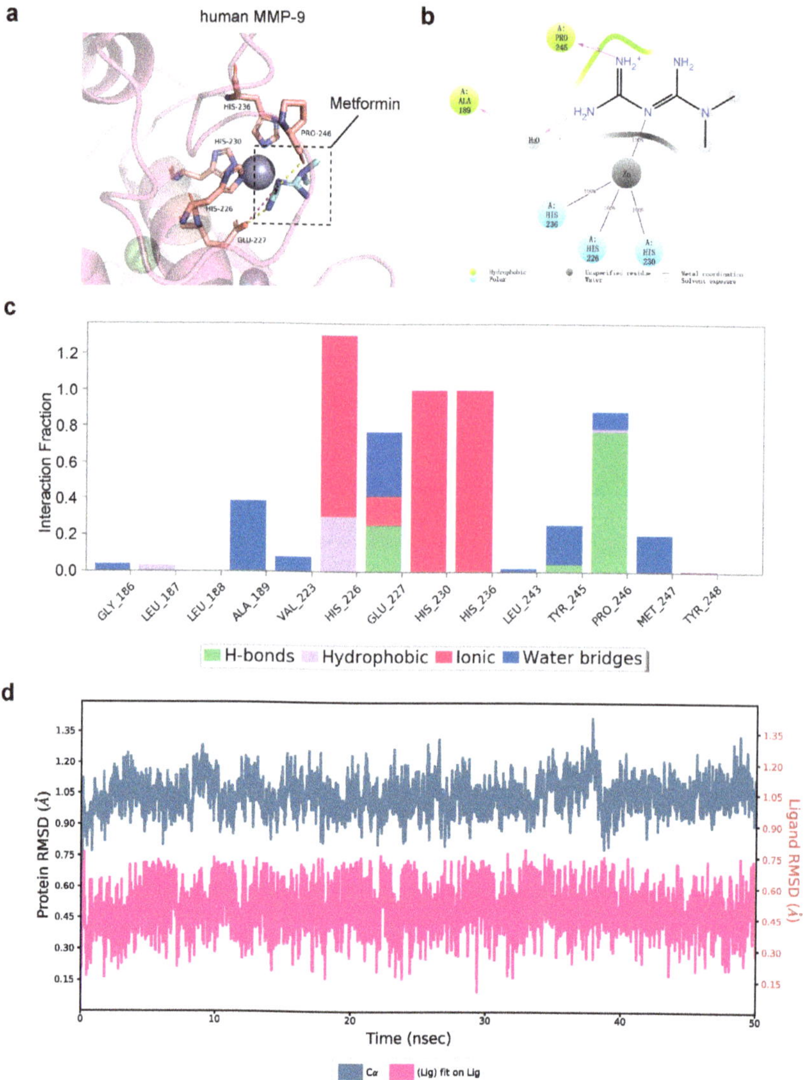

Figure 1. Matrix metalloproteinase-9 (MMP-9) is predicted to bind directly to metformin. (**a**) Predicted binding mode of metformin with MMP-9 (PDB id: 4wzv). (**b**,**c**) Protein–ligand contact histogram of metformin and the corresponding two-dimensional diagram predicted through MD simulations. A percentage value suggests that for X% of the simulation time, the specific interaction is maintained. (**d**) RMSD of the interaction between MMP-9 and the ligand metformin in MD simulations. MD, molecular dynamics; RMSD, root mean square deviation; MMP-9, matrix metalloproteinase-9.

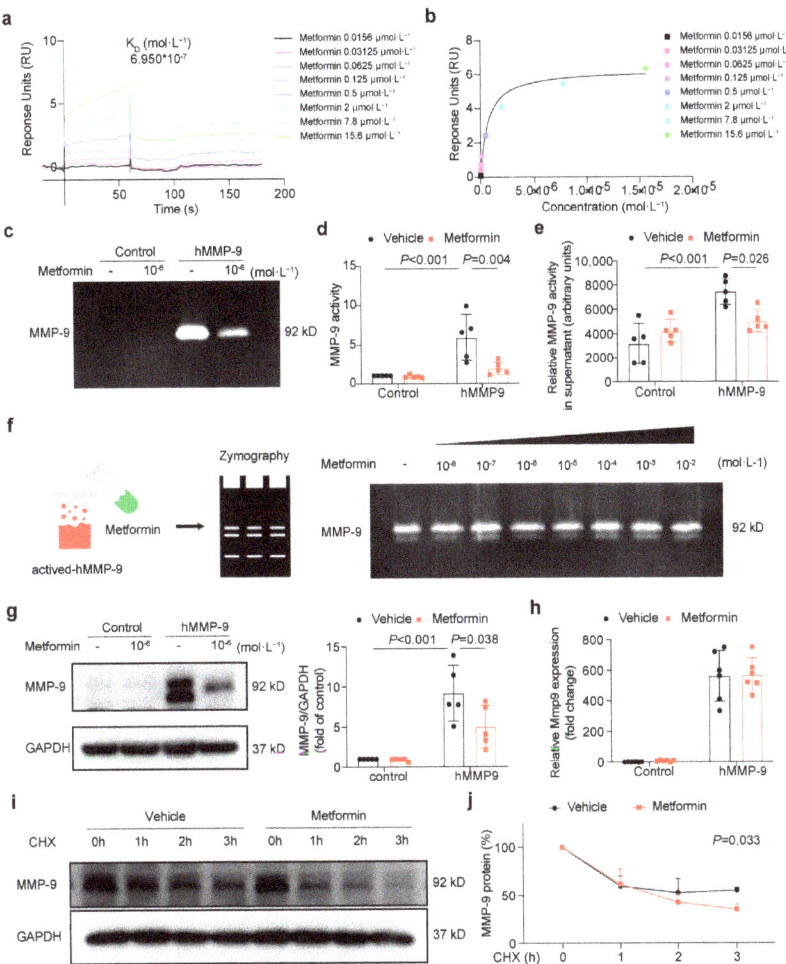

Figure 2. Metformin directly interacts with MMP-9 and attenuates its activity. (**a**,**b**) SPR analysis of the binding between metformin and MMP-9. Recombinant human MMP-9 protein was immobilized on an activated CM5 sensor chip, and metformin was then flowed across the chip. (**c**,**d**) Representative gelatin zymogram and the quantified values of the 92 kDa MMP-9 activity in the cultured supernatant. Data are shown as mean ± SD (two-way ANOVA followed by Tukey's test, n = 5). (**e**) MMP activity in supernatant from cultured HEK293A cells was measured using a Gelatinase Assay Kit (two-way ANOVA followed by Tukey's test, n = 5). (**f**) Representative gelatin zymogram of the recombinant human MMP-9 activity after incubation with different concentrations of metformin. (**g**) The exogenous MMP-9 protein level in HEK293A cells after incubation with metformin for 24 h (both of the 2 bands were quantified, two-way ANOVA followed by Tukey's test, n = 5). (**h**) MMP-9 mRNA expression level in HEK293A cells after incubation with metformin for 24 h (n = 6). (**i**,**j**) Exogenous MMP-9 degradation in metformin-treated HEK293A cells when protein synthesis was inhibited by 10 μM cycloheximide (two-way ANOVA followed by Tukey's test, n = 5).

To verify whether the inhibition of MMP-9 activity by metformin is a direct binding effect, we conducted a test tube experiment. The results showed that the activity of MMP-9 was not affected by metformin binding directly to MMP-9 (Figure 2f). However, Western blotting results suggested that metformin treatment could decrease the protein level of

MMP-9 (Figure 2g). Western blot analysis of MMP-9 in the total cell lysate consistently revealed two bands of apparent molecular masses of 85 and 92 kDa. It was previously shown that the 85 kDa band represents an underglycosylated precursor form of MMP-9 found intracellularly, whereas the 92 kDa band represents a fully glycosylated mature form that is secreted into the extracellular space [21]. Further, we detected the transcription level of MMP-9 by polymerase chain reaction and found that metformin did not change the mRNA level of MMP-9 (Figure 2h).

Accordingly, we became interested in establishing whether metformin downregulated the MMP-9 protein level by driving its degradation. To this end, we used eukaryotic inhibitor cycloheximide to inhibit protein synthesis in HEK293A cells to study the degradation of MMP-9 with or without metformin. We found that the exogenous MMP-9 protein was continuously degraded from 1 to 3 h, and metformin treatment effectively decreased MMP-9 protein expression by accelerating its degradation (Figure 2i,j).

3.3. Metformin Inhibits Local Plaque and Circulation MMP-9 Activity in ApoE-/- Mice

To further confirm whether metformin inhibits MMP-9 activity in vivo, we constructed a carotid artery plaque model in ApoE-/- mice (Figures S2 and S3) [19]. After a consecutive 14-day metformin treatment (200 mg·kg^{-1}) by intragastric gavage (Figures 3a and S4), we found that active MMP-9 and MMP-9 expression decreased in the plaque by immunofluorescence staining. However, metformin did not affect MMP-2/12 expression, which was reported to be related to plaque instability (Figure 3b,c). Moreover, the serum MMP-9 activity was detected using an MMP activity assay (Figure 3d). The results showed that metformin treatment inhibited local plaque and circulating MMP-9 activity.

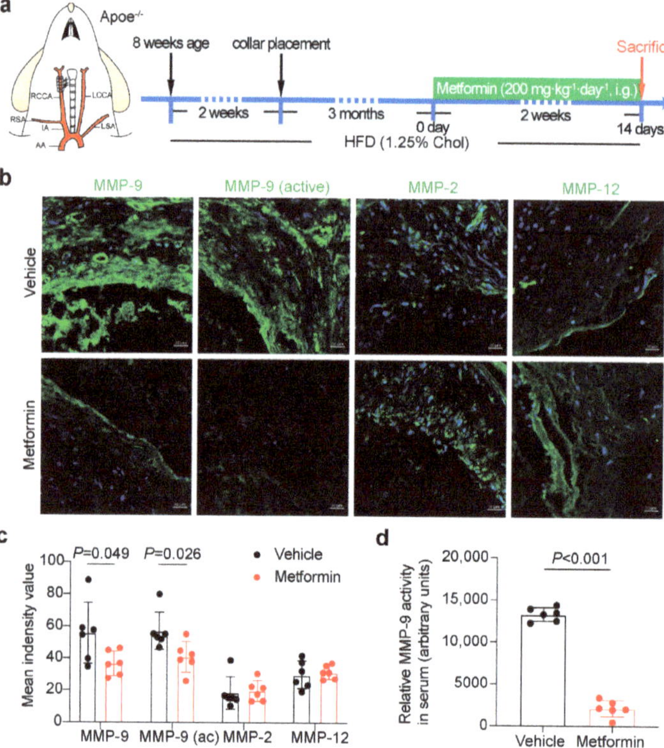

Figure 3. Metformin inhibits local plaque and circulating MMP-9 activity in ApoE-/- mice. (**a**) Flowchart illustrating the experimental procedure for actuating metformin treatment in a collar-induced carotid

atherosclerotic plaque model. (**b**) Representative images of immunofluorescence staining for active-matrix metalloproteinase (MMP)-9, MMP-2, and MMP-12 in plaque after metformin treatment. Scale bars represent 20 μm. (**c**) Quantification of immunofluorescence staining for MMP family in plaque after metformin treatment. Unpaired Student's *t*-test, n = 6 per group. (**d**) A Gelatinase Assay Kit was used to detect relative MMP activity in serum. Unpaired Student's *t*-test, n = 6 per group.

3.4. Metformin Improves Atherosclerotic Plaque Stability in ApoE-/- Mice

To determine the protective effects of metformin on atherosclerosis, we assessed the vulnerability index (VI) of the RCCA plaque using histology. The composition of plaques, including macrophages, collagen, lipids, and smooth muscle cells (SMCs) was demonstrated by CD-68, Sirius red staining, oil red O staining, and α-SMA immunostaining, respectively (Figure 4a,b). Sirius red staining results showed that the collagen content was preserved by the metformin treatment. Oil red O staining, α-SMA, and CD-68 immunofluorescence results suggested that there were no significant differences in lipid, SMCs, and macrophage content after metformin treatment. As each feature alone is insufficient for identifying high-risk plaques, the ratio between stable and unstable plaque components is often used to calculate the VI (macrophage content + lipid core content)/(SMC content + collagen content) in experimental studies [22]. The results showed that with the metformin treatment, plaque VI was significantly decreased, indicating that metformin had a beneficial effect on plaque stability (Figure 4c).

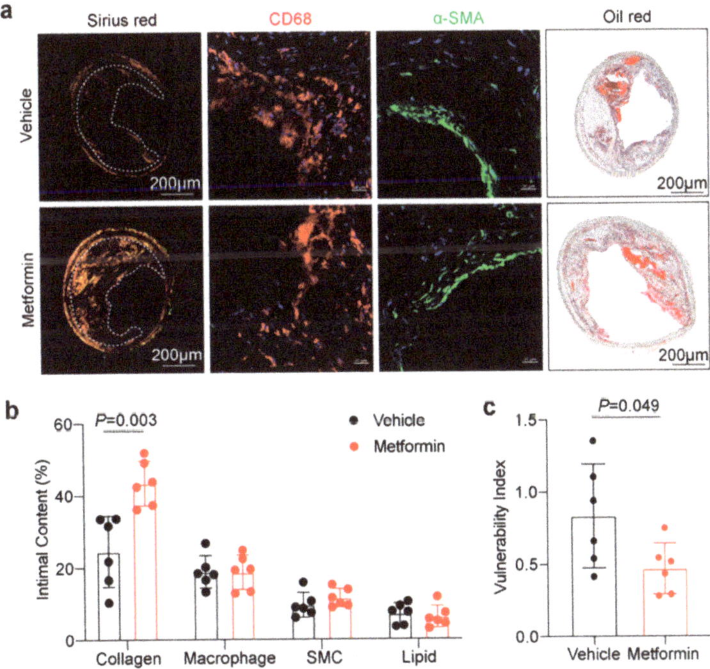

Figure 4. Metformin improves atherosclerotic plaque stability in ApoE-/- mice. (**a**) Representative images of Sirius red staining for plaque collagen, immunostaining for the macrophage marker CD-68, smooth muscle cell marker α-SMA, and oil red O staining for intimal lipid in plaque within the right common carotid artery. Scale bars for Sirius red staining and oil red O staining represent 200 μm and 20 μm for immunostaining. (**b**) Quantification of the positive area as a percentage of the whole plaque area. Unpaired Student's *t*-test, n = 6 per group. (**c**) The vulnerability index is calculated by dividing the area of macrophage+lipid by that of smooth muscle cells+collagen. Unpaired Student's *t*-test, n = 6 per group. Data are presented as the mean ± SD. HFD, high-fat diet; Chol, cholesterol.

4. Discussion

In this study, we demonstrated that metformin directly binds to MMP-9 and accelerates its degradation. Furthermore, we proved that metformin improved atherosclerotic plaque stability by inhibiting local plaque and circulating MMP-9 in ApoE-/- mice (Figure 5).

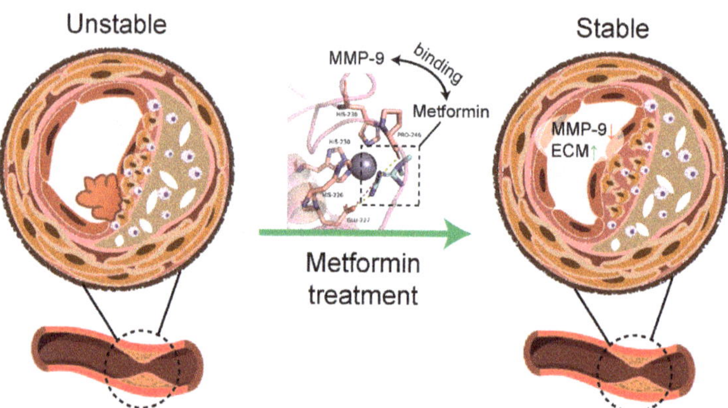

Figure 5. Schematic showing metformin directly binding to MMP-9 to improve plaque stability. MMP matrix metalloproteinase-9, ECM extracellular matrix.

Collagens are most abundant in the extracellular matrix, joined by elastin that confers elastic recoil to the artery [23]. Loss of collagen, which normally provides the main tensile strength of the artery wall, is an important cause of atherosclerotic plaque rupture, which underlies most cases of ACS [24]. MMPs have specific proteolytic activity against the ECM, which can result in the thinning of the fibrous cap and plaque instability [11,25]. MMP-9, also known as gelatinase B, is a widely investigated member of the MMP family. Studies have shown a strong relationship between MMP-9 and plaque instability [26,27], which indicates that MMP-9 may be a therapeutic target for preventing plaque instability. Currently, inflammatory pathways are the main therapeutic targets for plaque instability, such as the monoclonal antibody inhibiting interleukin-1β (called canakinumab) [28] and PCSK-9 inhibitors [29]. Both canakinumab and PCSK-9 inhibitors have anti-inflammatory effects. Moreover, PCSK-9 inhibitors also have an inhibitory effect on MMP-2, but cannot inhibit MMP-9 [30]. So, the mechanism by which canakinumab and PCSK-9 inhibitors stabilize plaques may be different from metformin. In addition, there are few plaque-stabilizing drugs targeting MMP-9. Metformin interferes with the pathophysiology of multiple cancers and diabetes by reducing MMP-9 expression [31–33]. However, there are still many studies showing that metformin can increase MMP-9 expression [34], including some clinical trials [35,36]. Whether metformin stabilizes plaque by modulating MMP-9 activity and expression remains unknown. Our results indicated that metformin directly binds to MMP-9, and significantly downregulated MMP-9 expression/activity levels in local plaque and circulation, which may explain the role of metformin in improving plaque stability.

It is generally accepted that metformin inhibits pro-inflammatory cytokine release, such as IL-1β, IL-6, and TNF-α, to have anti-inflammatory effects [37–41]. Destabilization of the atherosclerotic plaque is associated with increased inflammatory cytokine production [42,43]. To investigate whether metformin protects plaque stability by inhibiting inflammation, we measured plaque IL-1β, IL-6, and TNF-α levels. The immunofluorescence staining results suggested that metformin treatment did not affect the levels of IL-1β, IL-6, and TNF-α in plaque (Figure S5). H. Wu et al. found that macrophage infiltration was significantly reduced after 16 weeks of metformin treatment [44]. However, our immunofluorescence staining results suggested that as short as two weeks of metformin treatment had no significant anti-inflammatory effect. This may have been due to the short treatment time in our animal model. Additionally, metformin has been reported to promote macrophage cholesterol efflux, thus decreasing the lipid content of atherosclerotic plaques and increasing plaque stability [44]. In this study, after consecutive 14-day metformin treatment (200 mg·kg^{-1}) by intragastric gavage, we found that only the collagen content of the plaque was preserved, whereas intimal lipids, macrophages, and SMCs showed no significant difference, indicating that metformin improved plaque stability by reducing ECM degradation.

Metformin has protective effects by activating AMPK in intact cells and in vivo [45]. AMPK confers benefits in chronic inflammatory diseases, such as atherosclerosis, independent of its ability to normalize blood glucose levels. There was evidence that metformin inhibited TNF-α-induced MMP-9 upregulation in neutrophils, which might have been mediated via an AMPK-dependent pathway [46]. Metformin administration suppressed MMP-9/MMP-2 and mTOR expression and increased Akt and AMPK expression, indicating that metformin reduced the expression of MMPs via the AMPK signaling pathway [47]. In this study, we first found that metformin binds to MMP-9. The MMP-9 binding regions of metformin are situated in the active cavity and engage in several interactions with MMP-9. Moreover, the combination of metformin and MMP-9 significantly accelerated MMP-9 protein degradation, which may also account for the effect of metformin downregulating MMP-9 expression level and improving plaque stability.

Protein homeostasis is responsible for basic cellular functions, such as the regulation of the level of key enzymes and the removal of abnormal proteins [48]. Our results suggested that the combination of metformin and MMP-9 significantly accelerated MMP-9 protein degradation. Chang Y et al. reported that cells treated with MG-132, a proteasome inhibitor, exhibited a significant MMP-9 protein accumulation compared to its accumulation in the untreated controls, indicating that the degradation of the MMP-9 protein is in a proteasome-dependent manner. Moreover, SMURF1, an E3 ubiquitin ligase, binds MMP-9 to promote its degradation [49]. In this study, we first found that metformin binds to MMP-9. The MMP-9-binding regions of metformin are situated in the active cavity and engage in several interactions with MMP-9. Further, MMP-9 was shown to have two N-glycosylation sites, which seems to be important for MMP-9 protein structure stabilization and secretion, on asparagine residues at position 38 in the propeptide domain and in the catalytic domain at position 120 [50–52]. In subsequent research, we have two directions to further explore the potential mechanism of metformin regulation of MMP-9: (1) metformin affects the binding of MMP-9 to SMURF1, thus promoting MMP-9 ubiquitination and accelerating its degradation; (2) metformin affects the role of N-glycosylation in MMP-9 and decreases MMP-9 protein structure stabilization.

In conclusion, we have demonstrated that metformin directly binds to MMP-9 and accelerates its degradation, thus preserving the collagen content of plaque and improving atherosclerotic plaque stability. Further, these findings could significantly impact the development of the search for new drugs and pleiotropic mechanisms.

Supplementary Materials: The following supporting information can be downloaded at: https://www.mdpi.com/article/10.3390/jcdd10020054/s1, Figure S1: MMP-9 was successfully overexpressed in HEK293A cells. Figure S2: Serum triglyceride and total cholesterol levels increased in animal models. Figure S3: The carotid plaque model was successfully constructed. Figure S4: Metformin successfully activated AMPK. Figure S5: Metformin treatment had no significant anti-inflammatory effect in our model.

Author Contributions: Conceptualization, H.X., Y.Z., X.C. and S.W.; methodology and software, X.C. and S.W.; validation, formal analysis and data curation, X.C.; resources and technical supports, S.W., W.X. and M.Z.; writing—original draft preparation, X.C.; writing—review and editing, H.X. and Y.Z.; funding acquisition and project administration, H.X. and Y.Z. All authors have read and agreed to the published version of the manuscript.

Funding: This research was funded by the National Natural Science Foundation of China [81830009 to Y.Z., 82030072 to H.X.], Michigan Medicine-PKUHSC Joint Institute for Translational and Clinical

Research [BMU2019JI007 to Y.Z.], CAMS Innovation Fund for Medical Sciences to [No. 2021-I2M-5-003 to H.X.] and the Key Clinical Projects of Peking University Third Hospital [BYSYZD2019022 to H.X.].

Institutional Review Board Statement: The investigations conformed to the US National Institutes of Health Guide for the Care and Use of Laboratory Animals (NIH Publication No. 85-23, revised 1996). Animal experiments were approved by the Committee of Peking University on Ethics of Animal Experiments (LA 2018-112) and conducted in accordance with the Guidelines for Animal Experiments, Peking University Health Science Center.

Informed Consent Statement: Not applicable.

Data Availability Statement: The data presented in this study are available in this article.

Conflicts of Interest: The authors declare no conflict of interest. The funders had no role in the design of the study; in the collection, analyses, or interpretation of data; in the writing of the manuscript; or in the decision to publish the results.

References

1. Libby, P.; Buring, J.E.; Badimon, L.; Hansson, G.K.; Deanfield, J.; Bittencourt, M.S.; Tokgozoglu, L.; Lewis, E.F. Atherosclerosis. *Nat. Rev. Dis. Prim.* **2019**, *5*, 56. [CrossRef] [PubMed]
2. Roth, G.A.; Johnson, C.; Abajobir, A.; Abd-Allah, F.; Abera, S.F.; Abyu, G.; Ahmed, M.; Aksut, B.; Alam, T.; Alam, K.; et al. Global, Regional, and National Burden of Cardiovascular Diseases for 10 Causes, 1990 to 2015. *J. Am. Coll. Cardiol.* **2017**, *70*, 1–25. [CrossRef] [PubMed]
3. Salvatore, T.; Galiero, R.; Caturano, A.; Vetrano, E.; Rinaldi, L.; Coviello, F.; Di Martino, A.; Albanese, G.; Marfella, R.; Sardu, C.; et al. Effects of Metformin in Heart Failure: From Pathophysiological Rationale to Clinical Evidence. *Biomolecules* **2021**, *11*, 1834. [CrossRef] [PubMed]
4. Goldberg, R.B.; Aroda, V.R.; Bluemke, D.A.; Barrett-Connor, E.; Budoff, M.; Crandall, J.P.; Dabelea, D.; Horton, E.S.; Mather, K.J.; Orchard, T.J.; et al. Effect of Long-Term Metformin and Lifestyle in the Diabetes Prevention Program and Its Outcome Study on Coronary Artery Calcium. *Circulation* **2017**, *136*, 52–64. [CrossRef]
5. Zhang, K.; Yang, W.; Dai, H.; Deng, Z. Cardiovascular risk following metformin treatment in patients with type 2 diabetes mellitus: Results from meta-analysis. *Diabetes Res. Clin. Pract.* **2020**, *160*, 108001. [CrossRef]
6. Papaioannou, T.G.; Kalantzis, C.; Katsianos, E.; Sanoudou, D.; Vavuranakis, M.; Tousoulis, D. Personalized Assessment of the Coronary Atherosclerotic Arteries by Intravascular Ultrasound Imaging: Hunting the Vulnerable Plaque. *J. Pers. Med.* **2019**, *9*, 8. [CrossRef]
7. McDermott, M.M.; Guralnik, J.M.; Corsi, A.; Albay, M.; Macchi, C.; Bandinelli, S.; Ferrucci, L. Patterns of inflammation associated with peripheral arterial disease: The InCHIANTI study. *Am. Heart J.* **2005**, *150*, 276–281. [CrossRef]
8. Allison, M.A.; Criqui, M.H.; McClelland, R.L.; Scott, J.M.; McDermott, M.M.; Liu, K.; Folsom, A.R.; Bertoni, A.G.; Sharrett, A.R.; Homma, S.; et al. The effect of novel cardiovascular risk factors on the ethnic-specific odds for peripheral arterial disease in the Multi-Ethnic Study of Atherosclerosis (MESA). *J. Am. Coll. Cardiol.* **2006**, *48*, 1190–1197. [CrossRef]
9. Mallat, Z.; Corbaz, A.; Scoazec, A.; Besnard, S.; Leseche, G.; Chvatchko, Y.; Tedgui, A. Expression of interleukin-18 in human atherosclerotic plaques and relation to plaque instability. *Circulation* **2001**, *104*, 1598–1603. [CrossRef]
10. Young, J.L.; Libby, P.; Schonbeck, U. Cytokines in the pathogenesis of atherosclerosis. *Thromb. Haemost.* **2002**, *88*, 554–567. [CrossRef]
11. Sakakura, K.; Nakano, M.; Otsuka, F.; Ladich, E.; Kolodgie, F.D.; Virmani, R. Pathophysiology of atherosclerosis plaque progression. *Heart Lung Circ.* **2013**, *22*, 399–411. [CrossRef]
12. Holm Nielsen, S.; Jonasson, L.; Kalogeropoulos, K.; Karsdal, M.A.; Reese-Petersen, A.L.; Auf dem Keller, U.; Genovese, F.; Nilsson, J.; Goncalves, I. Exploring the role of extracellular matrix proteins to develop biomarkers of plaque vulnerability and outcome. *J. Intern. Med.* **2020**, *287*, 493–513. [CrossRef]
13. Galis, Z.S.; Sukhova, G.K.; Lark, M.W.; Libby, P. Increased expression of matrix metalloproteinases and matrix degrading activity in vulnerable regions of human atherosclerotic plaques. *J. Clin. Investig.* **1994**, *94*, 2493–2503. [CrossRef]
14. Gu, C.; Wang, F.; Zhao, Z.; Wang, H.; Cong, X.; Chen, X. Lysophosphatidic Acid Is Associated with Atherosclerotic Plaque Instability by Regulating NF-kappaB Dependent Matrix Metalloproteinase-9 Expression via LPA2 in Macrophages. *Front. Physiol.* **2017**, *8*, 266. [CrossRef]
15. Loftus, I.M.; Naylor, A.R.; Goodall, S.; Crowther, M.; Jones, L.; Bell, P.R.; Thompson, M.M. Increased matrix metalloproteinase-9 activity in unstable carotid plaques. A potential role in acute plaque disruption. *Stroke* **2000**, *31*, 40–47. [CrossRef]
16. Jiang, X.B.; Wang, J.S.; Liu, D.H.; Yuan, W.S.; Shi, Z.S. Overexpression of matrix metalloproteinase-9 is correlated with carotid intraplaque hemorrhage in a swine model. *J. Neurointerv. Surg.* **2013**, *5*, 473–477. [CrossRef]
17. de Nooijer, R.; Verkleij, C.J.; von der Thusen, J.H.; Jukema, J.W.; van der Wall, E.E.; van Berkel, T.J.; Baker, A.H.; Biessen, E.A. Lesional overexpression of matrix metalloproteinase-9 promotes intraplaque hemorrhage in advanced lesions but not at earlier stages of atherogenesis. *Arterioscler. Thromb. Vasc. Biol.* **2006**, *26*, 340–346. [CrossRef]

18. Fukuda, D.; Shimada, K.; Tanaka, A.; Kusuyama, T.; Yamashita, H.; Ehara, S.; Nakamura, Y.; Kawarabayashi, T.; Iida, H.; Yoshiyama, M.; et al. Comparison of levels of serum matrix metalloproteinase-9 in patients with acute myocardial infarction versus unstable angina pectoris versus stable angina pectoris. *Am. J. Cardiol.* **2006**, *97*, 175–180. [CrossRef]
19. von der Thusen, J.H.; van Berkel, T.J.; Biessen, E.A. Induction of rapid atherogenesis by perivascular carotid collar placement in apolipoprotein E-deficient and low-density lipoprotein receptor-deficient mice. *Circulation* **2001**, *103*, 1164–1170. [CrossRef]
20. Daugherty, A.; Tall, A.R.; Daemen, M.; Falk, E.; Fisher, E.A.; Garcia-Cardena, G.; Lusis, A.J.; Owens, A.P., 3rd; Rosenfeld, M.E.; Virmani, R.; et al. Recommendation on Design, Execution, and Reporting of Animal Atherosclerosis Studies: A Scientific Statement From the American Heart Association. *Circ. Res.* **2017**, *121*, e53–e79. [CrossRef]
21. Olson, M.W.; Bernardo, M.M.; Pietila, M.; Gervasi, D.C.; Toth, M.; Kotra, L.P.; Massova, I.; Mobashery, S.; Fridman, R. Characterization of the monomeric and dimeric forms of latent and active matrix metalloproteinase-9. Differential rates for activation by stromelysin 1. *J. Biol. Chem.* **2000**, *275*, 2661–2668. [CrossRef] [PubMed]
22. Tomas, L.; Edsfeldt, A.; Mollet, I.G.; Perisic Matic, L.; Prehn, C.; Adamski, J.; Paulsson-Berne, G.; Hedin, U.; Nilsson, J.; Bengtsson, E.; et al. Altered metabolism distinguishes high-risk from stable carotid atherosclerotic plaques. *Eur. Heart J.* **2018**, *39*, 2301–2310. [CrossRef] [PubMed]
23. Van Doren, S.R. Matrix metalloproteinase interactions with collagen and elastin. *Matrix Biol.* **2015**, *44–46*, 224–231. [CrossRef] [PubMed]
24. Hansson, G.K. Inflammation, atherosclerosis, and coronary artery disease. *N. Engl. J. Med.* **2005**, *352*, 1685–1695. [CrossRef]
25. Falk, E.; Nakano, M.; Bentzon, J.F.; Finn, A.V.; Virmani, R. Update on acute coronary syndromes: The pathologists' view. *Eur. Heart J.* **2013**, *34*, 719–728. [CrossRef]
26. Park, J.P.; Lee, B.K.; Shim, J.M.; Kim, S.H.; Lee, C.W.; Kang, D.H.; Hong, M.K. Relationship between multiple plasma biomarkers and vulnerable plaque determined by virtual histology intravascular ultrasound. *Circ. J.* **2010**, *74*, 332–336. [CrossRef]
27. Chen, F.; Eriksson, P.; Hansson, G.K.; Herzfeld, I.; Klein, M.; Hansson, L.O.; Valen, G. Expression of matrix metalloproteinase 9 and its regulators in the unstable coronary atherosclerotic plaque. *Int. J. Mol. Med.* **2005**, *15*, 57–65. [CrossRef]
28. Ridker, P.M.; Everett, B.M.; Thuren, T.; MacFadyen, J.G.; Chang, W.H.; Ballantyne, C.; Fonseca, F.; Nicolau, J.; Koenig, W.; Anker, S.D.; et al. Antiinflammatory Therapy with Canakinumab for Atherosclerotic Disease. *N. Engl. J. Med.* **2017**, *377*, 1119–1131. [CrossRef]
29. Kosowski, M.; Basiak, M.; Hachula, M.; Okopien, B. Impact of Alirocumab on Release Markers of Atherosclerotic Plaque Vulnerability in Patients with Mixed Hyperlipidemia and Vulnerable Atherosclerotic Plaque. *Medicina* **2022**, *58*, 969. [CrossRef]
30. Basiak, M.; Kosowski, M.; Hachula, M.; Okopien, B. Impact of PCSK9 Inhibition on Proinflammatory Cytokines and Matrix Metalloproteinases Release in Patients with Mixed Hyperlipidemia and Vulnerable Atherosclerotic Plaque. *Pharmaceuticals* **2022**, *15*, 802. [CrossRef]
31. Kimber-Trojnar, Z.; Dluski, D.F.; Wierzchowska-Opoka, M.; Ruszala, M.; Leszczynska-Gorzelak, B. Metformin as a Potential Treatment Option for Endometriosis. *Cancers* **2022**, *14*, 577. [CrossRef]
32. Tay, K.C.; Tan, L.T.; Chan, C.K.; Hong, S.L.; Chan, K.G.; Yap, W.H.; Pusparajah, P.; Lee, L.H.; Goh, B.H. Formononetin: A Review of Its Anticancer Potentials and Mechanisms. *Front. Pharmacol.* **2019**, *10*, 820. [CrossRef]
33. Malekpour-Dehkordi, Z.; Teimourian, S.; Nourbakhsh, M.; Naghiaee, Y.; Sharifi, R.; Mohiti-Ardakani, J. Metformin reduces fibrosis factors in insulin resistant and hypertrophied adipocyte via integrin/ERK, collagen VI, apoptosis, and necrosis reduction. *Life Sci.* **2019**, *233*, 116682. [CrossRef]
34. Shao, W.; Wang, D.; He, J. The role of gene expression profiling in early-stage non-small cell lung cancer. *J. Thorac. Dis.* **2010**, *2*, 89–99.
35. Goldstein, B.J.; Weissman, P.N.; Wooddell, M.J.; Waterhouse, B.R.; Cobitz, A.R. Reductions in biomarkers of cardiovascular risk in type 2 diabetes with rosiglitazone added to metformin compared with dose escalation of metformin: An EMPIRE trial sub-study. *Curr. Med. Res. Opin.* **2006**, *22*, 1715–1723. [CrossRef]
36. Hanefeld, M.; Pfutzner, A.; Forst, T.; Kleine, I.; Fuchs, W. Double-blind, randomized, multicentre, and active comparator controlled investigation of the effect of pioglitazone, metformin, and the combination of both on cardiovascular risk in patients with type 2 diabetes receiving stable basal insulin therapy: The PIOCOMB study. *Cardiovasc. Diabetol.* **2011**, *10*, 65. [CrossRef]
37. Hammad, A.M.; Ibrahim, Y.A.; Khdair, S.I.; Hall, F.S.; Alfaraj, M.; Jarrar, Y.; Abed, A.F. Metformin reduces oxandrolone- induced depression-like behavior in rats via modulating the expression of IL-1beta, IL-6, IL-10 and TNF-alpha. *Behav. Brain Res.* **2021**, *414*, 113475. [CrossRef]
38. Koh, S.J.; Kim, J.M.; Kim, I.K.; Ko, S.H.; Kim, J.S. Anti-inflammatory mechanism of metformin and its effects in intestinal inflammation and colitis-associated colon cancer. *J. Gastroenterol. Hepatol.* **2014**, *29*, 502–510. [CrossRef]
39. Docrat, T.F.; Nagiah, S.; Chuturgoon, A.A. Metformin protects against neuroinflammation through integrated mechanisms of miR-141 and the NF-kB-mediated inflammasome pathway in a diabetic mouse model. *Eur. J. Pharmacol.* **2021**, *903*, 174146. [CrossRef]
40. Mummidi, S.; Das, N.A.; Carpenter, A.J.; Kandikattu, H.; Krenz, M.; Siebenlist, U.; Valente, A.J.; Chandrasekar, B. Metformin inhibits aldosterone-induced cardiac fibroblast activation, migration and proliferation in vitro, and reverses aldosterone+salt-induced cardiac fibrosis in vivo. *J. Mol. Cell Cardiol.* **2016**, *98*, 95–102. [CrossRef]
41. Chung, M.M.; Nicol, C.J.; Cheng, Y.C.; Lin, K.H.; Chen, Y.L.; Pei, D.; Lin, C.H.; Shih, Y.N.; Yen, C.H.; Chen, S.J.; et al. Metformin activation of AMPK suppresses AGE-induced inflammatory response in hNSCs. *Exp. Cell Res.* **2017**, *352*, 75–83. [CrossRef] [PubMed]

42. Wen, H.; Liu, M.; Liu, Z.; Yang, X.; Liu, X.; Ni, M.; Dong, M.; Luan, X.; Yuan, Y.; Xu, X.; et al. PEDF improves atherosclerotic plaque stability by inhibiting macrophage inflammation response. *Int. J. Cardiol.* **2017**, *235*, 37–41. [CrossRef] [PubMed]
43. Ni, M.; Wang, Y.; Zhang, M.; Zhang, P.F.; Ding, S.F.; Liu, C.X.; Liu, X.L.; Zhao, Y.X.; Zhang, Y. Atherosclerotic plaque disruption induced by stress and lipopolysaccharide in apolipoprotein E knockout mice. *Am. J. Physiol. Heart Circ. Physiol.* **2009**, *296*, H1598–H1606. [CrossRef] [PubMed]
44. Wu, H.; Feng, K.; Zhang, C.; Zhang, H.; Zhang, J.; Hua, Y.; Dong, Z.; Zhu, Y.; Yang, S.; Ma, C. Metformin attenuates atherosclerosis and plaque vulnerability by upregulating KLF2-mediated autophagy in apoE(-/-) mice. *Biochem. Biophys. Res. Commun.* **2021**, *557*, 334–341. [CrossRef] [PubMed]
45. Hawley, S.A.; Gadalla, A.E.; Olsen, G.S.; Hardie, D.G. The antidiabetic drug metformin activates the AMP-activated protein kinase cascade via an adenine nucleotide-independent mechanism. *Diabetes* **2002**, *51*, 2420–2425. [CrossRef]
46. Zhang, D.; Tang, Q.; Zheng, G.; Wang, C.; Zhou, Y.; Wu, Y.; Xuan, J.; Tian, N.; Wang, X.; Wu, Y.; et al. Metformin ameliorates BSCB disruption by inhibiting neutrophil infiltration and MMP-9 expression but not direct TJ proteins expression regulation. *J. Cell Mol. Med.* **2017**, *21*, 3322–3336. [CrossRef]
47. Chen, Z.; Wei, H.; Zhao, X.; Xin, X.; Peng, L.; Ning, Y.; Wang, Y.; Lan, Y.; Zhang, Q. Metformin treatment alleviates polycystic ovary syndrome by decreasing the expression of MMP-2 and MMP-9 via H19/miR-29b-3p and AKT/mTOR/autophagy signaling pathways. *J. Cell Physiol.* **2019**, *234*, 19964–19976. [CrossRef]
48. Hershko, A. Ubiquitin-mediated protein degradation. *J. Biol. Chem.* **1988**, *263*, 15237–15240. [CrossRef]
49. Chang, Y.; Jin, H.; Li, H.; Ma, J.; Zheng, Z.; Sun, B.; Lyu, Y.; Lin, M.; Zhao, H.; Shen, L.; et al. MiRNA-516a promotes bladder cancer metastasis by inhibiting MMP9 protein degradation via the AKT/FOXO3A/SMURF1 axis. *Clin. Transl. Med.* **2020**, *10*, e263. [CrossRef]
50. Kotra, L.P.; Zhang, L.; Fridman, R.; Orlando, R.; Mobashery, S. N-Glycosylation pattern of the zymogenic form of human matrix metalloproteinase-9. *Bioorg. Chem.* **2002**, *30*, 356–370. [CrossRef]
51. Duellman, T.; Burnett, J.; Yang, J. Functional Roles of N-Linked Glycosylation of Human Matrix Metalloproteinase 9. *Traffic* **2015**, *16*, 1108–1126. [CrossRef]
52. Kumar, S.; Cieplak, P. Role of N-glycosylation in activation of proMMP-9. A molecular dynamics simulations study. *PLoS ONE* **2018**, *13*, e0191157. [CrossRef]

MDPI
St. Alban-Anlage 66
4052 Basel
Switzerland
Tel. +41 61 683 77 34
Fax +41 61 302 89 18
www.mdpi.com

Journal of Cardiovascular Development and Disease Editorial Office
E-mail: jcdd@mdpi.com
www.mdpi.com/journal/jcdd

www.ingramcontent.com/pod-product-compliance
Lightning Source LLC
LaVergne TN
LVHW070924170726
843515LV00005B/859

* 9 7 8 3 0 3 6 5 8 1 1 6 3 *